Memoirs of the Connecticut Academy of Arts and Sciences
Volume XXIII

ASTRONOMY AT YALE

1701–1968

by Dorrit Hoffleit

Oliver Winchester, Frontispiece

ASTRONOMY AT YALE
1701–1968

by

Dorrit Hoffleit
Senior Research Astronomer, Ret.
Department of Astronomy

Yale University
New Haven, Connecticut
The Connecticut Academy of Art and Sciences
1992

First published 1992 by the Academy
New Haven, Connecticut

Printed in the United States of America

Composed in ITC Garamond by Brevis Press,
Bethany, Connecticut.

Printed and bound by Thomson-Shore, Inc.,
Dexter, Michigan.

The paper used in this publication meets the
minimum requirements of American National Standard
for Information Sciences—Permanence of Paper for
Printed Library Materials, ANZI Z39.48–1984

Publication of this book was assisted by a gift
from the O. P. and W. E. Edwards Foundation, Inc.

ISBN: 1–878508–02–4

Library of Congress Catalog Card Number: 91–070522

Publications of the Connecticut Academy of Arts and Sciences
can be obtained from the Academy, Drawer 93A Yale Station,
New Haven, Connecticut 06520.

TO THE MEMORY OF
Kate *Sanio* Hoffleit
1879–1974
and
Beulah *Cole* Hibbert
1888–1984

CONTENTS

*Undergraduate Peter Wlasuk contributed appreciably to the literature and archival research for these chapters.

ILLUSTRATIONS

TABLES

PREFACE AND ACKNOWLEDGMENTS

The American Astronomical Society was founded in 1899. In conjunction with celebrations of its 75th anniversary in 1974 Professor Demarque asked me to present a short paper on the activities at the Yale Observatory in that era. At that time I mentioned that as a retirement project I wanted to research and write up the history of astronomy at Yale. He warmly encouraged this, suggesting I not wait until retirement. However, I had pressing commitments first to finish several star catalogues for which we had limited-time government support.

Many requests that came to the department asking for information on topics in the history of the Observatory were referred to me. Shortly before my retirement an exhibit had been set up in the Sterling Memorial Library depicting the history of astronomy at Yale. There I spotted a few errors. In particular, there was a photograph of a spectacular fireball, attributed to W. L. Elkin's experiments in meteor photography. I was sufficiently familiar with Elkin's work to realize that this photograph could not have been one of his. After some effort I was able to identify it. An amateur had serendipitously captured it while trying to photograph a faint comet. He turned the photograph over to H. A. Newton who published his analysis in the *American Journal of Science.* Most of the queries that had come to me could not be answered quickly but required archival research. These various circumstances clinched my determination to write the present account in order that the answers to many questions might in the future be found in one place. The book is written mainly for astronomers as a fact-finding reference. It is not intended as a discourse on the philosophy of science.

When in 1983 the time was finally ripe for pursuing this work, a patroness, Mrs. Harold Hibbert, widow of a former Yale chemistry professor who later spent the major part of his career at McGill University, offered Yale $5,000 specifically to help with this project. This enabled me to employ a "bursary student," Yale senior Peter Wlasuk, who helped very effectively in scrutinizing the archival materials that were made available to me. Had he been able to continue at Yale he would have been co-author of this volume. He did write quick drafts of several chapters, without however including either any references or direct quotations, to both of which I am particularly partial. Hence I repeated all of his research and independently wrote my own usually enlarged versions, then found his helpful for checking accuracy and content. In the Table of Contents an asterisk indicates the chapters for which his help is acknowledged.

As my investigations continued I became tempted to use as a subtitle, "The Stepchild Department of the University." There are numerous instances when, mainly for ever prevalent financial reasons, the needs of the Observatory were considered secondary to those of other departments, and in one instance, even to the needs of a private girls' school. For example, Professor Denison Olmsted (1791–1859) fought in vain for a proper observatory to house the 5-inch Dollond telescope, then the largest refractor in the States; and Dirk Brouwer (1902–1966) was prematurely evicted from the Observatory on Canner Street before even minimally desirable quarters were ready for occupancy. He once told me he believed the President of the University wished he would drop dead! Misfortunes, however, often serve as incentives, and for all its adversity the astronomy faculty has always demonstrated outstanding scientific leadership. The final chapter summarizes and attempts to trace the progress of teaching astronomy at Yale, from its early founding through the era of Dirk Brouwer.

Among the references at the end of each chapter, *Yale Obs. Archives* refers indiscriminately to archival materials, not just those deposited at the Yale University Archives at Sterling Memorial Library, but also to materials located in various astronomy faculty offices and offices of the secretaries and business manager of the Observatory, most of which will eventually be deposited at Sterling. I am grateful to Judith Schiff, Chief Archivist at Sterling, for giving me access to the materials I needed. The citations to the Annual Reports of the Department do not necessarily refer to the published reports. The carbon copies of the original versions that I consulted are frequently preliminary, containing some information omitted in the final printed versions.

In the Index I have given the dates of individuals wherever they were readily available. Some presented confusion from same names, which I have attempted to resolve. For example various biographical sources indicate ten men named James Hillhouse; seven named William Hillhouse; two A. P. Stokes; numerous H. Smith; and four Asaph Hall of whom the third was the U.S. Naval Observatory professor who had discovered the moons of Mars. The fourth, Professor Hall's son and the second astronomer named Asaph Hall, was usually but not always called Jr. A few of the dates may be off by one year when different sources have been discordant; the era, however, is more important here than the specific date of birth or death. This Index indicates by the letter Y those men and women who earned either bachelor or graduate degrees at Yale. The honorary degrees cited are Yale degrees awarded to faculty members who had not been students at Yale. The letters NAS cite the men who were members of the National Academy of Sciences.

What I have written are my own impressions of the progress of astronomy at Yale from the inception of the college through the directorship of Dr. Brouwer. This appeared an appropriate stopping place; for within a very few years of his decease the character of research in the department changed abruptly from strongest emphasis on celestial mechanics to the broader fields of astrophysics and cosmology. Undoubtedly I must have overlooked numerous items others might have included; and inserted quotations some may consider trivial but which to me help portray the spirit of the times or the character of the personnel. In general I have tried to keep my own opinions to a minimum but have enjoyed quoting the opinions of others, especially of contemporaries in each era. I have attempted to be as complete as possible in citing references in order that others may check anything they question. To readers who find mistakes, relevant omissions, or disagreements, I shall be grateful if they will report them to me.

I submitted the first draft of the manuscript to two astronomers and two historians of science for comments. In instances where I did not acquiesce in suggested changes by any one of these referees I checked if the others had made similar recommendations, and in some cases asked still others for opinions. One critic was adamant on two issues. He insisted that more reasons should be stated as to why certain investigations were carried out; and he wanted all my references to obituaries or biographical memoirs deleted as "hagiographic" or "far from objective sources." On the first of these points, I have sometimes been irritated by historians of science who do not clearly differentiate between documentable facts and their own personal inferences. If an investigator or his contemporaries have specifically stated why they did what they did, I am happy to quote them. Hindsight, say some four generations later, may well indicate reasons why something might or should have been investigated. But this educated hindsight may not necessarily coincide with the pioneer's motives. Serendipity and occasionally perhaps even idle curiosity could have played a role. Let the documented facts speak for themselves. On the second point, I recall one historian as saying he wished more obituaries had been written as they are often the only sources of information on the scientists. The eulogies I quote (and all eulogies are not obituaries) come from authorities who would hardly have sunk to mere sanctifying and whose opinions I would expect to be unbiased and sincere, e.g. President Dwight or H. A. Newton. And if a Harvard astronomer eulogizes the work of a Yale astronomer some fourteen years before the death of the latter (Elkin), that I feel fully merits citation. My disagreements with others reminded me of a quotation I had once extracted from a small town newspaper. Two politicians were disputing their policies in the newspaper when one wrote, "Those that disagree with me are entitled to their own rediculous opinions." After weighing to the best of my understanding the comments of my critics, I hope I shall be forgiven for clinging to my own "rediculous opinions."

Finally, I dedicate this volume to the memory of the two ladies who have played major roles in my career. My mother, Kate *Sanio* Hoffleit (1879–1974), a highly cultured but impoverished refugee from discrimination in Germany, sold her cherished home in Pennsylvania to enable her to give her two children the highest education of which they were capable. Beulah *Cole* Hibbert (1888–1984), a friend I first met at the Yale Faculty Club in 1957, took an interest in my career and gave support first to my summer work at the Nantucket Maria Mitchell Observatory, later to the initial stages of this project.

Carol Runlett has graciously prepared the manuscript on computer disks. I am especially grateful for her patience in making innumerable insertions and revisions after the first drafts were completed. Our departmental librarian, Pauline DiGioia, has given considerable help in tracing hard-to-find references, not only at Yale but from other institutions. I have frequently been privileged, through the courtesy of its Director, Janet A. Mattei, to use the library of the American Association of Variable Star Observers and to scan its archives, particularly for the era when Yale's E. W. Brown was President of the AAVSO. Horace Smith, Yale Ph.D. 1980, called my attention to an O'Keefe reference to the meteoric procession of 1913. David DeVorkin graciously sent me a copy of the part of Rothenberg's Ph.D. dissertation dealing with the work of Denison Olmsted. Amateur astronomer Vincent Anyzeski alerted me to Lee De Forest's interest in meteors. Professor C. E. Gilbert of the Yale Fine Arts Department kindly identified for me Luther Terry as the artist H. A. Newton had cited as having made a copy of Raphael's Madonna di Foligno. With my friend Barbara Narendra, Archivist at Yale's Peabody Museum, I enjoyed many stimulating discussions and she voluntarily provided me with numerous helpful references. Professor Savage of the Connecticut Academy suggested my writing an Appendix on women astronomers at Yale.

For all of the direct quotations from publications 1965 or later I have obtained the permission to quote from the publishers or authors. In particular I am indebted to Pergamon Press for permission to use a high percentage of the material in my paper, "Yale Contributions to Meteoric Astronomy," published in *Vistas in Astronomy,* **32**, 117–43, 1988. While there is considerable overlap between that article and the present text, each contains material not included in the other. The same holds for a paper, "Yale and USNO Cooperation," published by Gordon and Breach Science Publishers, Inc., in *Comments on Astrophysics,* **16**, No. 1, 17–30, 1992; and also for two papers published by the American Association of Variable Star Observers in their *Journal,* "Elias and Francis Loomis on Variable Stars, 1866–69," **13**, 67–77, 1984, and "The Earliest Infrared Light Curves," **16**, 29–33, 1987. Figure 17.2 was available only as a dark low-contrast Xerox copy of an old newspaper, supplied by the Yale University Archives. With his electronic computer David Guenther was able to make a clear high-contrast copy suitable for reproduction. From J. Maynard of the Yale Printing Service I have received numerous helpful hints on how to prepare the manuscript for publication. I have benefited from helpful comments and additional references from Drs. David DeVorkin, John Lankford, Brian Marsden, and Janet A. Mattei. Most importantly I have enjoyed and much appreciated the encouragement of all the successive chairmen of the Yale Astronomy Department while this work was in progress: Professors P. Demarque, W. van Altena, R. Larson, and A. Oemler. Authors are frequently notorious for failing to catch their own typographical and spelling mistakes, even when they know better. Hence I am indebted to copy editor Anne F. Wilde for correcting many of mine. Finally I am particularly grateful to the Connecticut Academy of Arts and Sciences for undertaking the publication of this book, and to Marlene Demarque for seeing it through completion; and we are all grateful to the anonymous donor who contributed generously to the costs of publication.

D.H.

1
OVERVIEW OF TWO AND A HALF CENTURIES

The following letter was written by Frank Schlesinger, Director of the Yale Observatory from 1920 to 1941, to Burton E. Livingston, then President of the American Association for the Advancement of Science, dated September 20, 1920. It concerned a proposal to establish a History of Science Section in the AAAS.[1]

> At the request of Dr. Brasch [F. E. Brasch of the Library of Congress] I give you herewith a brief word in statement of my opinion regarding the importance of the study of the history of science. I have always been of the opinion that in this country we have neglected too much this side of science. As in other fields of human endeavor, the history of the subject has a most important function in teaching us what mistakes to avoid, and what is the best attitude to maintain regarding new ideas and new theories. The study of biographical works is, in my opinion, especially important for the young scientist. It tells him better than anything else can what rewards he can hope for and what discouragements and obstacles he must expect to encounter. I have no doubt that a great many bitter disappointments might be averted if such information as this could be put into the hands of beginners as early as possible.

These arguments apply equally to the purpose of this book. Few astronomers, even those trained at Yale, are aware of the past astronomical activities at this institution.

When New Haven Colony was founded in 1638 its leader, the Rev. John Davenport, strongly advocated establishing a college there similar to Harvard. After innumerable obstacles, Yale College was finally founded in 1701, the third college in the Colonies after Harvard and William and Mary. Typically of the times, its major goal was the training of young men for the ministry, in this case exclusively the Puritan or Congregational. One of the earliest topics taught, aside from theology, was astronomy with, however, pointed allusions to its exemplification of the power and bounty of God. With the acquisition in 1714 of scientific books from England, especially Isaac Newton's *Principia* and *Opticks,* the tutors began learning his new branch of mathematics, "fluxions," now known as the calculus. This knowledge was essential to the understanding of both gravitation and the principles of optics.

The apparition of a spectacular fireball in 1742 gave rise to President Clap's amazing theory of *Terrestrial Comets.* The impending expected return of the comet, later named Halley, further stimulated the early study of celestial mechanics, mainly for the determination of orbits. The great shower of meteorites at Weston, Connecticut, in 1807, prompted investigation of its extraterrestrial origin, established by B. Silliman and J. Kingsley at Yale. Comets and meteors and their concomitant application of celestial mechanics appear to have been the major objects of research for nearly two centuries.

With the acquisition of the 5-inch Dollond telescope in 1829, the gift of Sheldon Clark, Yale achieved the enviable distinction of having the largest refractor in America. This reputation was enhanced when Professor Denison Olmsted and Tutor Elias Loomis in 1835, using the telescope, were the first in America to spot Halley's comet upon its expected return that year. In 1833 the magnificent Leonid meteor shower was observed by Olmsted, and at West Point by A. C. Twining, resulting in the first acceptable cosmic theory of the origin of meteor showers. Later, H. A. Newton, comparing the observations of 1833 with much earlier displays, was the first correctly to predict the return of the Leonids in 1866, and showed how perturbations by Jupiter had changed the original course of the swarm.

Unfortunately, Yale's eminence dwindled as Harvard in 1847 acquired its famous 15-inch "great refractor" and the Yale administration slackened its interest in astronomy. Although William Hillhouse in 1856 bequeathed instruments (especially a transit instrument), and in 1858 Mrs. James A. Hillhouse and her daughters gave six acres of choice land in memory of her father-in-law, James Hillhouse, Yale Treasurer 1782–1832, especially for the erection of an observatory, nothing came of these gestures and the able astronomers struggled along as best they could without adequate support, nevertheless achieving noteworthy results based on such observations as their meager equipment allowed. In 1871 Oliver Winchester gave 32 acres of land for the erection of an observatory, most of the land to be sold to provide funding for the proposed buildings. The national financial crisis of 1873 frustrated his plans. On a scale much reduced from the original architect's drawings, the observatory on the corner of Prospect and Canner Streets was finally erected in 1882. Largely to meet operating expenses, the horological and thermometric bureaus were set up by Leonard Waldo. Very successful at first, they, too, ultimately were discontinued.

Meanwhile, new and exciting projects were undertaken. The observatory (for a time known as the Winchester Observatory) under the direction of H. A. Newton, ordered a heliometer, a divided objective telescope such as F. W. Bessel in Königsberg, Germany, had used for the determination of the first reliable published parallax of a star, 61 Cygni, in 1838. The Yale heliometer was the first and only such telescope in America and at that time the largest in the world, with a diameter of 6 inches. The first scientific use made of the new instrument was for the observation of the transit of Venus in 1882. In 1884 W. Lewis Elkin was appointed astronomer in charge of the heliometer. He had spent the previous two years in South Africa working with David Gill at Capetown and learning the techniques of that most experienced expert on the determination of stellar parallaxes with the heliometer. At Yale, Elkin is reputed to have determined more and better parallaxes than anyone before the advent of photography made the visual technique obsolete. Elkin himself was well aware of the potential advances in astrophotography, having analyzed Gould's reductions of Rutherfurd's famous pioneering plates of the Pleiades, and measured some Lick photographs.

At the same time, Elkin made significant strides in the study of meteors. Acting upon a suggestion made much earlier by J. Homer Lane, Elkin built two batteries of cameras equipped with rotating shutters for obtaining the velocities as well as the heights of meteors. His pioneering efforts provided the model for more precise programs instituted at Harvard decades later.

Thus, by the onset of the twentieth-century comets and meteors, celestial mechanics as applied chiefly to orbital calculations, and finally astrometry were the main subjects of interest at Yale. After Elkin's retirement in 1910 until the appointment of Schlesinger in 1920, the Observatory was inactive. In this interval E. W. Brown perfected his lunar theory and worked on his massive lunar tables. No further work of note on meteors has been carried out at the Yale Observatory except for a paper on meteoritic dust by R. Wildt and an undergraduate, P. Hodge, in 1956. After more than a century of preeminence in studies of meteors, this field became effectively nonexistent at Yale.

Schlesinger brought new light and new subjects. He has been described as "the father of modern astrometry." It was mainly he who perfected the photographic methods for parallax determination, and who initiated the programs for "zone catalogues"—the determination of precise positions of tens of thousands of stars in discrete belts around the sky, and their proper motions obtained by intercomparing positions measured at large time intervals, usually on the order of 20 years. With the collaboration of Jan Schilt, first at Yale, later at Columbia, the magnitudes for a high percentage of the stars in the zone catalogues were determined.

While Schlesinger concentrated his own research efforts on astrometric problems, he did not fail to encourage others in different aspects of astronomy. Notably, during his regime, graduate student John Hall was the first to do pioneering work on infrared photometry of variable stars, inventing his own equipment. Following him, Dr. Arthur Bennett improved the equipment and carried out many more observations.

In 1925 Schlesinger established a Southern Station at Johannesburg, South Africa, where he erected a 26-inch Grubb long-focus refractor for the determination of parallaxes of southern stars. This telescope was in operation at Johannesburg until 1952 where 66,456 plates had been taken. It was then moved to Mt. Stromlo where the total number of plates was increased to 69,950. In 1963 Yale left Mt. Stromlo and donated the 26-inch telescope to the Australian observatory. Also as early as 1928 a "catalogue camera" was erected at Johannesburg for extending the zone catalogue work far into the southern hemisphere. Clearly, astrometry reigned supreme during Schlesinger's directorship.

Meanwhile E. W. Brown and Dirk Brouwer were deeply immersed in the problems of theoretical mechanics. When Brouwer took over the directorship in 1941, celestial mechanics was soon to catapult to worldwide preeminence at Yale. This tremendous achievement was, of course, given most significant impetus by the advent of the "space age"; celestial mechanics is vital to the navigation of space vehicles. Astrometry, however, was not neglected. While the work on the parallaxes continued, it did so at a reduced pace. The zone catalogue work, on the other hand, was pushed, although not for the sake of studying the stars as such, as Schlesinger did, but because Brouwer considered the astrometry of the zone catalogues as "the handmaiden of celestial mechanics." The positions and a knowledge of their changes of position provided an essential reference frame for the determination of precise positions of solar-system objects, particularly for the asteroids, and of space vehicles. Stellar astronomers, on the other hand, still viewed the zone catalogue data as most essential to the analysis of galactic structure, galactic rotation, and the assignment of stars to the various population groups.

For a relatively short time Yale indulged in a radio-astronomy program especially concerned with radiations from the planet Jupiter. Although preliminary results were encouraging, the University policy of "publish or perish" was detrimental to projects of long duration, especially the innovative ones. No matter how worthwhile, they do not yield definitive results quickly, so the investigators had to leave. They took their equipment with them to Texas.

During Brouwer's regime, photometry (especially of variable stars) and astrophysics (under R. Wildt) were by no means neglected, although they played a comparatively minor role. After Brouwer's death in 1966 Wildt, as Acting Director, discouraged the continuance of most of the theretofore predominant projects. When Dr. Pierre Demarque came as Director in 1968, the remaining active projects were again encouraged; but now theoretical astrophysics, studies of external galaxies, and cosmology became the major topics of investigation.

Throughout the time span of this survey, a change in directorship generally brought about a change in emphasis to different sub-branches of astronomy. In each era, however, Yale proved itself a leader, at least in America, in whatever field it specialized in at the time.

REFERENCE

1. F. Schlesinger to B. E. Livingston, September 20, 1920, *Yale Obs. Archives.*

2
A COLLEGE IN THE MAKING

Recounting bits of the history of the college seems appropriate for visualizing the setting, not always auspicious, in which astronomy at Yale was first seeded, sometimes neglected, and finally flourished.

Like most early American colleges, Yale was founded, in 1701, with a major goal of preparing young men for the ministry. However, the impetus for founding a college in New Haven came much earlier, as has been described in detail by Oviatt in *The Beginnings of Yale*[1] and by W. L. Kingsley in his voluminous *Yale College, A Sketch of Its History*.[2] More recently G. W. Pierson, official Yale Historian, has researched the history of the founding of Yale College and noted inconsistencies in the available literature.[3] Doubt is thus cast on some of the oft-quoted details. The questioned sequences are traced largely to historical accounts published by Thomas Clap, Rector from 1740 to 1745 and then as first President of Yale (1745 to 1766). He is suspected of having altered facts to suit his own convenience. Clap clung more tenaciously than most of his Congregational contemporaries to early convictions that "Colleges are Religious Societies" for training men for the ministry.[4] Hofstadter,[5] describing sectarianism at Yale during the regime of Clap, remarked that Clap

> provides us with a case study in illiberalism, for there all the animus of sectarian passion was unleashed and there the restrictive measures that were being taken informally, partially, and sometimes rather apologetically at other colleges were made a formal part of the college laws and were openly defended. There an attempt was made to reduce one of the major colonial colleges to the status of a severely sectarian agency.

As intolerant a despot as Clap proved to be in his capacity as administrative head of the college, he did play an important role in the early history of astronomy at Yale (*see* Chap. 3). Except for Clap, the questioned accounts of the founding of Yale have little bearing on the history of astronomy at Yale.

John Davenport (Fig. 2.1), an Episcopalian turned Puritan, had escaped religious persecution in England and come to Boston in 1637 with a group led by Theophilus Eaton, who was to become the first governor of the New Haven Colony. In Boston, Davenport became influential in plans for the young college that a year later was to be named Harvard. Davenport, however, not only wanted to escape from the dictates of the Church of England, he wanted to establish an independent church-state of his own. In April 1638 he and his followers therefore founded the New Haven Colony and settled at the mouth of the Quinnipiac River in what is now New Haven. Among his earliest plans was the founding of a college, but financial and political crises kept educational problems in the background. Nevertheless, Davenport's personal interest in establishing the equivalent of Harvard in New Haven never flagged. Land had in fact been set aside for the support of a college. Finally in about 1657, Davenport wrote Edward Hopkins (distantly related by marriage to Governor Eaton and Elihu Yale's grandmother). Davenport outlined his hopes for a college and hinted at his financial needs. Hopkins' response was favorable but he unfortunately died before he could make funds available. However, his will provided for funds to establish both a college and a grammar school. New Haven finally succeeded in establishing only the school, Hopkins Grammar School, which, after a checkered early existence, has survived to the present day. In 1668 the discouraged Davenport accepted a call to the First Church in Boston, last stronghold of his cherished theological principles, where he died only two years later. It would be nearly another third of a century before his dream for a college in New Haven could be realized.

In 1700, Harvard graduate James Pierpont revived the project of creating a college in New Haven. As the General Assembly of Connecticut Colony was to meet for the first time in New Haven in 1701, a consortium of about ten ministers from throughout Connecticut felt that this would be an opportune time to persuade the Assembly to grant a charter for founding a college. They met in Branford, where

Fig. 2.1. John Davenport (1597–1670). Founder of the New Haven Colony. Among his earliest hopes was the founding of a college like Harvard in the new colony. (Courtesy Yale University Archives).

each is reputed to have brought books for the college they were founding.[3] After much discussion of conflicting opinions, a charter was indeed granted by the General Court of Connecticut and seven ministers were selected as Trustees of the College. On November 11, 1701, the first organizational meeting of the *Colony Collegiate School* was held in the nearby small town of Saybrook. James Pierpont opened the meeting by reading the provisions of their charter, granting the Trustees

> Liberty, and privilege, for the founding, suitably endowing and ordering a Collegiate School, within his Majesty's Colony of Connecticut, wherein Youth may be instructed in the Arts & Sciences, who through the blessing of Almighty God may be fitted for Publick employment both in Church & Civil State.[6,7]

The primary concerns to be discussed were the site of the school, who should be the Rector, and what the rules for the students should be. Admission to the college was to depend wholly upon a reading knowledge of Latin and Greek, to be tested by the Rector and a second minister. The head of the college was to be called Rector, not President. He was to teach theological divinity "but in no other system than the Trustees permitted."[8] In general the curriculum was to follow that at Harvard. At a later meeting Harvard graduate Abraham Pierson of the village of Killingworth (now Clinton) was selected Rector.

While a student at Harvard, young Pierson (B.A. 1668) had studied "Calvinistic theology and a quaintly unsophisticated study of natural laws called 'Physicks.'"[9] Physics (from the 17th through the 19th centuries more generally known as *natural philosophy*) is said to have been Pierson's favorite subject, on which he wrote a textbook, *Physicks*. From this, in their senior year his students were taught the rudiments of mathematics and physics. No copy of the text seems to have survived, but it presumably closely followed what Pierson had been taught at Harvard, which itself was about a century behind what was being taught in England. As Oviatt comments, "But for the first decade and more of the Collegiate School's career, it continued to give the orthodox education of the times, untouched by what was going on in the outside English speaking world, and remote from its cultural influences."[10] Pierson's students were still being taught that the sun revolves around the Earth.[11] This was more than a century and a half after the death of Copernicus, three-quarters of a century since Kepler's, and two decades after the first publication of Newton's *Principia*. But as historian of science, T. S. Kuhn, comments, "Not until 1822 did the [Catholic] Church permit the printing of books that treated of the earth's motion as physically real, and by then all but the most rigidly orthodox Protestant sects had been long persuaded."[12] However, he added, "By mid-seventeenth century it is difficult to find an important astronomer who is not Copernican; by the end of the century it is impossible."[13]

Finding the college again in serious financial straits, Pierpont, shortly before his death in 1714, had taken it upon himself to write the Massachusetts Colony Agent in London, Jeremiah Dummer (Harvard B.A. 1700), on the Collegiate School's behalf.[14] The socially prominent Dummer then contacted wealthy and influential friends, including Edmund Halley, Isaac Newton, and even Elihu Yale. Dummer succeeded in collecting or having sent to the college some 1,000 volumes, mostly theological, many literary, and some scientific. Among the latter were personal copies from Isaac Newton of his *Opticks* and the second edition of his *Principia* published in 1713.

The arrival of the books in 1714 brought to a head the necessity for permanent housing not only for the valuable library but also for the students and tutors. The college had been loosely organized, students being permitted to go for their training to whatever Congregational parsonages they preferred. When the General Assembly voted funds for a building, much haggling among the Trustees ensued, with some bitterness. The three factions favored respectively, Hartford, Saybrook, and New Haven, for which each attempted fund drives to enhance its own interest. New Haven not only raised the largest amount but also provided land for the proposed building. At long last, in 1717, a majority vote among the Trustees established the permanent site of the Collegiate School in New Haven. As funds were insufficient for the completion of a building, the Trustees again turned to Dummer and to Yale. Largely at the suggestion of Cotton Mather at Harvard, Elihu Yale's contributions resulted in the naming of the college at its Commencement in September 1718 as *Yale College.*

When the cartons of books from Dummer had arrived in Saybrook in 1714, Samuel Johnson (Yale 1714) was still a student there. Later (1716–19) he became a tutor at Yale, and ultimately the President of Kings College, now Columbia University. (This was not the famous English lexicographer Samuel Johnson, whom James Boswell eulogized, although they became acquainted through the American's son, Wm. Samuel Johnson, in the late 1770s.[15]) The books were of tremendous fascination to the young student. In particular he was drawn to Newton's *Principia,* a treatise for which his previous training ill prepared him and he assiduously aspired on his own to make up for his deficiencies. According to Oviatt,

> He had, so he says, been graduated from Saybrook with a very large conceit of his intellectual attainments. He had drawn up an elaborate system "of all the parts of learning within his reach," and considered himself a learned man. But his first plunge into Sir Isaac Newton had shattered this high opinion of himself, and with an intellectual energy that came from the needs of his starved mind, he had reconstructed his notions of the universe and therefore his theology. He seemed to himself, immediately this great horizon dawned upon him, "suddenly emerging out of the glimmer of twilight into the full sunshine of the open day."[16]

This last sentence his biographer, E. Beardsley, attributed to Johnson's having discovered and devoured Francis Bacon's *Instauratio Magna*[17] before his graduation.[18] In any case, not only was Johnson beginning to comprehend modern science, but his religion, also, as might have pleased Episcopalian Elihu Yale, was beginning to deviate from his inherited beliefs. Nor was he alone: a classmate and future Yale tutor, Daniel Browne, had like sentiments. The Trustees at first either overlooked these departures from earlier teachings, or, in the case of the sciences, felt they were not important. But when they became aware of the theological deviations they finally dismissed Johnson.

In 1744 Johnson published a pamphlet, *An Introduction to the Study of Philosophy* (Fig. 2.2), in which he described philosophy as "the Study of Truth and Wisdom, in the Pursuit of Happiness" and in which he outlined three divisions, Rational, Natural, and Moral Philosophy. *Natural* he described as partly speculative and partly practical, and he elucidated: (1) it treats in general with mathematics and mechanics of quantity, bodies and motion; and (2) explains and accounts for all the phenomena in both heaven and earth in all parts of physics and astronomy.[19] To this pamphlet was appended a three page catalogue of relevant literature by Yale's President Clap[20] who probably used it in his own classes.

For all of Johnson's achievements, President Ezra Stiles in his diary wrote of Johnson, "He was a very indifferent Writer, especially of Sermons. But a very considerable Reader all his Days. . . . He loved to see what was going forward in the learned World, but was not himself very learned."[21] Most damaging was Stiles' opinion, "Dr. Johnson was always of the opinion of the last Book he read." In his pursuit of science this was probably fortunate.

[23]

The TABLE.

PHILOSOPHY is the Study of Truth and Wiſdom, in the Purſuit of true Happineſs.

And it is divided into THREE PARTS.

I. RATIONAL; which teaches us to cultivate our *Rational Powers*; viz.

1. *Reaſon*, in *Logic*; to which are referred *Ontology* or *Metaphyſics*, and *Dialectic*.
2. *Speech*, in *Grammar* and *Rhetoric*; to which are refered *Oratory*, *Hiſtory*, and *Poetry*; and, to all of them, the *Art of Criticiſm*.

II. NATURAL; which teaches the Knowlege of the *Natural World*, and the Practices thereon depending; being therefore *partly Speculative*; and partly *Practical*; and is,

1. *General*; which treats of *Quantity*, *Bodies*, and *Motion*, in general, in *Mathematics* and *Mechanics*.
2. *Special*; which explains and accounts for all the *Phænomena*, both in *Heaven* and *Earth*, in all the Parts of *Phyſics* and *Aſtronomy*.

III. MORAL; which teaches the Knowlege of the *Moral World*, and the Practices thereon depending; being partly *Natural*, and partly *Revealed*; and is,

1. *Speculative*; which contemplates the Nature of *created Spirits*, and of GOD, the *Creator* and Lord of all Things, in *Pneumatology* and *Theology*.
2. *Practical*; which teaches the Practice of what is neceſſary to our complete Happineſs, both here and for ever, in *Ethics*, *Oeconomics*, and *Politics*, both Civil and Eccleſiaſtical.

A

Fig. 2.2. A page from Samuel Johnson, *An Introduction to the Study of Philosophy,* London, 1744.

REFERENCES

1. E. Oviatt, *The Beginnings of Yale,* 1701–1726, New Haven, 466 pp., 1916.
2. W. L. Kingsley, *Yale College, A Sketch of Its History,* Vol. **1**, 504 pp.; Vol. **2**, 533 pp., 1879.
3. G. W. Pierson, *The Founding of Yale, The Legend of the Forty Folios,* xxi + 275 pp., New Haven and London, 1988.
4. R. F. Butts and L. A. Cremin, *A History of Education in American Culture,* p. 81, 1953.
5. R. Hofstadter, *Academic Freedom in the Age of the College,* N.Y., 1955, 6th prtg., p. 163, 1969.
6. Ref. 1, p. 197.
7. *Historical Register of Yale University, 1701–1937,* p. 7, 1939.
8. Ref. 1, p. 199.
9. *Idem,* p. 266.
10. *Idem,* p. 275.
11. *Idem,* p. 422.
12. T. S. Kuhn, *The Copernican Revolution,* Cambridge, Mass., 1957, 4th prtg., p. 109, 1971.
13. *Idem,* p. 227.
14. Ref. 1, p. 290.
15. E. E. Beardsley, *Life and Correspondence of Samuel Johnson, D.D.,* 3rd ed., pp. 361–65, 1887.
16. Ref. 1, p. 398.
17. F. Bacon, *Instauratio Magna, The Advancement of Learning,* 1605.
18. Ref. 15, p. 6.
19. S. Johnson (no author indicated on title page), *An Introduction to the Study of Philosophy: Exhibiting a General View of All the Sciences, for the Use of Pupils,* vi + 26 pp., 1744.
20. Ref. 15, p. 123, footnote.
21. F. B. Dexter, *Biographical Sketches of Graduates of Yale College,* N. Y., **1**, 125, 1885.

3
FIRST CENTURY OF ASTRONOMY AT YALE

At first the curriculum at Yale was strictly classical, with only minor attention to astronomy. Students were permitted to choose whether they would learn the astronomy of Ptolemy or of Copernicus.[1] Yale was founded 41 years after Harvard stopped teaching that the sun revolved around the Earth. As most of the early curriculum at Yale was modeled after Harvard, and most of the instructors were Harvard graduates, one might have expected the Copernican system to have been in vogue at Yale from the beginning.[2] However, real interest in the sciences at Yale started only after the acquisition of the Dummer books for the library in 1714, when Samuel Johnson and Timothy Cutler first looked into Isaac Newton's *Principia* and *Optics*.

F. E. Brasch[3] writing of the Newtonian era in the American Colonies points out that Cotton Mather (1663–1728) was "the first scholar in the colonies to catch the real spirit of Newton's philosophy," even though neither Harvard nor any of its professors at the time possessed any copy of the *Principia,* first published in 1687. However, a Harvard man, Thomas Brattle, had in 1680 provided Newton with observations of comets which Newton used for demonstrating the gravitational influence of the sun. This, and his being a member of the Royal Society with many contacts with British savants, accounts for Mather's early knowledge of the gravitational theory. Yale's was the first copy of the *Principia* to arrive in America.

While still the Rector (1719–22), Cutler had taught some mathematics and elementary physics. "Common arithmetic and a little surveying was the *ne plus ultra* of mathematical requirements."[4] In order that the students should understand Newton and Locke, he had added geometry and astronomy. For Freshmen and Sophomores, logic, Greek, Hebrew, and Latin were required. For Juniors, besides these, physics was also required and, in their Senior year, metaphysics and mathematics. But Johnson and Cutler, the two teachers who were the earliest to advance an understanding of physical sciences at Yale, were quickly severed from the young college for their religious deviations.

For fully understanding the theory of gravitation, Newton's new mathematics, fluxions (now called the calculus), had to be mastered. As all of the tutors (the title of professor was not used until 1755) at that time were primarily theologians, they were more concerned with reconciling the new theory with accepted religious principles. As in the case of Einstein's theory during the first half of the twentieth century, those few who did comprehend the new theory were held in awe. Not until Thomas Clap became Rector of Yale in 1739, and President in 1745, a position he held until 1766, did anyone after Johnson and Cutler show real enthusiasm for mastering the subject of gravitation, and even Clap does not appear to have stressed it with his students. However, Ezra Stiles (1727–95), who had been Clap's student from 1742 to 1746, stated in a eulogy,[5,6] "I have known him to elucidate so many of the abstrusest theorems and ratiocina of Newton that, I doubt not, the whole *Principia* of that illustrious philosopher was comprehended by him; a comprehension which, it is presumed, very few mathematicians, of the present age, have attained."

When Clap first became Rector he revised the students' curriculum, decreasing the amount of time to be spent on logic and introducing more natural philosophy and mathematical science. He gave instruction in conic sections, fluxions, surveying, navigation, and the calculation of eclipses.[7]

An early text discovered in the Yale Library, one that might possibly have been used in Yale classes, E. Wells's, *The Young Gentleman's Astronomy*[8] presents in its preface the philosophy of education at that time.:

The most proper method to make Young Gentlemen *Learned,* is this; to teach them at first only such Elements

> of the liberal Arts and Sciences, as are *most useful* in the common Affairs of Life, and withal *most easy* to be known. They have a competent Apprehension of the Usefulness of such Things as occur in the common Concerns of Life; and consequently hereby that Question frequently put by Young Students, *of what Use is this?* will be answered aforehand, and so they will be rendered *willing* to understand what they apprehend the Use of. And when they find that the Understanding thereof carries in it no Difficulty, then they will be also *encouraged* to proceed. And when they have thus gone through, and become Masters of the *most useful* and *easy* Elements of the liberal Arts and Sciences, they will thereby be enabled with much more Ease to conquer the more *difficult* Parts of Learning, if their own Inclinations shall lead them thereto hereafter, when they are come to riper Years, and so can judge more rightly of the worth of Learning.

Curiously, the names of Kepler and Newton are nowhere mentioned, even though Kepler's laws and gravity are invoked in support of the "Copernican Hypothesis" as it was still known. Interestingly, at that early date, the chapter on comets states,

> It is not known yet, whether Comets belong only to this our Solar System, or whether they may not also pass into other of the Mundane Systems, which have the fixed Stars for their respective Suns . . . It is supposed most probable by the Learned in *Astronomy,* that they move in some conic Section, which has the Sun in one of its Focus's.

One text that was used by Clap[9] (and also as late as 1793 by Stiles[10]) was by the English theologian and poet, Isaac Watts (1674–1748), *Knowledge of the Heavens and Earth Made Easy; or The First Principles of Astronomy and Geography by the Use of Globes and Maps* (London 1726). In 1738 Watts had presented to Yale both a celestial and a terrestrial globe which Clap then used for teaching students how to solve problems such as the positions of the sun in the ecliptic, and for predicting times of sunrise and sunset.

The first attempts by the Trustees to acquire "philosophical apparatus" were in 1734 when the Trustees and other gentlemen subscribed to the purchase of a reflecting telescope, microscope, barometer, air pump and sundry other articles. Joseph Thompson of London donated a complete set of surveying instruments.[11] As a good teacher, Clap appreciated the value of models. He built a 7-foot diameter primitive but effective orrery in 1743 for illustrating the motions of the planets and, as an opportune innovation, added a parabola for Halley's comet, and later another, probably for the bright comet that appeared in 1743–44. Clap's orrery, or planetarium, is reputed to have been the first built in America.[12] It was ruined during the war in 1779.

As elsewhere, research was slower than teaching in getting started at Yale. The earliest publications seemed restricted to publications of almanacs. Between 1720 and 1809 over fifty were published by students, tutors, and associates. Teaching predominated for more than a century. However, Thomas Clap showed a keen interest in mathematics and astronomy. This interest was aroused when, on November 24, 1742, a brilliant daylight detonating fireball was observed from places as widely separated as Norwalk, Connecticut and Braintree, Massachusetts, a span of some 130 miles. Although he, himself, only heard but did not see the fireball, his interest was strongly stimulated. He took great pains to collect all the observations he could throughout New England. This led to his interpretation of bright fireballs as "terrestrial comets," a theory first apparently formulated about 1756, but not published until 1781, by Stiles.[13,14] Three subsequent brilliant daylight fireballs, on May 10, 1759, May 10, 1765, and September 18, 1765, added further fuel to his theory. In brief, Clap supposed that detonating meteors were solid bodies, about half a mile in diameter, traveling in elliptical orbits around the Earth, at minimum altitudes of 20 or 30 miles. As such a body entered the Earth's atmosphere friction caused rumbling, explosive noises, and luminous effects. Yet all this was not considered sufficient to slow the meteor to the extent of preventing its escape, and it supposedly continued to circumnavigate the Earth.

Original as Clap believed his theory to be, he found reason to suspect another (John Pringle in England[15,16]) of stealing it; and he was himself suspected of plagiarism when Stiles in 1766 discovered a reference in the October 1755 *Gentleman's Magazine* to a fireball of August 1755. The author "B.J." suggested that if the object had been moving parallel to the horizon at an initial velocity of 4.95 miles per second it would assume the motion of a satellite revolving around the earth in a circular orbit. Tactfully, Stiles wrote Clap, "I imagine that Mr. B. J. and yourself hit upon the same hypothesis by separate efforts of genius." Nevertheless, in an obituary Stiles wrote of Clap,[17] "In mathematics and natural philosophy, I have no reason to think he was equalled by any man in America, except the most learned professor Winthrop of Harvard." Neither Clap nor Stiles seems to have commented on a letter from Winthrop to John Pringle published in the *Philosophical Transactions*[18] in 1764. There Winthrop pointed out that an Earth-circling satellite at a height of 40 or 50 miles would have a velocity of only 7 to 10 miles a second whereas the fireball with an estimated velocity of 30 miles a second would have had to be orbiting the sun.

The original manuscript of Clap's dissertation, *Terrestrial Comets,* was lost when a trunkful of his manuscripts was carried off by the British when their troops ransacked New Haven during the Revolution; they dumped it into the harbor at Fairfield. Clap had probably written it as one of his lectures for students. Fortunately it had been copied by the Rev. Ebenezer Baldwin,[19] a tutor in the College, and this copy, discovered after Baldwin's death, is probably what Stiles later had published.

Dubious as Clap's theory may seem today, it has recurred repeatedly in the intervening two centuries, not, however, by interpreting fireballs as comets but some of them as Earth satellites. In the mid-nineteenth century several fireballs were reported as circling the Earth, merely transiting through the Earth's atmosphere. But inevitably inaccurate estimates of their velocity, and neglect of air resistance in the orbital computations, led to wrong conclusions.[20] For a long time the famous meteoric procession of February 9, 1913,[21] appeared to be a convincing case, but it too proved controversial. Wylie[22] in 1939 re-examined the available data and found the Earth-circling assumption untenable. Twenty years later, in 1959, O'Keefe[23] made a critical investigation of the old data and Wylie's computed orbits, and reached the surprising conclusion that the meteors of February 9, 1913, were indeed traveling in a nearly circular satellite orbit around the Earth. Extensive systematic searches by Clyde Tombaugh[24] in the mid-twentieth century again failed to reveal any small earth satellites.

Clap was particularly intrigued by the predictions for the return of the comet of 1682, later known as Halley's. Halley had only indicated that the comet would return sometime by the end of 1758 or the beginning of 1759. Clap and Chauncy Whittelsey, a tutor, recomputed the orbit, predicting perihelion on September 14, 1758, in contrast to Clairault's prediction of April 15, 1759.[25] Brian Marsden's[26] recent *Catalogue of Cometary Orbits* tabulates the perihelion date as March 13, 1759. The comet was first seen in Europe on Christmas, 1758; Winthrop at Harvard saw it on April 3, 1759, but to Yale's disappointment, Clap did not see it until April 29.

In 1743 Clap had acquired from the London spectacle-maker, Edward Scarlett, what was at the time described as a large telescope, diameter not stated, but presumably a Gregorian type speculum reflector of 2.5-foot focal length. It was provided with "a glass to be screwed on to look at the sun."[27] Clap taught the students to use this, and a theodolite he acquired later. It was probably with this telescope that Clap observed the transit of Mercury on October 25, 1743, as well as a faint fast-moving comet in January 1760.[28] Early in the 1750s there was no observing with the telescope as it had been returned to England for "new specula." During the course of the Revolution the telescope deteriorated badly and was declared useless. Ezra Stiles was unable to use it for observing the transit of Mercury on November 5, 1789. During his presidency, the College in 1789 purchased a new larger reflector, 3.5 feet long, with magnifying powers of 50 to 400, and a micrometer. This telescope was unfortunately lost in 1822 in a shipwreck off the coast of Ireland when Prof. Alexander M. Fisher (1794–1822) was en route to England to return the telescope for repairs. For the next eight years Yale was without a telescope.

Nehemiah Strong (Yale 1755), who had been one of Clap's pupils and then a tutor (1757–60), was Professor of Mathematics and Natural Philosophy from about 1771 to 1781. Plans for funding the badly needed endowed professorship had been originated and carried through the legislature by James Hillhouse, without whom, W. L. Kingsley comments, "nothing would have been or could have been done."[29] Strong was only the second man at Yale to be given the title Professor; the first was Naphtali Daggett in 1755, Professor of Sacred Theology, subsequently President of Yale. Although scientifically highly qualified, Strong was unpopular among the students, both because of his manners and later for his reputed Tory loyalty during the Revolution. According to Kelley[30] the Yale Corporation had a policy of getting rid of those it disliked, and hence pushed Strong out of office in 1781. The ostensible reason for his termination, however, was lack of funds. The position was not filled again until Josiah Meigs (Yale 1778) was appointed Professor in 1794. Meanwhile, Strong in 1784 published a short treatise of 53 pages, *Astronomy Improved, or a new theory on the harmonious regularity observable in the mechanism or movements of the planetary system,* of which Fulton[31] states, "Here for the first time in an American text, one finds a clear statement of the laws of gravitation as they operate in the planetary system."

Ezra Stiles (1727–95) was President of Yale from 1778 until 1795. His successor, Timothy Dwight, eulogized Stiles as "probably the most learned man in

America, at the time of his death; and was probably excelled by few in the world."[32] Among Stiles' varied interests was always an enthusiastic attention to astronomy. He had entered Yale as a student in September 1742 at age 15. As an under-privileged but able student he was befriended by Clap. During his undergraduate years he observed the comet of 1743–44 and investigated the number, periods, distances, velocities, and other properties of comets.[33] He received his A.B. and M.A. degrees in 1746 and 1749, and then was appointed a Tutor at Yale. He served from 1749 until 1755 when he was called to Newport, Rhode Island, as pastor of its Second Congregational Church, a post he held with love and distinction until he was persuaded to return to Yale as successor to President Daggett.

While in Newport, Stiles evidently had kept in touch with Clap, sharing interests and observations. Stiles, like Clap, took great interest in the predictions and search for Halley's comet, being particularly interested in the inequality of successive cycles, attributed to perturbations by planets, according to Newton's law. On this he corresponded with Winthrop at Harvard,[34] who replied, "You have very happily conjectured, at what will probably turn out to be the truth, that the same cause which contracted the last, will probably prolong the present period."

On June 3, 1769, Stiles, still in Newport, observed the transit of Venus with a sextant he had ordered to be made especially for the occasion. The major purpose of observations of the transit of Venus was for the determination of the parallax of the sun. Unfortunately, the observations by Stiles and his associates were subject to large uncertainties[35] and did not serve their intended purpose.[36,37]

Described as "one of the most learned American divines,"[38] Stiles stimulated and encouraged students in whatever fields interested them most and, like Clap, befriended those in need. Among his students was Josiah Meigs, class of 1778, an effective and popular orator on many subjects, whom Stiles appointed Tutor in Mathematics, Natural Philosophy and Astronomy. Meigs held the post from 1781 to 1784, the first appointment specifically for professing astronomy, and as Professor of Mathematics and Natural Philosophy from 1794 until 1801, when he became president of the new college at Athens, Georgia.[39]

During the Revolution much of the so-called philosophical apparatus either disappeared or fell into disrepair. Stiles tried raising funds for replacements but it was several years before enough was contributed, and globes, a microscope, the 3.5-foot telescope, and a new orrery were ordered from England. Meanwhile a student, Joseph Badger (Yale 1785) in 1784 had built a 6-foot diameter orrery with gearing to show the motions of the planets (including Uranus, discovered in 1781) and the orbits of three comets.[40]

The American Revolution and the difficult postwar years encroached heavily on Stiles's administrative duties, and astronomy could not be vigorously pursued. Until the end of Yale's first century we find reports of observations (in addition to those already mentioned) of the fireball of October 17, 1788 (observed by Amos Bassett, Yale 1784, later also a tutor), which reawakened interest in Clap's theory. There were observations of the comets of January 6, 1760, August—September 1768, June—July 1770, and December—May 1783–84, and a report on a futile search for one suspected by Maskelyne in 1789. Also observed were the total eclipse of October 27, 1780, and the partial of August 27, 1783, for which Elizur Goodrich, Secretary of the Yale Corporation, had computed the expected times of contact for New Haven. But no theoretical work was published.

The eclipse of October 27, 1780, could not have been seen as total from New Haven, according to Oppolzer's *Canon der Finsternisse.*[41] The records of observations by Nehemiah Strong and Stiles have evidently been lost.[42] In contrast, Harvard astronomers who likewise failed to see totality, nevertheless made a noteworthy early discovery. Because the eclipse occurred during the Revolutionary War, Harvard observers had obtained special permission from the British to observe from Penobscot Bay where it was predicted to be total. Unfortunately their predictions were not sufficiently accurate and the eclipse was not total even that far north. However, the Harvard observers were so close to the edge of totality that they observed Baily's beads 56 years before Baily described them.[43]

The next eclipse of interest in Yale history did not occur until the annular eclipse of September 17, 1811. Jeremiah Day and J. L. Kingsley[44] observed it with the 3-foot focal length reflector for the purpose of obtaining the as yet undetermined longitude of the College, at the site of the Lyceum. From twelve observations of the altitude of the sun at meridian passage they determined the latitude as +41 18′. Their observations of the times of contact at the beginning and end of eclipse yielded the longitude as $4^h\ 51^m\ 49.8^s$. They commented that small errors might be present, arising from uncertainties in the available lunar tables. E. Loomis later estimated that these coordinates could be in error by some minutes of arc in latitude and 10 to 15 seconds of time in longitude. He himself, in 1835, used a sextant and small transit instrument to determine the coordinates of the Athenaeum[45] (about 250 feet northerly from the Lyceum) and obtained $4^h\ 51^m\ 40^s\ +41^\circ\ 18'\ 28''$. Still later H. A. Newton[46] described this longitude as differing by less than 2^s from the best determination available in 1890.

Toward the end of his life Stiles attempted to initiate an investigation into ancient oriental astronomy. Believing "the Zodiac and much of astronomy to be antediluvian"[47] and at the same time being anxious to ascertain if the missing tribes of Israel had wandered into India and China, he wrote, "I have great hopes from the Astronomers, who, last year, accompanied the Ambassador from the King of England to the Emperor of China." He hoped that contact with the Calcutta Society would clarify the problem. His correspondents, however, had died before his letter and supporting manuscript could reach them; and the reply by the President of the Calcutta Society came shortly after Stiles himself had died.

The Life of Ezra Stiles was published in 1798 by the Rev. Abiel Holmes, Stiles' son-in-law and father of the more famous *Amiable Autocrat,*[48] Oliver Wendell Holmes, Sr. Abiel Holmes (Yale 1783) had been a student under Stiles, later a Tutor at Yale, and ultimately Pastor of the First Church of Cambridge, Massachusetts. Stiles had willed Holmes the use of his cabinet of manuscripts, comprising forty volumes, from which Holmes quoted freely. Of Stiles, Holmes says,

> While a resident graduate and a tutor, together with his pursuit of the various branches of literature, and of theology, he paid particular attention to the study of philosophy and astronomy; and his experiments in the one science, with his calculations in the other, comprise a manuscript quarto volume. . . . He delighted equally in the acquisition and the communication of science.[49]

In the funeral oration for Stiles, another of his former students, Josiah Meigs, remarked, "He was well versed in most branches of mathematical knowledge. In natural philosophy and astronomy he excelled." With the exception of sacred literature, Meigs commented, astronomy was Stiles' favorite science.[50]

From the foundation of Yale in 1701 through 1795, Yale graduated 2,372 students of whom 616 (26%) were students during Stiles's 17 years as President.[51] All were presumably taught some astronomy for cultural reasons, though few made a career of it (J. Meigs, B.A. 1778, and J. B. Day, B.A. 1795, being the outstanding exceptions, as well as College administrators).

Timothy Dwight (1752–1817) was Ezra Stiles's successor as President of Yale, taking office in September 1795. At that time there were 110 students in the College. While himself a student, Dwight (Yale 1869) had shown a keen interest in the sciences. From 1771 to 1777, as Tutor, he stressed sciences as essential to a well-rounded education, and he carried out mathematical and scientific studies, but from textbooks, not laboratory experiments. As McKeehan[52] points out, upon becoming President, Dwight was interested in science as an administrator who wished for Yale to be strong in many fields. Several new professorships were therefore established, among them Professor of Chemistry and Mineralogy, to which Benjamin Silliman was appointed in 1802. Dwight himself attended Silliman's lectures as assiduously as did the students.[53]

Jeremiah Day, who took over the chair of Mathematics and Natural Philosophy after the resignation of Meigs in 1801, found a significant shortage of adequate textbooks. He therefore assiduously wrote his own, covering algebra, mensuration, plane trigonometry, and navigation and surveying. Some of these were used at other colleges even fifty years later.[54]

REFERENCES

1. J. F. Fulton, *Bull. Hist. Medicine,* **20**, 100, 1946.
2. *cf. Yale Alumni Weekly,* Feb. 5, 1937.
3. F. E. Brasch, *The Newtonian Epoch in the American Colonies* (1680–1783), Worcester, Mass., 1940; reprinted from *Proc. Amer. Antiquarian Soc.,* n.s. **49**, 314, 1939.
4. E. Oviatt, *The Beginnings of Yale, 1701–1776,* New Haven, p. 423, 1946.
5. E. Baldwin, *Annals of Yale College from Its Foundation to the Year 1831,* 2nd ed., New Haven, p. 92, 1838.
6. A. Holmes, *The Life of Ezra Stiles,* Boston, p. 394, 1798.
7. W. L. Kingsley, *Yale College, A Sketch of Its History,* N. Y., **1**, 66, 1879.
8. E. Wells, *The Young Gentleman's Astronomy,* London, 2nd. ed., 1718.
9. B. M. Kelley, *Yale, A History,* New Haven, p. 79, 1974.
10. E. S. Morgan, *The Gentle Puritan, A Life of Ezra Stiles,* New Haven, p. 52, 1962.
11. Ref. 7, p. 62.

12. H. M. Fuller, *The Philosophical Apparatus of Yale College,* in *Papers in Honor of Andrew Keogh, Librarian,* New Haven, p. 168, 1938.
13. L. W. McKeehan, *Yale Science—The First Hundred Years, 1701–1801,* New York, 82 pp., 1947.
14. H. A. Smith, *Sky and Telescope,* **53**, 420, 1977.
15. Ref. 13, p. 35.
16. J. Pringle, *Phil. Tans.,* **51**, 259, 1759.
17. J. L. Kingsley, *Amer. Quart. Register,* **8**, 13, 201, 1835.
18. J. Winthrop, *Phil. Trans.,* **54**, 185, 1764.
19. Ref. 5, p. 88.
20. J. Ashbrook, *Sky and Telescope,* **14**, 334, 1955.
21. C. A. Chant, *Jour. Roy. Ast. Soc. Canada,* **7**, 145, 438, 1913; Abstract by G. J. Burns, *Jour. Brit. Ast. Assoc.,* **24**, 109, 1913.
22. C. C. Wylie, *Pop. Ast.,* **47**, 291, 1939.
23. J. A. O'Keefe, *Jour. Roy. Ast. Soc. Canada,* **53**, 59, 1959.
24. C. Tombaugh, J. C. Robinson, B. A. Smith, and A. S. Murrell, *The Search for Small Natural Earth Satellites,* New Mexico State Univ., 1959.
25. *cf.,* C. P. Olivier, *Comets,* Baltimore, pp. 97–100, 1930.
26. B. Marsden, *Catalogue of Cometary Orbits,* Smithsonian Astrophys. Obs., 4th ed. 1982.
27. Ref. 12, p. 170.
28. Ref. 13, p. 18, 71.
29. Ref. 7, p. 109.
30. Ref. 9, p. 100.
31. Ref. 1, p. 102.
32. Ref. 5, p. 130.
33. Anon., *Amer. Quart. Register,* **8**, 195, 1836.
34. Ref. 6, p. 67.
35. Ref. 9, p. 153–57.
36. H. Woolf, *The Transits of Venus,* Princeton, p. 79, 1959.
37. A. E. Lownes, *Sky and Telescope,* **2**, 3, 1943.
38. J. L. Kingsley, in Jared Sparks' *Library of American Biography,* Ser. II, **6**, 1–79, 1845.
39. *See also* Chap. 26.
40. Ref. 10, p. 379.
41. T. Oppolzer, *Canon der Finsternisse,* Vienna, 1887; Dover, N.Y., lax + 375 pp., + 160 charts, 1962.
42. Ref.13, p. 49, 74.
43. S. A. Mitchell, *Eclipses of the Sun,* N.Y., 4th ed., p. 129, 1935.
44. J. Day and J. L. Kingsley, *Mem. Conn. Acad. Arts and Sci.,* **1**, 353, 1812.
45. E. Loomis, *Amer. Jour. Sci.,* **34**, 309, 1838.
46. H. A. Newton, *Elias Loomis, LL.D.,* Memorial Address, New Haven, p. 13, 1890.
47. Ref. 6, pp. 331–38.
48. E. M. Tilton, *Amiable Autocrat,* N.Y., 470 pp., 1947.
49. Ref. 6, p. 30.
50. ———, *Idem,* p. 354.
51. ———, *Idem,* p. 432.
52. Ref. 13, p. 57.
53. Ref. 5, p. 138.
54. Ref. 7, p. 115.

4
STONES FROM HEAVEN

Fireballs depositing meteorites on the surface of the Earth had occasionally been observed from time immemorial. Their origin was superstitiously attributed to either a benevolent or a wrathful deity, depending, especially in times of war, upon whose territory the object happened to fall. For example, the meteorite that fell in Ensisheim, Alsace, in 1492 was proclaimed by Emperor Maximilian of Germany as a sign of God, directed against his enemies the Turks. This 55-kilogram stone is displayed in the town hall of Ensisheim[1,2]; it is the earliest fall that has actually been preserved. After W. Herschel in 1787 supposed that he saw active volcanoes on the Moon, some imagined that meteorites came from the Moon.[3] Other early attempts to account for meteorites assumed that somehow they were the effects of atmospheric lightning. In 1810 Jeremiah Day, Professor of Mathematics and Natural Philosophy from 1801 to 1817 and subsequently President of Yale, summarized the "very liberal supply of hypotheses" that had from time to time been offered to explain the origin of meteoritic stones:[4]

(1) "The materials of which meteoric stones are composed, are raised into the air in a state of exhalations or gases—that, in the upper regions of the atmosphere, they are occasionally collected in great abundance—that, some of them being inflammable, a combustion takes place—and the particles of the whole, by their mutual attractions, rush together, and form a mass, which descends by its weight, to the ground."
(2) "The meteoric stones are masses of matter thrown from volcanoes."
(3) Supposed to be sent from lunar volcanoes.
(4) President Clap's theory of terrestrial comets.

Day discusses objections to all of these, stating,

The inquiry has commenced, with a number of suggestions, which may be true; but which must be left, to be confirmed or refuted, by subsequent observations . . . The first step toward an important discovery, is often an ingenious conjecture . . . It must be granted, that but little progress has as yet been made, in explaining the origin, nature, and use, of the bodies, from which the atmospheric stones proceed . . . Whether the mysteries of the subject will be unveiled, upon a farther investigation, time must determine.

However, it was not until mid-nineteenth century that meteorites were generally accepted as stemming from extraterrestrial sun-orbiting bodies.[3,5]

The first meteorite fall recorded in the United States occurred at Weston, Connecticut, on December 14, 1807. (The date is sometimes erroneously given as Dec. 13, 17, or 26.) A brilliant detonating fireball, about half the apparent diameter of the moon, was seen rising in the north and moving more slowly than ordinary meteors. Passing intermittently behind clouds, it flashed and scintillated, leaving a conical train of maximum width about 10 or 12 diameters greater than the head. Fading gradually, it disappeared about 15° short of the zenith. The total flight duration was estimated at 30 seconds; then 30 or 40 seconds later, detonations as from cannon shot, followed by rumbling sounds were heard. The fireball was observed from Canada to New York, from Salem, Massachusetts, to the Delaware River. At the time it was estimated to be from several hundred feet to a quarter of a mile in diameter. Stones fell in three areas over a span of ten miles. The largest stone weighed about 200 lbs, but it was broken on impact.[6–8]

This fall was thoroughly investigated by Benjamin Silliman (who had not actually seen the fireball but heard of it only three days later) and J. L. Kingsley, at Yale. Silliman included a chemical analysis of the meteorites,[9] indicating that their composition differed from known terrestrial rocks; in particular, their nickel—iron content was remarkable.

Yet incredulity ran rampant. To no less an authority than Thomas Jefferson, himself a credible scientist, is attributed the oft-quoted cryptic statement, "It is easier to believe that two Yankee Professors would lie than to admit that stones could fall from heaven."[7,8] The origin of this precise quotation is obscure. S. L. Mitchill, in a eulogy on Jefferson in 1826,

Fig. 4.1. Jeremiah Day (President 1817–46), Theodore Woolsey (President 1846–71), and Benjamin Silliman. Commencement 1860. (Courtesy Yale University Archives).

said that when Jefferson was told of the fall and shown a specimen, he immediately declared, "It is a lie."[10,11] The news had first reached Mitchill, then a member of Congress, in a letter from his wife in New York.[12] Mitchill also corresponded with H. Holley and J. Bronson who had investigated the sites of fall[13] and sent Mitchill the specimen that was shown Jefferson. C. U. Shepard, in a memorial address at the unveiling of the bronze statue of Silliman at Yale on June 24, 1884, stated,[14] "The fall of the Weston meteorite . . . enabled Professor Silliman, in conjunction with Professor Kingsley, to describe a phenomenon whose reality was until then universally doubted, the great Jefferson even having said of the reported fall that 'it is more natural for men to lie than for stones to fall from the sky.' "

Spontaneous oral statements are almost always impossible to verify and are subject to future embellishment. Thus Jefferson was known to dislike Yankees, and having been recently fooled by a reported discovery of a salt mountain about 1,000 miles up the Missouri, which proved to be false, he was naturally suspicious of any report of something unusual.[15,16] (In his explorations of the Louisiana Purchase, William Dunbar[17,18] had recorded many saltlicks; and some areas of the Louisiana Purchase are now known to contain many saltdomes, but no such feature large enough to have been designated as a mountain.)

That Jefferson should have made the statement imputed to him is more difficult to understand when one remembers that in 1800 he presented at a meeting of the American Philosophical Society, of which he was President, a paper by William Dunbar[19] on a spectacular fireball which landed at Baton Rouge, Louisiana, on April 5, 1800. Described as being the "size of a large house," the object was seen flying over Baton Rouge from SW to NE, followed by loud noises, a crash, and an earthquake. The ground where it fell was broken up and surrounding vegetation badly scorched. The area was presumably searched only superficially; there seems to have been no excavation for buried meteorites, as none would have been expected. E. Howard[20] in 1803 deplored the fact that the search had not been thorough and expressed hope that a meteorite from this fall might crop up in the future. Although at that time such occurrences were more apt to be attributed to lightning effects,[21] one may nevertheless wonder why Jefferson did not sense some resemblance between the Weston and the Baton Rouge falls. In a later letter to Daniel Simon, with his more customary calm judgment, Jefferson recommended that the subject of the Weston fall needed to be more thoroughly investigated by competent scientists.[22]

The eminent navigator, Nathaniel Bowditch, like Silliman, also collected all the information he could on the Weston fireball, from its first sighting from Rutland, Vermont, until its final explosion. After four years of study he gave apparently convincing evidence that it had been moving at a velocity of 3.5 miles per second, at an altitude of about 18 miles, in a path nearly parallel to the surface of the earth, and that its original mass must have been six million tons.[23,24] This seemed to put to rest all arguments opposed to the theory that the Weston stones had fallen from heaven, and supported the theory of their extraterrestrial origin. Moreover, from a comparison between the mass collected and his estimate of the total mass involved, Bowditch ventured the opinion, reminiscent of Clap's, "that by far the greater part of the mass continued in its course without falling to the Earth."[23] Two generations later, H. A. Newton reexamined the available data and disagreed with Bowditch on the path of the Weston fireball. Newton found that instead of heading 7° west of south it traveled east of south, passing 50 miles above the Helderbergs, New York, and finally exploded at a height of 7 or 8 miles above Weston.[25,26] (Newton stated "I make it to have moved from a point N.40°W., 35° high,"[25] this being the position as seen from New Haven.) Unfortunately, Silliman and Kingsley had not published a path for the fireball with which the Bowditch and Newton determinations might have been compared.

As early as 1794, E. F. F. Chladni in Germany had collected reports on meteorites which led him to suggest that they were of extraterrestrial origin.[27] Later chemical analyses by E. Howard[28,29] in England and J. L. C. de Bournon[30,31] in France (1801–08) revealed that even stony meteorites contained nickel—iron enclosures not found in ordinary rocks. Such chemical curiosities were also found in the stones that fell in l'Agile, France, on April 26, 1803, a fall of some 3,000 stones. J. B. Biot declared them to be extraterrestrial,[32,33] and the French Academy of Sciences in Paris was at last convinced. Although skepticism such as Jefferson's still persisted, the investigations of Silliman and Kingsley were ultimately persuasive in America.[5]

The largest unbroken fragment of the Weston fall weighed 36.5 lbs.[34] Silliman had been anxious to obtain this, but when it was put up for sale it was purchased by Colonel George Gibbs of Newport, R.I., for his large and already famous mineral collection, consisting of 10,000 specimens, including a few meteorites. One-half of this large collection Gibbs deposited on loan at Yale in 1810, the remainder in 1812. In 1824, citizens of New Haven and officers at Yale purchased the entire collection, known as the "Gibbs Cabinet,"[35] for $20,000. Thus the Weston fall provided the nucleus of the Yale collection of meteorites, the oldest such collection in America. At that time the Weston specimen was also the world's largest known stony meteorite.

The largest iron meteorite in the Yale collection, known as the "Gibbs Iron" or the "Red River, Texas" or the "Louisiana" meteorite is approximately 3.4 × 2.3 × 1.3 ft and weighs 1,635 lbs. In 1808 Captain Anthony Glass was trading among Pawnee and Hietan Indians near Red River, Texas, when he was informed of a curious mineral lying on the bank of a creek.[36–38] It was so heavy that it took six Indians just to tilt it on end. In 1810 the finders, thinking it was valuable platinum, worked under great difficulties to transport it nearly 2,000 miles, first through wilderness and hostile savage terrain to the Mississippi River, to New Orleans, and thence by ship to New York. There Gibbs noted that it was iron, not platinum, and bought it from the disappointed speculators for far less than they had expected. Gibbs first displayed it at the New York Historical Society. Later it was set up in Central Park where some workers, considering it an unsightly rock, began to bury it. Mrs. Gibbs fortunately detected them and the meteorite was saved. She then presented it to Yale in 1835 as a memorial to her husband, who had died in 1833.[39] At that time it was the largest in any collection; it is still the largest at Yale although it is now housed in the Astronomy Department. (It was temporarily exchanged in 1968 for Mrs. H. A. Loop's 1882 portrait of Elias Loomis, which had been willed to the Astronomy Department.[40]) A small segment, about 6 × 8 inches, had been sawed off one end of the meteorite. On this polished surface is engraved the legend,

IN MEMORY OF GEORGE GIBBS
A LOVER OF SCIENCE AND HIS COUNTRY
PRESENTED TO YALE COLLEGE BY LAURA W. GIBBS 1835
METEORIC IRON OF TEXAS
WEIGHT 1365 LB

Incidentally, it was George Gibbs who in 1817 suggested to Silliman that he found the *American Journal of Science and Arts,*[41] which was then first published in 1818.

In 1810 another large meteorite, estimated as weighing 1,700 lbs (actually 1,350) was discovered near Santa Rosa, New Grenada, or Colombia.[42] Its origin then unknown, it was used (1818–24) by a local blacksmith as an anvil; later as a revered monument in the Santa Rosa marketplace; and it is now at the Bogota National Museum.[43] C. U. Shepard (part-time lecturer in Natural History at Yale from 1833 to 1847) in 1829 had noted the similarity of the two massive irons in appearance, size, and composition. He found the composition of the Red River to be 90.02% iron, 9.67% nickel;[42] and for Santa Rosa he quoted 91.41% iron, 8.59% nickel.[44] A later determination by W. S. Clark[45] gave 91.23% iron and 8.21% nickel. (Gibbs and Silliman had earlier found nickel in the Red River meteorite,[46] but Shepard's was a more exact determination.) Even though they fell some 3,000 miles apart, Shepard speculated that "they were derived from one and the same meteorite, which traversed the atmosphere of our planet in a direction lengthwise of the American continent."[42] This interesting speculation has subsequently been discarded.[47]

Shepard (1804–86), an Amherst graduate, was an avid collector of meteorites. By 1886 his collection, then the largest in America, became the property of Amherst College, where Shepard taught from 1844 to 1877. In Harrison Brown's *Bibliography on Meteorites,*[48] Shepard has 71 references for 1827–88, more than any other American of that era except J. Lawrence Smith, with 75 between 1854 and 1884. Of Shepard's, eleven stem from the fifteen years he was associated with Yale.

The first catalogue of the Yale collection of meteorites was compiled in 1869 by George J. Brush, curator of the Gibbs Mineral Cabinet.[49] (Brush was a member of the first class to graduate from Yale's Sheffield School in 1852.) His catalogue listed 46 stony and 56 iron meteorites. Among numerous later additions this was augmented by an 826-pound Canyon Diablo (i.e. Meteor Crater) specimen donated in memory of Elias Loomis by his sons and friends; and by H. A. Newton's private collection containing samples from 100 different localities.[34] These included several hundred of the thousands of mostly iron nodules from the Estherville, Iowa, fall of May 10, 1879, a large conglomerate of stone and metal that fractured upon impact with our atmosphere.[50] By 1978 the Yale collection had grown to thousands of specimens from 386 localities.[34]

Newton had been perhaps the first to point out that more meteorites had been seen to fall between noon and midnight than between midnight and noon. This he rightly attributed to the fact that those meeting the Earth head-on, in its revolution around the sun, burn up faster than those catching up with the Earth. Thus most of the head-on retrograde meteorites do not survive their passage through the atmosphere as meteorites. Newton also deduced in 1888 that meteorites had previously had orbits more nearly like those of short-period comets than the comets with parabolic orbits,[25,51] a conclusion that seemed to indicate that the meteorites had their origin within the solar system. W. H. Pickering, however, in 1910 asserted[51,52] that the majority of stony meteorites had orbits close to the orbit of Earth, not reaching out as far as the asteroid belt. Giovanni Schiaparelli in 1908, on the other hand,[51,53] inferred from his own computations that the meteorites came from the distances of the fixed stars. In general, Newton's conclusions come closest to fulfilling the test of time, in the sense that most meteorites are now believed to have had orbits closely resembling the asteroidal—in fact that the meteorites are actually asteroidal debris.[54]

REFERENCES

1. F. Heide, *Meteorites,* (Trans. by E. Anders of *Kleine Meteoritenkunde,* 2nd ed. Berlin, 1957), Chicago, pp. 64, 118, 1964.
2. P. Merian, *Ann. d. Physik u. Chemie,* **198**, 182, 1864.
3. D. W. Sears, *The Nature and Origin of Meteorites,* N.Y., p. 8, 1978.
4. J. Day, *Mem. Conn. Acad. Arts and Sci.,* **1**, 163, 1810.
5. J. R. Eastman, *Bull. Phil. Soc., Washington,* **11**, 275, 1890.
6. B. Silliman and J. L. Kingsley, *Trans. Amer. Phil. Soc.,* **6**, 323, 1808; *Mem. Conn. Acad. Arts and Sci.,* **1**, 141, 1810.
7. O. C. Farrington, *Meteorites, Their Structure, Composition, and Terrestrial Relations,* Chicago, p. 11, 1915.
8. J. F. Fulton and E. H. Thomson, *Benjamin Silliman,* N.Y., p. 76, 1947.
9. B. Silliman, *Trans. Amer. Phil. Soc.,* **6**, 335, 1808; *Mem. Conn. Acad. Arts and Sci.,* **1**, 151, 1810.
10. H. Shapley, *Proc. Amer. Phil. Soc.,* **87**, 234, 1944.
11. S. L. Mitchill, *A Discourse on the Character and Services of Thomas Jefferson, More Especially as a Promoter of Natural and Physical Sciences,* N.Y., p. 35, 1826.

12. S. L. Mitchill, *Harper's Monthly Mag.,* **58**, 752, 1870.
13. J. Bronson, *The New York Spectator,* Dec. 19, 1807.
14. C. U. Shepard, 1884, *Memorial Addresses at the Unveiling of the Bronze Statue of Professor Benjamin Silliman at Yale College,* New Haven, p. 8, 1885.
15. Ref. 11, p. 27.
16. J. G. Burke, Private communication, March 30, 1985; *Cosmic Debris: Meteorites in History,* Berkeley, p. 56, 1986.
17. T. Jefferson and W. Dunbar, *Documents Relating to the Purchase and Exploration of Louisiana,* Part II., Boston and N.Y., 1904.
18. Mrs. Dunbar Rowland, *Life, Letters and Papers of William Dunbar,* p. 134, 1930.
19. W. Dunbar, *Trans. Amer. Phil. Soc.,* **6**, 25, 1804.
20. E. Howard, *Ann. d. Physik u. Chemie,* **13**, 315, 1803.
21. E. M. L. Patrin, *Idem,* p. 332.
22. J. P. Foley, *The Jefferson Cyclopedia,* N.Y., p. 549, 1900.
23. N. Bowditch, *Jour. Nat. Phil. Chem. and Arts* (also called *Nicholson's Jour.*), **28**, 89, 206, 1811.
24. ———, *Mem. Amer. Acad. Arts and Sci.,* **3**, 213, 1815. Quoted by M. Wilson, *American Science and Invention,* N.Y., p. 44, 1954.
25. H. A. Newton, *Amer. Jour. Sci.,* (3), **36**, 1, 1888; *Nature,* **38**, 250, 1888.
26. C. S. Lyman, *The New Englander,* **27**, 134, 1868.
27. E. F. F. Chladni, *Ueber den Ursprung der von Pallas gefundnen und andern ähnlichen Eisenmassen,* Riga, 63 pp., 1794.
28. E. Howard, *Phil. Mag.,* **12**, 181, 1801.
29. ———, *Phil. Trans.,* **92**, Pt. I, 168, 1802.
30. J. L. C. de Bournon, *Idem.,* p. 180.
31. ———, *Encycl. Methodique,* **5**, 548, 1808.
32. J. B. Biot, *Phil. Mag.,* **16**, 224, 1803.
33. ———, *Jour. d. Mines,* **14**, 485, 1804.
34. B. Narendra, *Discovery,* **13**, 10, 1978.
35. B. Silliman, Ed., *Amer. Jour. Sci.,* **25**, 214, 1834.
36. C. H., *Ibid.,* **8**, 218, 1824.
37. R. Cox, *Ibid.,* **33**, 257, 1838.
38. B. Silliman, Jr., and T. S. Hunt, *Ibid.,* (2), **2**, 370, 1846.
39. Anon., *Ibid.,* **27**, 382, 1835.
40. R. Brown, *Report* for 1889–90. Presented by the Board of Managers of the Observatory of Yale University to the President and Fellows, p. 4, 1890.
41. Ref. 8, p. 117.
42. C. U. Shepard, *Amer. Jour. Sci.,* **16**, 217, 1829.
43. V. F. Buchwald, *Handbook of Iron Meteorites,* Berkeley, **3**, 1075, 1975.
44. M. de Rivero and J. B. Boussingault, *Ann. Chem. et Phys.,* **25**, 438, 1824; *Edinburgh Phil. Jour.,* **11**, 120, 1824.
45. W. S. Clark, *Amer. Jour. Sci.* (2), **15**, 7, 1853.
46. Ref. 36, p. 225.
47. H. H. Nininger, *The Moon, Meteorites and Comets,* B. M. Middlehurst, Ed., Chicago, p. 162, 1963.
48. H. Brown, *A Bibliography on Meteorites,* Chicago, viii + 686 pp., 1953.
49. G. J. Brush, *Catalogue of Meteorites in the Mineralogical Collection of Yale College,* New Haven, 4 pp., 1869.
50. H. H. Nininger, *Out of the Sky,* Denver, p. 310, 1952.
51. Ref. 7, p. 213.
52. W. H. Pickering, *Pop. Ast.,* **18**, 262, 1910.
53. G. Schiaparelli, *Riv. Fis., Math. e Sci. Nat.,* **18**, 523, 1908.
54. T. Gehrels, Ed., *Asteroids,* Tucson, pp. 61, 822, 849, 892, 926; 1979.

5
SHELDON CLARK AND THE FIRST YALE OBSERVATORY

An early benefactor of Yale College was Sheldon Clark (1785–1840),[1] (Fig. 5.1), a farmer of modest means but with a craving for education. His father, who had intended a college education for his son, died when Sheldon was but a small boy. The parsimonious grandfather who adopted him considered a formal education a waste of time and money and largely confined the boy to rural activities. When the grandfather died in 1811, he left his estate to Sheldon who, then age 26, consulted Benjamin Silliman on the possibility of access to Yale's educational facilities without having had the required formal preparation. Impressed by the young man, Silliman obtained for him permission for his use of the library and attendance at lectures and science classes as a special student. In the ensuing years Sheldon was mainly occupied with farm labor and teaching at the rural school of his district in Oxford, Connecticut. In 1822 he visited Silliman again, with a proposal to deposit $5,000 with Yale College, to be invested at compound interest for 24 years, after which the fund was to be used to establish a professorship of moral philosophy and metaphysics. (From 1847 to 1892 Noah Porter, a classicist and President of Yale, was the first to hold this professorship.) Clark later donated $1,000, to be similarly invested to establish one or more scholarships. At the time of his death, Clark's contributions to Yale, with accumulated interest, amounted to $30,000, three times as much as any other individual had donated up to that time.

Having heard of the loss of the college telescope in the shipwreck of the *Albion* in 1822, Clark determined in 1828 to replace it with a larger and better telescope. This he did, not by encroaching upon his inherited funds but by occasional contributions, ranging from $50 to $200, depending on proceeds of his farm produce, until the requisite amount of $1,200 had accumulated (August 26, 1829). The telescope was ordered from Dollond and Sons of London, at the time probably the best lens maker in the world and the one most influential in the development of achromatic telescope lenses. (Achromatic telescopes were first built by Chester Moor Hall as early as 1733 and perfected by John Dollond in 1758.[2,3]) Although Clark stipulated that the telescope be delivered within two years (or the money returned to him), it arrived in November of the same year, 1829. The maker declared it to be "perfect, and such an instrument as he was pleased to send as a specimen of his powers." It was a 5-inch, 10-foot focal-length achromatic refractor, the largest refractor then in America. It was much appreciated by both students and professors. Members of the senior class in November, 1832, wrote a letter of appreciation to Clark, which stated in part:[1]

> After enjoying the rare privilege of looking through the large telescope, some acknowledgment to the liberal and high-minded donor was deemed proper and requisite . . .
>
> Actions distinguished for generous munificence, and designed to accomplish a useful and exalted object, claim the admiration of all; but from those who receive the benefit, they demand not only admiration but gratitude. And considering the circumstances, where shall we look for one which displays this noble quality in a stronger light? Where for one intended to effect a nobler object than the extension of scientific knowledge?

The telescope was mounted in the Athenaeum (Fig.5.2), a building originally erected in 1761 as Yale's first chapel, ostensibly designed by President Thomas Clap.[4] The steeple was replaced in 1829 by an octagonal tower "built in imitation of the Tower of Winds in Athens."[5] In remodeling the Athenaeum, the architectural purity of surrounding buildings was taken into account, to the detriment of the use of the telescope. Although the telescope had come with a sturdy alt-azimuth mount, it had to be put on castors enabling it to be rolled from one window to another. Still the telescope could not point to objects more than 30° above the horizon. In 1870 a revolving cylindrical dome was added above the octagonal

Fig. 5.1. Sheldon Clark (1785–1840). (Courtesy Yale University Archives)

Fig. 5.2. The Athenaeum, showing tower added in 1829 to house the Dollond refractor.

room; it looked awkward but provided better use of the telescope. However, by this time the college had obtained a larger telescope, a 9-inch Alvan Clark from Sheffield, which naturally commanded greater attention.

In its heyday "Clark's Telescope," as the 5-inch Dollond was then called, despite the handicaps of its location, brought Yale into the limelight of American astronomy. In 1835 a return of Halley's Comet was expected. Astronomers worldwide were searching for it, each hoping he would be the first to spot it. On the night of August 31, Professor Denison Olmsted and Tutor Elias Loomis detected a faint fuzzy speck only about half a degree from the predicted position. This was three days before William Mitchell on Nantucket found it, thinking he was the first. Ultimately, news from Europe indicated that Dumouchel in Italy had seen it as early as August 6. (Communication was incredibly slow in those days; Alexander Bell, inventor of the telephone, was not born until 1847; the first telegraph line in the U.S. was not laid until 1844 and the first trans-Atlantic cable not until 1858.) Meanwhile, in America, the discovery by the Yale astronomers was widely hailed.

Sheldon Clark died tragically in 1840 in a fall from a scaffold while repairing his barn. He left most of his property to Yale; his telescope put Yale at the forefront of American astronomy for a generation or more. Silliman wrote of him,[1]

> He kept his money always at work—loaned all the cash he did not need, (and his personal wants were few)—required his interest and payments at the day—but was exactly just in his dealings—prompt to give his advice when desired, and kind in his treatment of all. His hoarding was not for himself; wife and children he had none, and he laid by his thousands—the results not of traffic or speculation, but of laborious thrifty industry—to furnish the means of a superior education to the children of others, and to generations yet unborn.

Many students and tutors made use of the telescope, among them several who later became famous in astronomy or academia. Frederick A. P. Barnard (1809–99), Yale 1828, Tutor 1828–30, later became President of Columbia University where he advocated the admission of women. Six months after his decease Barnard College was opened, named in his honor. William Chauvenet (1820–70), Yale 1840, as Professor of Mathematics in the Navy, was instrumental in the establishment of the U.S. Naval Academy where he later became head of the Department of Astronomy and Navigation. From 1855 until 1862 he became President, then Chancellor of Washington University, St. Louis, Missouri. Another astronomer, Joseph Stillman Hubbard (1823–63), Yale 1843, served at the U.S. Naval Observatory 1845–63 where he was concerned with the parallax of α Lyrae, the determination of the orbit of the famous comet of 1843, and the identification and orbit of the recently discovered planet Neptune. Yale students later associated with Yale as professors or administrators included E. Loomis, 1830; C. S. Lyman, 1837; and E. C. Herrick, hon. M.A., 1838, college Librarian and Treasurer. Two undergraduate classmates (1839), E. P. Mason and H. L. Smith, with enthusiasm for observing, were stimulated to build their own telescope. At that time the largest reflector in America appears to have been an 8.5-inch built by Amasa Holcomb of Southwick, Massachusetts. These young students ground, polished, and mounted a 12-inch, 14-ft focal-length speculum mirror of the Herschelian type.[6] With it and the 5-inch Dollond, Mason and Smith made careful observations of some diffuse nebulae that Sir William Herschel had observed some fifty years before. Mason,[7] hoping to detect changes, made precise measurements of positions of the stars embedded in the nebulosities relative to which he carefully drew the contours or isophotes of the nebulosities. This is probably the first instance of the use of isophotes in astronomy. No less an authority than John Herschel praised the accuracy of Mason's drawings.[8]

Mason is also credited with having been the first American to observe meteors systematically through a telescope.[9] Actually there is a very early record of a single serendipitous observation by J. H. Schroeter[10] in 1795; and Olbers[11] remarked in 1837 that he had, over a span of fifty years, used a comet seeker to observe meteors, expecting to see many faint ones. All he saw were actually naked-eye objects. Hence he concluded that the faint meteors which other observers had claimed to see must at least in part have been confused with other types of objects, presumably terrestrial. Mason, on the other hand, observed the Perseids at their maximum in 1839, when he saw fifty in an interval of four or five nights, which, on the average were much fainter than those observed by naked eye. Erroneously, but reasonably for his era, he assumed that they were fainter because they were farther away. Thus he reached the puzzling conclusion that the telescopic meteors occurred at an elevation of some 4,000 miles, in contrast to only 50 to 60 miles for those seen by the naked-eye. This youthful contribution was nevertheless a step in the right direction—namely the acquisition of new data essential to future theories. Moreover, he speculated on the possibilities of using simultaneous observations of meteors from various localities for obtaining the difference in longitude between observing stations, particularly from New York to New Haven and to Philadelphia. Similarly he proposed occultation observations for the same purpose, for which he carried out numerous computations and observations, and solicited observations by others.

During his sophomore year Mason, at Olmsted's suggestion, kept a journal of a remarkable group of sunspots during March of 1837, observed with the 12-inch. Mason accompanied his journal with seven drawings showing the variation of the spots. Olmsted described the composite poster, unfortunately now lost, as "a magnificent drawing, which, for beauty of penmanship and extreme finish in the delineation of the minutest changes of figure or dimensions of the spots, has never been surpassed by any astronomical drawings with the pen, which I have ever seen."[12]

On September 18, 1838, there occurred an annular eclipse of the sun. Contacts were observed by Olmsted at the 5-inch Dollond, and by two of his students, H. L. Smith with a 3-ft f.1. Gregorian reflector, and E. P. Mason with a 7-ft f.1. Newtonian. (Both telescopes had been built by the students themselves.) Mason, in addition to timing the contacts, used a divided object-glass micrometer to measure the separation of the cusps during the partial phases. From fifty such measurements he computed the diameter of the sun as 31′ 54.81″ ± 0.19″, differing by only 0.6″ from the ephemeris predicted value, 31′ 54.2″. He commented that due to atmospheric turbulence, the diameter determined from measurements of the sharply defined cusps was apt to be more accurate than direct measurements of the solar diameter. A crude value from four such measurements 20 minutes before first contact yielded 31′ 53.7″.[13]

Another observation with the 5-inch that brought distinction to Yale was the discovery by Francis Bradley (assistant in practical astronomy, 1845–47) and Herrick in 1845 of the splitting of Bela's Comet, described in the next chapter.

We have dignified the location of the Dollond telescope atop the Athenaeum as Yale's "first observatory." Olmsted did not consider this awkward arrangement an observatory. He tried repeatedly to raise funds for a proper building but the College apparently felt a greater need for faculty salaries and other wants. Unable to use the telescope for objects at higher than 30° altitudes, Olmsted concentrated on work that could be done without a telescope.

After the installation of the revolving tower on the Athenaeum in 1870, the 5-inch Dollond unfortunately remained mostly idle. The building itself was demolished in 1893.[4] Much later the telescope was on public exhibition at the Franklin Institute in Philadelphia, from 1932 until 1946, when it was returned to Yale for display during the celebrations of the centennial anniversary of the Sheffield School. For a time it was at the Smithsonian National Museum in Washington. With the expected return of Halley's Comet in 1985 the famous telescope was returned to Yale for exhibition at the Peabody Museum.

REFERENCES

1. B. Silliman, *Amer. Jour. Sci.,* **41**, 217, 1841.
2. H. C. King, *The History of the Telescope,* London, p. 144, 1955.
3. I. S. Bowen, "Telescope," in *Encyclopaedia Britannica,* **21**, 909, 1966.
4. R. C. Carroll, Ed., *Buildings and Grounds of Yale University,* New Haven, pp. 44, 83, 95, 1979.
5. W. E. Decrow, *Yale and The City of Elms,* Boston, pp. 13, 27,1882.
6. J. Ashbrook, *Sky and Telescope,* **44**, 366, 1972.
7. E. P. Mason, *Trans. Amer. Phil. Soc.,* n.s., **7**, 165, 1841.
8. J. F. W. Herschel, *Amer. Jour. Sci.* (2), **5**, 91, 1848.
9. E. P. Mason, *Trans. Amer. Phil. Soc.,* n.s., **8**, 122, 1843.
10. W. Valentiner, *Handwörterbuch d. Ast.,* **2**, 120, 1898.
11. H. W. M. Olbers, *Schumacher's Jahrbuch,* p. 37, 1837.
12. D. Olmsted, *Life and Writings of Ebenezer Porter Mason,* N. Y., p. 93, 1842.
13. D. Olmsted, *Amer. Jour. Sci.,* **35**, 174, 1839.

6
THE OLMSTED ERA

The achievements of Denison Olmsted (1791–1859) at Yale fall mainly into three categories: as a teacher, a writer of textbooks, and a pioneer in the study of meteor showers, especially their radiant points and problems of their orbital motions.[1,2] He graduated from Yale in 1813, and after teaching a year in a private boys' school, returned to Yale as a tutor in 1815. On receiving his master's degree in 1816, he delivered an oration on the state of education in Connecticut, proposing the creation of a state-supported "seminary for school masters," a suggestion antedating by some seven years the first classes for teacher training by Samuel Reed Hall in Concord, Vermont, in 1823. Subsequently several such programs were started in various academies in New York and Massachusetts, some state supported. However, the first state supported Normal School, designed primarily for teacher training much as Olmsted had suggested, was opened by Horace Mann, first secretary of the Massachusetts Board of Education, in Lexington, Massachusetts, in 1839 with Cyrus Peirce as head.[3–5] (From its beginning this school was intended for women only. It was moved to Framingham in 1852 and later became Framingham Teachers College; it is now the co-educational Framingham State College.)

Meanwhile James Bates Thomson (1808–83),[6] a Yale graduate of the class of 1834, was undoubtedly influenced by Olmsted in regard to methods of teaching and the necessity for training prospective teachers how to teach. Until 1842 Thomson was in charge of an academy on Nantucket Island where Cyrus Peirce (first Principal of the Nantucket High School) presumably influenced him even more. Thomson was also a writer of mathematical texts. In 1842 Yale's President J. B. Day assigned Thomson the task of writing an abridgment of Day's treatise on algebra. Entitled *Elements of Algebra, being an abridgment of Day's algebra, adapted to the capabilities of the young, and the method of instruction, in schools and academies,* this text went through twelve editions between 1842 and 1849, followed by his *New Practical Algebra* in 1877. Thomson also devoted five or six years to the organization and extension of teachers' institutes. In recognition of his efforts Hamilton College awarded him an honorary Doctor of Laws degree in 1853, and the University of Tennessee awarded another in 1882.

From 1817 to 1825 Olmsted was Professor of Chemistry (including geology) at the University of North Carolina. He prepared himself for this post by studying under Benjamin Silliman. At North Carolina he collected minerals and sent duplicates to Silliman at Yale.[7] (The Gibbs Cabinet contained primarily European samples.) In 1825 Olmsted was recalled to Yale as Professor of Mathematics and Natural Philosophy, the title later (1836) changed to Professor of Natural Philosophy and Astronomy.[1] It has been surmised[8] that Olmsted dropped the mathematics course because he was weak or ill prepared in the subject. The same might have been said of his early preparation in natural philosophy. Certainly in mathematics he was not as gifted as his predecessor, Alexander Metcalf Fisher.[9] As a teacher, however, Olmsted assiduously mastered whatever was needed for his classes. It would appear then that Olmsted had assumed more responsibilities than one professor alone could easily handle. He had to curtail somewhere. So when a new professorship in mathematics was established and a full-time mathematician needed, Olmsted retained the appointments that most interested him: science, especially astronomy and meteorology. Silliman described Olmsted as "among the most distinguished of my professional associates."[10]

Finding a dearth of suitable textbooks, Olmsted set about writing his own. His first, *Elements of Natural Philosophy* (including properties of matter, mechanics, hydrostatics, pneumatics, meteorology, acoustics, electricity, magnetism, and optics) was published in 1831 (Vol. I) and 1832 (Vol. II). It met with great success and continued to be used both at

Fig. 6.1. Denison Olmsted (1791–1859) (Courtesy Yale University Archives).

Yale and in other institutions right up to the time of his death. This was followed by an abridgment, *School Philosophy* (1832) for high schools, and *Rudiments of Natural Philosophy and Astronomy* (1842) for use in elementary schools, which went through more than fifty editions, including one in raised type for the blind.[1] In the Introduction he wrote, "Some years since, I announced to the public an intention of preparing a series of text-books in Natural Philosophy and Astronomy, adapted, respectively, to Colleges, Academies, and Common Schools." In the Introduction to the 1859 edition he stated, "No one should write a text-book who has not already had experience teaching students of the age-group for which the text is intended." He himself had wide experience in teaching all age groups. His first *Introduction to Astronomy* appeared in 1839, soon followed by an abridged version for schools, and one called *Letters on Astronomy Addressed to a Lady* in 1840. In all, he published six different texts, some going through several editions. It is estimated that the sale of his books amounted to some 200,000 copies.

Olmsted was also an early advocate of informing the public of advances in science. As a motto on a frontispiece of his *Letters on Astronomy* he quotes, ostensibly from the *Edinburgh Review,*

> The public now demand of those professionally devoted to the sciences, that they shall not confine the knowledge they have such favored opportunities of acquiring, to the lecture-room, but shall render it, as far as practicable, available to the well-informed of all professions, and to the more intelligent, at least, of the other sex.

He begins each chapter of the *Letters* with a verse from the literature: quotations from Virgil, Milton, Pope, Addison, and others. This, more than the astronomical contents, distinguishes this text from his treatises designed for college students, and seems consistent with the presumably greater interest of upper-class ladies. Interestingly, a great lady astronomer of a century later, Cecilia Payne-Gaposchkin, began the chapters in her *Introduction to Astronomy* (1954), and in her *Variable Stars* (1938) similarly with quotations from Shakespeare, Milton, Tennyson, Spenser, and others. Her one quotation from Milton's *Paradise Lost* is the only one that Olmsted and Payne-Gaposchkin had in common:

> . . . . like a comet burned
> That flies the length of Ophiuchus huge
> In th'arctic sky, and from his horrid hair
> Shakes pestilence and war.

Olmsted reputedly considered that stimulating interest in science by teaching was more important than the pursuit of research as such. In a biographical sketch, C. S. Lyman wrote,[1]

> In forming our estimate, on the whole, of Prof. Olmsted's scientific character, we must bear in mind that he himself always regarded it as his more appropriate sphere of effort, in the circumstances in which he was placed, not so much to *cultivate* science as to *teach* and *diffuse* it. Teaching, indeed, was his chosen and ever cherished work, and the one for which by temperament, talents, training and attainments, he was peculiarly fitted.

Among Olmsted's ablest students were Elias Loomis (his eventual successor) and Ebenezer Porter Mason (1818–40) a young man of exceptional abilities and accomplishments[11,12] who died at the early age of 21. Olmsted had encouraged Mason to write a supplement on practical astronomy to Olmsted's *Astronomy*. This fine text[13] Mason completed just a month before he succumbed to tuberculosis. T. R. Treadwell in 1943 in his *Notes on a Forgotten Episode* described Mason's *Practical Astronomy* as an "extraordinary accomplishment, a work of genius exerting an influence for many years on the teaching of astronomy."[14]

While Olmsted was highly respected by most of his students and colleagues, not all of the students who were required to take his courses were happy. The following sketch by D. G. Mitchell[15] has been interpreted by Yale historian A. P. Stokes[16] as referring to Olmsted:

> I saw a long-faced, dark-haired man, who looked as if he were always near some explosive, electric battery, or upon an insulated stool. He was, I believe, a man of fine feelings; but he had a way of reducing all action to dry, hard, mathematical system, with very little poetry about it. I know there was not much poetry in his problems in physics, and still less in his half-yearly examinations. But I do not dread them now.

While lecturing, professors usually sat on the same level as their students. Olmsted is reputed[17] to have been the first to elevate the professor to a slightly higher platform, so that he could more readily observe all the students. Some of them resented this departure from the previous informality, considering it undemocratic and haughty. For teaching the upper classes Olmsted gave three courses of lectures, one on natural philosophy and optics, one on astronomy, and the third on meteorology. Lyman remarks that the one on meteorology was "perhaps, on the whole, the most attractive and useful." On all of these subjects Olmsted published profusely, over fifty research articles primarily in *Silliman's Journal,* as the *American Journal of Science* was then affectionately known. Olmsted's theory (1830) on the formation of hailstones was probably the most nearly correct of any up to that time.[18]

Stimulated by the intense displays in 1835 and 1837, aurorae claimed Olmsted's attention, and he was perhaps the first to reject a purely terrestrial origin for them. Finding that the displays extended over thousands of miles and at heights of hundreds of miles, he suggested that they might be caused by the passage of the Earth through "some nebulous body" in interplanetary space. Solar winds were, of course, unknown then, but Olmsted's suggestion comes as a fairly suggestive forerunner. However, he believed that the zodiacal light might be the source of both the aurorae and ordinary meteors. He indicated a period of 60–65 years for auroral occurrences. Elias Loomis was the most progressive analyst of aurorae in this era, although Brekke and Egeland considered him the most neglected auroral physicist in the 19th century.[19] From a compilation of observations for 1763–1854 by New Haven savants,[20] supplemented by others from the Boston area for 1742–1839, Loomis derived a provisional cycle of 58 years, but found the total time-span of the observations still too short to yield a definitive period. Most of the observations stemmed from Stiles for 1763–94, Herrick 1827–54, and Bradley 1842–54. In the *Smithsonian Report* for 1865, Loomis[21] analyzed data from 1685 to 1864 and again deduced a 59-year period, noting, "a subordinate period of about 10 years with a *maximum maximorum* every 59 or 60 years." He assumed there might be a correlation with the magnetic needle and noted that telegraph lines became generally unmanageable during aurorae. He also noted the correlation with sunspots. To account for the apparent periodicities, Loomis assumed that a 10–12 year cycle corresponded with Jupiter's period. If Jupiter influenced the sun, then Saturn should also, but to a lesser extent, and in a period of 30 years. A combination of these two periods, five revolutions of Jupiter corresponding to two of Saturn, should then account for the resulting period of approximately 59.5 years. Between 1795 and 1830 very few aurorae had been observed. If the period inferred by Olmsted and Loomis had been valid, then another minimum should have occurred about 1870 when, in fact, unusually many aurorae were seen.[22] The period has never been substantiated.

Edward C. Herrick, Librarian (1843–58) and later (1852–62) Treasurer of Yale, and an able, avid amateur astronomer, remarked at the time of the brilliant 1837 display,[23]

> It is greatly to be desired that careful contemporaneous observations on the Aurora Borealis should be made by persons stationed at many different places. Within the last two hundred years, a vast multitude of isolated accounts have been recorded, most of which are of comparatively little value to science. A tenth part of the labor which they have cost, had it been spent in well concerted contemporaneous observations in different parts of the world, would long ere this have contributed important data for a satisfactory theory. The most probable opinion is, that the aurora is in some way a result of magneto-electric action, but the laws which govern its capricious appearances have thus far eluded all investigation.

Further, he asked,

> Is it too much to hope that some of the many American ships which traverse the Indian Ocean, where opportunities for making the necessary observations most often occur, will hereafter bring home the desired information?

These various researches and speculations Olmsted incorporated into his lectures for the students, stimulating their cooperation.

Outlines of Olmsted's astronomy lectures for the Senior classes of 1850 and 1858 have come to hand. The excellent introductory lecture for 1850, describing the scope of the course, laws or reasoning in astronomy, and the advantages and pleasures of studying astronomy, is omitted in the 1858 version; newly introduced is a lecture on practical astronomy. On the whole, much of the material is somewhat rearranged and expanded. Of interest is the relative increase in coverage of stars and stellar systems relative to the solar system during the years of Olmsted's professorship. His 1839 *Astronomy* text devoted 25 pages to stars compared with over 240 on the solar system, or about 9 percent (an 1846 text for schools showed, 12%; the 1850 outline, 25%; and the 1858 outline 30%, indicating a steady increase in knowledge about the stars and the galactic system). Yale, however, contributed little to the accumulation of stellar and galactic research until after 1880 when it acquired the Yale heliometer. The one major exception seems to have been E. P. Mason's observational work on some of Sir William Herschel's nebulosities, described in the previous chapter.

Published outlines of Olmsted's lectures for the Senior class in meteorology for the years 1836(?), 1850, and 1858 are extant. In all of them meteors are the subject of the final lectures of the course, whereas meteors are not mentioned at all in his outlines of the lectures in astronomy. This seems puzzling in view of the fact that he himself was among the first to ascertain that meteors are probably sun-circling objects of cosmic origin. Chladni (1756–1827) had previously come up with a similar conclusion about aerolites, but not for ordinary "shooting stars." Olmsted's *Astronomy* mentions meteors only under zodiacal light, as a suggested source for meteors. In his *Letters on Astronomy Addressed to a Lady,* however, he does devote a whole chapter to "Meteoric Showers."

In the 1836 version of his meteorology outline, Olmsted lists four hypotheses respecting the cause of meteors: electricity, magnetism, hydrogen gas, and terrestrial comets, all of which he rejects as inadequate. In the latest version he again lists three of these same hypotheses: electricity, found inadequate; magnetism, "no proof of its agency—inadequate to account for the facts"; terrestrial comets, "the theory of falling stones . . . The meteors of November could not be of the same nature—differently constituted—velocity too great." None of his outlines mentions ordinary comets as a possible origin, even though Isaac Newton in the 17th century had proposed that meteors proceed from the tail of a comet, or Boubeé that they are fragments of an exploded comet,[24,25] and he, himself, had already considered a cometary origin in his analysis of the Leonids in 1834.[26] However, in his 1839 text-book *Astronomy,* in the chapter on comets, Olmsted wrote,

> Considering a comet (as most of them doubtless are) as a mass of exceedingly light nebulous matter, it is not probable, even though the earth were to make its way directly through a comet, that a particle of the comet would reach the earth. The portions encountered by the earth, would be arrested by the atmosphere, and probably inflamed; and they would perhaps exhibit, on a more magnificent

scale than was ever before observed, the phenomena of shooting stars, or meteoric showers.

In attempts to account for the origin of the meteors he discussed the proofs that they came from outside the Earth's atmosphere: their heights, parallel paths, high velocities, estimates of their size, and cause of their light and heat. In the later versions of his outline he added luminous trains, modes of existence before they fell to earth, relation to the solar system, why the radiant points of the November meteors were always in Leo, those of the August meteors always in Perseus, and why showers occur precisely at these times of the year. In both the earlier and the later versions he indicated the period of revolution at a year or some aliquot part thereof, but unlikely less than half a year. He suggested the zodiacal light as a possible origin. These stimulating lectures showed great advance over pre-1833 knowledge of meteors but relatively insignificant advances between 1835 and 1858. Had Olmsted but lived to witness the 1866 return of the Leonids, he would probably have discovered the 33 year period and resolved the quandary of the relation between the real period and the, to him, apparent period of a year or half year arising simply from the distribution of the meteoroids along their orbit.

If one were to select just one item for which Olmsted is best remembered by astronomers, it would probably be his discovery and interpretation of the radiant point of the spectacular and unexpected meteor shower of November 13, 1833. In New England one observer in a 15-minute interval from 5:45 to 6:00 A.M. counted 650 meteors in an area less than 10 percent of the sky. H. C. Twining at West Point estimated that there must have been 10,000 an hour after 5 A.M. Southern observers reported seeing shooting stars from 9 P.M. increasing in frequency to a maximum between 2:30 and 4:00 A.M. Although some half dozen observers noted that the meteors appeared to emanate from a point in Leo, it is Denison Olmsted at Yale and Twining who are generally credited with having been the first to discover the Leonid radiant, the first radiant to be discovered at all. This credit is justified by the fact that they were the ones to carry out further research, explaining the reason for a radiant and leading to the theory that meteors are of extraterrestrial origin. Olmsted, in particular, collected and analyzed the observations of others and hence does merit the greater credit.

Olmsted, as well as Twining, correctly interpreted the radiant as indicating that the meteors moved in parallel paths in space. That the radiant remained fixed relative to the stars throughout the night meant that the meteors must be cosmic and could not be earth-circling. Earlier it had generally been supposed that meteors were atmospheric phenomena (analogous to lightning) and were not related to bright fireballs and aerolites.

In 1834 Olmsted found that the position of the radiant as observed from various widely separated localities differed somewhat. He analyzed the discrepancies, assuming them to be real (not the consequence of observational uncertainties), in an effort to find the altitude of the radiant point itself; this would then enable him to compute the free-fall velocity from that altitude to the observed altitude of the meteors. Thus, comparing the coordinates of the radiant as seen from New Haven, with the position noted at a site 113 miles distant, he found an altitude of 1,572 miles; from 85 miles away, 2,955 miles, or an average of 2,263.5 miles; from observations from a location at the Gulf of Mexico, 2,027 miles; and a location at Lake Huron, 2,424 miles. He then accepted the mean of the three determinations, 2,238 miles, which would yield a velocity of only 4 miles per second (in contrast to modern determinations of 44 miles per second) when the meteors reached an average altitude of about 50 miles. He did, however, indicate that the differences in radiant positions could arise purely from errors in their determinations, which were merely eye-estimates and not accurate measurements, and that the true distance was probably much greater. The figure, 2,238 miles (also quoted in Olmsted's *"Letters Addressed to a Lady,"* 1840), apparently inspired a monody of disrespect attributed to a possibly inebriated poet (Edgar Allan Poe[27]) for a professor who would give a figure implying four-digit accuracy to a mean value derived from input data differing by nearly 1,400 miles:

> That devilish simper which he frequent wore
> When, with the glibness of a saint, he swore,
> That, 'His last work was plainer far than dirt'—
> As tainting too, thou muddy-headed squirt!

In 1836 Olmsted wrote that the remarkable exhibition of shooting stars on November 13 occurred six years in succession, having been observed as early as 1831 in Ohio, with the numbers first increasing to the magnificent display of 1833, and then decreasing steadily. Olmsted also collected data from all possible sources, from which he concluded that the intensity of the shower increased from northern to southern states. He argued,[28]

Does not the recurrence of this phenomenon for six successive years, at the *same period of the year,* plainly show its connection with the progress of the earth in its orbit? and does not the fact, that the greatest display occurs every where in places differing widely in longitude at the *same hour of the day,* as plainly indicate its connection with the motion of the earth on its axis? The supposition of a body in space, consisting of an immense collection of meteors stretching across the earth's orbit obliquely, so that the earth passes under it in its annual progress, while places on its surface lying westward of each other are successively brought, by the diurnal revolution, to the point of nearest approach, will satisfy both these conditions.

Moreover, he added,

It would not be at all inconsistent with the known extent of astronomical bodies, to give to the body in question a breadth of thousands, and a length of millions of miles. It was an accidental observation, made after the conclusion was formed, which ascribes the origin of meteoric showers to a revolving nebulous body, that first led me to suspect the *zodiacal light* to be the body in question. This, according to Laplace, *is* such a nebulous body, revolving round the sun in the plane of the solar equator.

Finally, he concluded,

According to this view of the subject, *the zodiacal light would no longer be regarded as a portion of the sun's atmosphere, but as a nebulous or cometary body, revolving round the sun within the earth's orbit, nearly in the plane of the solar equator, approaching at times very near to the earth, and having a periodic time of either one year, or half a year, nearly.*

Such, I affirm, would be the fact should the zodiacal light be proved to be the body which affords the meteoric showers.

Believing the Leonids to derive from a nebulous swarm between the Earth's orbit and the sun, Olmsted attempted to correlate the observations of the meteors with an assumed diffuse comet traveling in the same 182-day orbit, nearly in the plane of the ecliptic, inferior to the Earth's orbit but tangent to it at aphelion. Remarkably, he found in the meteorological records observations of auroral-like luminous effects resembling the zodiacal light but not coincident with it, and seen at appropriate times and places before and after sunrise or sunset to satisfy his hypothetical comet.[29] Later he did not pursue this suggestion.

Although he did not expect a rich display in 1837, he and others nevertheless observed as a matter of scientific interest. They found that the apparent radiant moved from Gemini in October to Leo on November 13. Again Olmsted invoked the passage of the earth through the zodiacal cloud as a likely interpretation.[30] The following year he and his associates likewise observed and collected observations from others.[31] His report for the 1838 display no longer mentions the zodiacal light. In fact, in the preceding article in the same issue of the *American Journal of Science,* Herrick pointed out that the Earth is in a situation more favorable for collision with this body early in December than in the middle of November. Olmsted concluded his report by noting that "several of the most eminent astronomers in Europe" were occupying themselves with the theory of shooting stars, adding ironically in parentheses, "which some of our own astronomers have supposed beneath their attention." He hoped that the difficulties of interpretation of the origin of meteors would soon be removed, "and we shall know whether to regard them as atmospheric concretions, or as visitants from another sphere." Was the reference to "atmospheric concretions" a step backward? The Leonid shower of 1833 was the most spectacular observed in Olmsted's generation, except possibly for the Leonids of 1799 and the Lyrids of 1803–04 which occurred while Olmsted was a child. In any case, these earlier showers had not inspired scientific investigations in America at those times. Olmsted did note in 1834 that the November 1799 meteors and those of 1833 represented the same swarm but, believing in his orbital period of 182 days, he did not infer the correct period of 33 years. Another generation was to pass before H. A. Newton finally resolved the mystery upon the 1866 return of the Leonids.

In 1846 Herrick wrote,[32]

Since the year 1838 no decided return of the meteoric display of November has been announced; nor indeed does the theory proposed by Prof. Olmsted, require the annual occurrence of this phenomenon. It is nevertheless quite important that observations should be maintained, in order, either to determine the negative, or else to secure the first symptoms of a return of the grand meteoric shower, which sooner or later may be expected in this period.

He and Yale undergraduate J. Homer Lane therefore observed on the night of November 13, 1845, between 4:10 A.M. and 5:10 A.M. and recorded six meteors of which five were conformable with the radiant in Leo; and F. Bradley, observing in a different part of the sky saw nine of which eight were conformable. This curious coincidence could at that time hardly have seemed compatible with the ultimately verified period of 33.25 years. However, as early as

1838, but only in a footnote, Herrick had stated,[33] "The cycle of the November shower seems to be, without much doubt, 33 or 34 years; that of the April shower is perhaps 27 years." Yet possibly in deference to Professor Olmsted's greater expected understanding, Herrick's confidence in Olmsted's annual period prevailed. He did advance for the shower meteors the asteroidal theory that was much later accepted for meteorites. Meteors, he stated,

> are encountered by the earth's atmosphere probably every hour in the year, but in much greater numbers at certain parts of the earth's orbit, than at others. The distribution of these bodies throughout the solar system can not yet be determined. The majority of them probably move in groups, and may be supposed to constitute one or more broad zones or rings, in some parts of which meteors are exceedingly numerous. When, at the return of certain periods, the earth traverses these dense parts, great meteoric showers occur . . . It may be supposed that in other years, the earth passes through a part where the meteors are less numerous, and then only a *sprinkling* of meteors is seen. Whether there are more zones than one, and if so, how they are situated, are problems which will probably long remain unsolved. Whether these meteors are the fragments of the supposed exploded planet, of which the four *asteroids* may have formed a part, or whether they were originally independent bodies, will perhaps be determined at a much later day.

The twentieth-century expert, Charles P. Olivier,[34] states that meteoric astronomy really began with the 1833 shower, and pays tribute to Olmsted "as being the only American, except his later successor, H. A. Newton, who during the nineteenth century deserves permanent recognition for his eminent contributions to meteoric astronomy." He might well have included E. C. Herrick in this eulogy. Further, noting the advantages to astronomers after the publication of Schiaparelli's book[35] on the theory of meteors in 1871, Olivier adds, "All consideration should, however, be given to those earlier pioneer observers on whose labors depended future advances and whose very mistakes showed to their successors the better way for advancing."

Meanwhile Edward C. Herrick interested himself as well in the occurrences of meteors in other seasons of the year. John Robie Eastman said of him,[24] "From 1838 [*sic*] when E. C. Herrick began his work he labored with untiring industry as an observer and compiler of observations and other data until his death in 1862, and no one in this country did so much as he in promoting the observation and investigation of the August meteors." From 1837 Herrick searched the literature and compiled catalogues of single meteors and fireballs, and in 1841 published a *Chronological List of Star-Showers* including 39 entries.[36] The first three were for observations of 1786 B.C., 686 B.C., and A.D. 7; the others ranged from A.D. 532 to 1833. In retrospect, four of these corresponded to the Leonids but were not yet identified as such. None was unambiguously a Perseid shower; presumably it had been assumed that because meteors were always seen in great numbers in August they were not necessarily considered as belonging to showers.

The preponderance of meteors in mid-August had been noted at least as far back as A.D. 830. Yet it is believed that the first to discover a radiant point in Perseus near Algol was John Locke in Cincinnati in 1834, as reported only in the local newspaper,[37] the *Cincinnati Daily Gazette,* August 11, 1834. Herrick,[38] unaware of Locke's discovery, observed the meteors in 1836 and 1837 and concluded that the radiant was near the cluster χ Persei. From other observations culled from the literature, especially 1833–37, but also earlier ones as far back as 1781, he found that maximum displays always occurred on or about August 9. By analogy with the Leonids, he concluded that the August meteors should also be periodic. Subsequently he continued his literature search and found at least eleven years between 1799 and 1837 when showers occurred. He commented that nearly all of the observations came from before midnight, whereas higher frequencies would have been expected after midnight when the earth meets the meteors more nearly head-on, and the radiant is higher in the sky and therefore more obvious. In 1839 his own observations were hampered by clouds, but he recruited observations from more fortunate observers from Georgia to Geneva, New York, and from Ohio to Middletown, Connecticut. From these, for the first time, he deduced that the radiant "lies fifty degrees north of the point towards which the earth is at the time tending. This fact may perhaps intimate that the meteoric zone does not lie in the plane of the ecliptic."[39] Herrick has been credited with having contributed more to the knowledge of meteors, especially the August meteors, than anyone else before any news of similar discoveries reached America from Europe. For example, both Herrick and Quetelet independently discovered the January Quadrantids in 1839.

By 1838 Herrick felt that three showers were well established, those of November (Leonids), August (Perseids), and April (Lyrids).[40] Of the last, however, little was as yet known and he advocated a careful lookout within ten days of April 30. Unusual displays had been seen at the end of April 1803 (one observer counting 167 in 15 minutes), and much earlier in A.D. 1095 and 1122. On evaluating the old records, Herrick made comments reflecting then current suppositions and misconceptions:[41]

> As hypotheses which I do not credit, are often interwoven with the testimony cited, I take occasion here to express my entire dissent from the suppositions that shooting stars, whether single or in showers, are connected, either as cause or effect, with earthquakes, pestilence, electrical charges, winds, seasons of heat or cold, or any particular sort of weather; or that the movements of these meteors have any correspondence with the direction of the wind, or with lines of magnetic dip or declination. That they are connected with the Aurora Borealis, is quite doubtful, yet it is well worthy of notice, that very brilliant displays of the latter have often occurred about the 13th of November.

Herrick could find no relevant records for other years than those cited, but his own observations in April 1839 appear to have clinched the shower relationship. In 1838 he had considered it improbable that other periodic showers would be found, as it seemed none was indicated in his search of "many hundred volumes" in the mainly meteorological literature. However, some uncertain statements still encouraged hope and vigilance. Very soon he was rewarded by the discovery of a new radiant in Perseus, December 5–8, 1838, confirmed by observations from over nearly all of the northern hemisphere (including China, England, France, and the North Cape), indicating a meteoric cloud of great extent.[42] This shower was later associated with the Bielid meteor stream, but Herrick at this time did not have sufficient data to infer any periodicity.

The Austrian Count von Biela, an amateur comet hunter, discovered a faint comet in 1826, one actually found twice before, in 1772 and 1805, but not until much later were the three identified as one and the same, with a period of 6.6 years. Upon its 1845–46 return, first seen on November 26, 1845, a remarkable discovery was made by Herrick and Bradley, observing with the 5-inch Dollond. On December 29 they noticed that the comet had split: the primary was accompanied by a faint fuzzy secondary at a separation of one minute of arc. Matthew Fontaine Maury in Washington, using a 9-inch refractor, independently rediscovered the doubling on January 13, the separation having increased to two minutes. Because he, unlike Herrick and Bradley, was quick to publish his discovery and make extensive follow-up observations, Maury has generally been given credit as the discoverer. Bradley and Herrick also continued to observe the pair of comets through March 31 when the companion, now separated by 16 minutes, was barely visible. The primary disappeared after April 16.

Upon their next return in 1852 the two comets had separated by about 1.5 million miles. Thereafter they have never been seen again. The world had witnessed the disintegration of a comet, the disintegration having started centuries earlier. However, years later it was found that the Andromedid meteors, seen as early as 1741, moved in the same orbit as Biela's comet.[43] Predictions by Weiss of Vienna of showers in 1872 and 1879 were brilliantly verified and the meteors have been renamed the Bielids. Unfortunately, both the comet and the shower appear to have been lost, the meteors in any case much depleted, if sparse observations at the correct times do indeed still represent this shower. There appear to be no records that Yale astronomers pursued further studies of the comet or the Bielids after the year of their epoch-making discovery, until much later when H. A. Newton examined all the available information.

Herrick also concerned himself with records of meteorites, and was particularly interested in very bright fireballs that reminded him of the Weston fall of 1807. Thus a fireball seen from Wilton, Connecticut, on December 14, 1837, estimated to be one-quarter to three-quarters the brightness of the full Moon, and a meteorite fall in Missouri on February 13, 1839, stimulated his thoughts on Weston: "It remains only to inquire whether it is more probable that the Weston meteorite was a satellite of the Earth, or a primary body moving around the sun."[44] Estimating that bodies opposed to the motion of the earth must have a velocity over 10 miles a second, and that Weston had a velocity of at least 20 miles a second, he concluded that it must have revolved around the sun.

Olmsted and Herrick, with their associates and students, were real pioneers in meteoric astronomy. They successfully laid the groundwork for Yale's future achievements in this field, and thereby assured the preeminence of Yale in America in the study of meteors. Much of what they discovered was inevitably also discovered independently in Europe. But news in those days traveled slowly. Here we are concerned only with Yale contributions as such, regardless of the possibility of slightly prior European discoveries about which the Yale astronomers could not have had any knowledge until much later. Suffice it to point out one reference to a presentation by A. S. Herschel (son of John and grandson of William Herschel) at a session of the Royal Institution of Great Britain, on May 25, 1866, "On the Shooting-stars of the years 1865–6, and on the probability of the Cosmical Theory of their Origin."[45] Here Herschel lauded Olmsted's work and deductions on the Leonid shower of 1833 and Herrick's on the Perseid meteors. Their European contemporaries or predecessors that Herschel mentioned are Quetelet on his discovery (then unknown to Herrick) of the Perseid radiant; Littrow's[46] deduction, from Biot's[47] catalogue of ancient Chinese records of meteors, of a 103-year period for the Perseids; and Humboldt's observations of the great Leonid shower of 1799.

REFERENCES

1. C. S. Lyman, *Amer. Jour. Sci.* (2), **28**, 109, 1859.
2. T. D. Woolsey, *The New Englander,* **17**, 575, 1859.
3. Funk and Wagnalls, *New Standard Encyclopedia,* **19**, 109, 1931.
4. E. Bridgwater and S. Kurtz, Eds., *The Columbia Encyclopedia,* p. 2102 (teacher training), 1963.
5. E. F. Guba, *Nantucket Odyssey,* Waltham, Mass., p. 294, 1965.
6. Anon. *Yale Necrology*: Obituary records for Yale men deceased in the year ending June, 1884, p. 182, 1884.
7. J. F. Fulton and E. H. Thomson, *Benjamin Silliman,* N. Y., p. 150, 1947.
8. M. Rothenberg, *The Educational and Intellectual Background of American Astronomy, 1825–1875,* Bryn Mawr Thesis, 1974.
9. T. Dwight, *Memories of Yale Life and Men, 1845–1889,* N.Y., p. 141, 1903.
10. Ref. 7, p. 268.
11. D. Olmsted, *Life of Ebenezer Porter Mason; Interspersed with Hints to Parents and Instructors on the Training and Education of a Child of Genius,* N.Y., 1842.
12. D. Hoffleit, *Historic Nantucket,* **33**, 13, 1986.
13. E. P. Mason, *Introduction to Practical Astronomy, Designed as a Supplement to Olmsted's* Astronomy; *Containing Special Rules for the Adjustment and Use of Astronomical Instruments, together with the Calculation of Eclipses and Occultations, and the Methods of Finding Latitude and Longitude,* N.Y., 141 pp. + VI Plates, 1841.
14. T. R. Treadwell, *Popular Astronomy,* **51**, 500, 1943.
15. D. G. Mitchell, *Reveries of a Bachelor,* N.Y., 214 pp. 1852.
16. A. P. Stokes, *Memorials of Eminent Yale Men,* New Haven, **1**, 162, 1914.
17. W. L. Kingsley, *Yale College—A Sketch of Its History,* N.Y., **2**, 504, 1879.
18. D. Olmsted, *Amer. Jour. Sci.,* **18**, 1, 1830.
19. A. Brekke and A. Egeland, *The Northern Lights,* N.Y., p. 108, 1983.
20. E. C. Herrick, F. Bradley, and E. Loomis, *Trans. Conn. Acad. Arts and Sci.,* **1**, 1, 1866.
21. E. Loomis, *Smithsonian Report,* p. 208, 1865; also *Arch. Sci. Phys. Nat.,* **31**, 273, 1868; and *Harper's New Monthly Mag.,* **39**, 1, 1869.
22. L. Harang, *The Aurorae,* N.Y., p. 10, 1951.
23. E. C. Herrick, *Amer. Jour. Sci.,* **33**, No. 2, 1837.
24. J. R. Eastman, *Bull. Philosophical Soc. Washington,* **11**, 275, 1890.
25. H. Shapley and H. Howarth, *Source Book in Astronomy,* N.Y., p. 88, 1929.
26. D. Olmsted, *Amer. Jour. Sci.,* **25**, 363; **26**, 132, 1834.
27. T. O. Mabbott, *Collected Works of Edgar Allen Poe,* Cambridge, Mass., Vol. **1**, Poems, p. 500, 1969. Quoted by permission of Harvard Univ. Press.
28. D. Olmsted, *Edinburgh Philosophical Jour.,* **23**, 89, 1837.
29. ———, *Amer. Jour. Sci.,* **31**, 386, 1837.
30. ———, *Ibid.,* **33**, 379, 1838.
31. ———, *Ibid.,* **35**, 368, 1839.
32. E. C. Herrick, *Ibid.,* (2), **1**, 86, 1846.
33. ———, *Ibid.,* **35**, 367, 1839.
34. C. P. Olivier, *Meteors,* Baltimore, p. 26, 1925.
35. J. V. Schiaparelli, *Sternschnuppen,* Stettin, viii + 268 pp. + IV plates, 1871.
36. E. C. Herrick, *Amer. Jour. Sci.,* **40**, 349, 1841.
37. J. Locke, *Trans. Amer. Philosophical Soc.,* n.s., **8**, 139, 1843; *also* Ref. 30 p. 43.
38. E. C. Herrick, *Amer. Jour. Sci.,* **33**, 176 and 401, 1838.
39. ———, *Ibid.,* **35**, 167, 1839.
40. ———, *Ibid.,* **33**, 360, 1838.
41. ———, *Ibid.,* **36**, 358, 1839.
42. L. A. J. Quetelet, *Ibid.,* **41**, 403, 1841.
43. A. C. B. Lovell, *Meteor Astronomy,* Oxford, p. 349, 1954.
44. E. C. Herrick, *Amer. Jour. Sci.,* **37**, 130, 1839.
45. B. Lovell, Ed., *Astronomy* (being the Friday Evening Discourses in Physical Sciences held at the Royal Institution: 1851–1939), **1**, 55, 1970.
46. C. L. v. Littrow, *Ast. Nachrichten,* **18**, 373, 1841.
47. E. Biot, *Catalogue general des étoiles filantes et des autres météores observés en Chine pendant vingt-quatre siècles,* 225 pp., and *Supplement,* 8 pp., Paris, 1846.

7
ELIAS LOOMIS AND SON

Olmsted died in 1859 and Herrick in 1862. Elias Loomis[1] (1811–89) was appointed Olmsted's successor and Professor of Natural Philosophy and Astronomy in 1860 (Fig. 7.1). Loomis had been Olmsted's student, 1826–30, and a tutor at Yale, 1833–36. Then he was called to Western Reserve University where he erected an observatory[2] and made many observations. These included, especially, 260 lunar culminations and 16 occultations, which he later compared with similar observations from abroad for the determination of accurate longitudes. Back at Yale after teaching at New York University from 1844 to 1860, Loomis followed admirably in Olmsted's footsteps in regard to textbook writing, teaching, and meteorology, and especially auroral investigations. However, after his return to Yale he contributed relatively little to the advancement of meteor studies except by keeping the subject very much alive in his lectures and treatises. Great strides in research in this area were left to a younger colleague, H. A. Newton, who as a student at Yale from 1846 to 1850 had presumably also come under the influence of Olmsted.

Yale's President Timothy Dwight (1828–1916, grandson of the first President T. Dwight), in the funeral services for Loomis[3] summarized his achievements:

> He delighted in original research, and had the patience and painstaking earnestness which are essential to the highest results in science.... He had another gift which belongs to the best mathematical minds—the power of clear statement. Among all the lecturers whom I have ever heard, I know of no one who surpassed him in this respect.... He was indeed no ordinary man ... He accomplished in the department of meteorology what no other scholar in our country has accomplished.... His pupils ... found in him a man who was as competent to help and to guide them as he was willing to do so, and as willing as he was competent.

Loomis in his devotion to teaching was an even more prolific textbook author than Olmsted. Between 1846 and 1878 he wrote no fewer than 23 treatises: 15 in mathematics (ranging from grade-school arithmetic through college calculus), 5 in astronomy, one on meteorology, one in both astronomy and meteorology, and one in natural philosophy (physics). Among these, some 600,000 copies were sold during his lifetime, providing him with a substantial income. Of these texts, half were written after his return to Yale in 1860. He also published nearly 150 articles in both popular and professional scientific journals. Before 1860 their subjects were about equally distributed among astronomy (30 papers), meteorology (25), and terrestrial magnetism (26). After 1860, however, his bibliography cites only 9 references in astronomy, in contrast to 46 in meteorology (including 13 on aurorae). In 1870 Loomis was the first to show a connection among terrestrial magnetism, the aurorae, and sunspots, all having the same period. He is credited with being the first, in connection with the storm of 1842,[1] to construct weather charts showing lines of equal barometric pressure, like those that were much later issued daily and systematically by the U. S. Signal Service when it was established in 1871, by the U. S. Weather Bureau since 1891, and by weather bureaus in many European countries.

While at Western Reserve, Loomis made significant efforts to alert the public to astronomical phenomena, and in particular he collected such observations on bright fireballs as members of the public were able to contribute. The fireball of November 15, 1859, especially interested him as it seemed to resemble the descriptions of the great aerolite of 1807 that dropped meteorites in Weston, Connecticut. Loomis's own computations indicated that the fireball of 1859 should have scattered meteorites along an east—west line at about latitude 35°13′N and longitude 75°W. Later he wrote wistfully[4] that the main mass should be found slightly west of Hughesville (Cape May), New Jersey, adding, "It is probable that fragments may yet be found. Will nobody look for them?" However, from H. A. Newton's account[5] that the fireball was observed from Massachusetts all the

Fig. 7.1. Elias Loomis (1811–89). (Courtesy Yale University Archives)

way to Virginia, the intermediate resting place indicated by Loomis seems very improbable, unless fragments had been separated much earlier in the path.

In the 1866 edition of his *Treatise on Astronomy* Loomis extrapolated backward the observed separations of the two components of Comet Biela which had first been observed as double by Herrick and Bradley in 1846. Loomis inferred that the comet must have split as early as 1844 and he surmised that "This separation may have been caused by the operation of some internal force like that which causes the tails of comets, or perhaps by collision with some other body like one of the asteroids." Later, in his 1869 version of the *Elements of Astronomy,* he invoked the late November (Andromedid) meteors as the culprits:

It has been found by computation that near the close of December, 1845, Biela's comet passed extremely near and probably *through* the stream of November meteors. It has been conjectured that this collision may have produced the separation of this comet into two parts: and that by subsequent encounters in 1859 and 1866, it may have been further subdivided and dissipated, so as to be permanently lost to our view.

On the dissipation of the comet he was right; but the meteors were mainly the end product rather than the cause of the disruption.

In his successive treatises on astronomy Loomis enumerated the various hypotheses that had, throughout history, been cited for the origin of meteors. These theories included purely atmospheric phenomena, both terrestrial and lunar volcanoes, and hypotheses that they were either Earth- or sun-orbiting bodies. Loomis himself strongly favored the assumption that the meteoroids move in orbits around the sun, and concluded that the only difference among ordinary shooting stars, detonating fireballs, aerolites, comets, and planets was in their size and density. Moreover, he continued,[6]

We hence conclude that the interplanetary spaces, instead of being absolutely void, are filled with a countless number of minute bodies, whose aggregate mass must be very great. The comets, like the earth, must encounter an immense number of these bodies, and a part of their motion must be thereby destroyed. This effect may be appreciable in the case of the earth and the other planets.

Loomis also remarked that although meteorites contained the same elements as are found in the crust of the earth, "The manner in which these elements

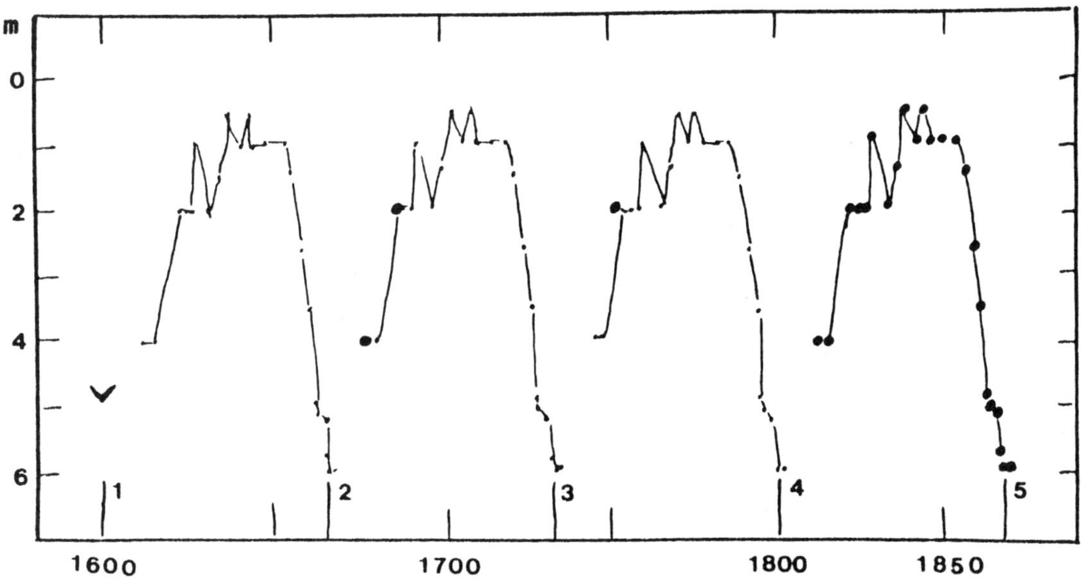

Fig. 7.2. Observations of η Carinae fitted to the Loomis 67-year period. Large dots represent the actual observations. The final cycle is retraced for epochs back to 1600. Predicted mimina are numbered 1–5.

are combined is peculiar, so that the general aspect of aërolites is sufficient to distinguish them from all other terrestrial minerals," a conclusion already reached at least half a century before. In his textbooks and lectures Loomis thus contributed much to the collection and dissemination of knowledge on meteors. However, his original research contributions in this field were comparatively minor.

Among his many publications Loomis apparently wrote only two papers on variable stars (based on data culled from the literature, not his own observations). One was on Algol[7] published in 1866, the other on η Carinae[8] (then called Argus) in 1869. He does, however, mention other variables in his textbooks. For Algol he collected all the published dates of minimum from which he deduced that the period changes. For η Carinae, a remarkably unusual nova-like star, he commented that Wolf[9] had deduced a period of 46 years, which Loomis found too short, and he derived a new period of 67 years. This was based only on the duration of the maximum cycle around 1843 and two isolated early observations, one by Halley in 1677, the other by Lacaille in 1751. Fig. 7.2 shows how he fitted the observations to this period.

Loomis had two sons, one of whom, Francis E. Loomis, obtained a Ph.D. from Yale in 1866 for a thesis on variable stars. He then went to Göttingen for more advanced training. There he wrote an expanded version of this thesis,[10] for which he was awarded another Ph.D., from Göttingen University in 1869. The thesis dealt mainly with the four variables, Algol, Mira, η Carinae, and T Coronae Borealis. For Algol he reached essentially but not exactly the same conclusions as Elias Loomis on the changing period, having added more observations that had been published since 1866. Fig. 7.3 shows the range of periods (dots joined by jagged lines) deduced by Francis Loomis, and for comparison, as horizontal segments, the periods given in the most recent *General Catalogue of Variable Stars*[11] (GCVS). The irregular deviations from a constant period are not even yet understood and it seems unlikely that any smooth curve can be found to represent all of the observations.

For Mira, the most often observed of all variables, then as well as now, Francis Loomis discussed the variations in cycle lengths. The intervals between successive maxima ranged from 228 to 369 days, and between minima from 251 to 381 days. Yet the mean period of 331.5 days was remarkably maintained. Loomis also inferred secondary periods of 40 to 160 years, which were only dubiously confirmed by more modern observations, mainly by the AAVSO[12,13] (the American Association of Variable Star Observers).

In the case of η Carinae, not only did the thesis contain exactly the same data the elder Loomis used in his analysis (published the same year) but both used exactly the same wording (except for the plural of the word magnitude) in reaching their conclusion that the star is probably periodic, "whose changes of brightness vary from the first to the sixth magnitude(s), and whose period is about seventy years." As neither author mentions or gives credit to the other, it is not clear who initially inspired the other.

Müller and Hartwig in the *Geschichte und Literatur des Lichtwechsels*[14] give some early references evidently not available to the Loomises. η Carinae was not included in Houtman's catalogue of 1603, which showed stars of the 4th and 5th magnitude, observed in the last decade of the 16th century. Therefore the variable was probably fainter than the 5th magnitude. This satisfies the Loomis 67-year period perfectly. η Carinae is reported to occur in the *Bayer Atlas* of 1603 as one of the stars marked *d,* not η. These are stars of about 3rd or 4th magnitude. It is not known whether the epoch of Bayer's observations preceded or followed those by Houtman. In view of the distortions of the configuration of Argo on the Bayer chart, the identification is dubious at best, although the magnitude would fit the descending branch of the predicted light curve. Also as pointed out by A. Winnecke,[15] Father F. Noël observed η Carinae at about second magnitude sometime between 1685 and 1689. This too fits the Loomis period.

In 1974 Feinstein and Marraco[16] (while reporting on a three-year period with a photoelectric amplitude of only 0.2 magnitude) compiled the light curve of η Carinae through 1973. At minimum between about 1900 and 1940, the star reached approximately 8th visual magnitude. Curiously there had been an increase of about one magnitude in 1889 (from 7.6 to 6.7) which seemed like the beginning of an ascending branch, a vestigial confirmation of the Loomis period but which failed to materialize, decreasing to low minimum within ten years, in contrast to the over twenty years' duration of the bright primary 1843 maximum. The star receded to its lowest recorded brightness by 1940 when it began a slow steady increase from 8 to 6 mag by 1973 (Fig. 7.4). Later observations by the AAVSO indicate it reached its maximum about 1980. The subsequent decline in mean annual magnitude is, however, well within the observational uncertainties.

As the inferred minima, as depicted in Fig. 7.2, appear to be more sharply defined than the maxima, the larger tick-marks at the bottom of both Figs. 7.2 and 7.4 indicate the predicted times of minimum reckoned from 1800 on the basis of Loomis' 67-year period. This seems to suggest that the maxima of η

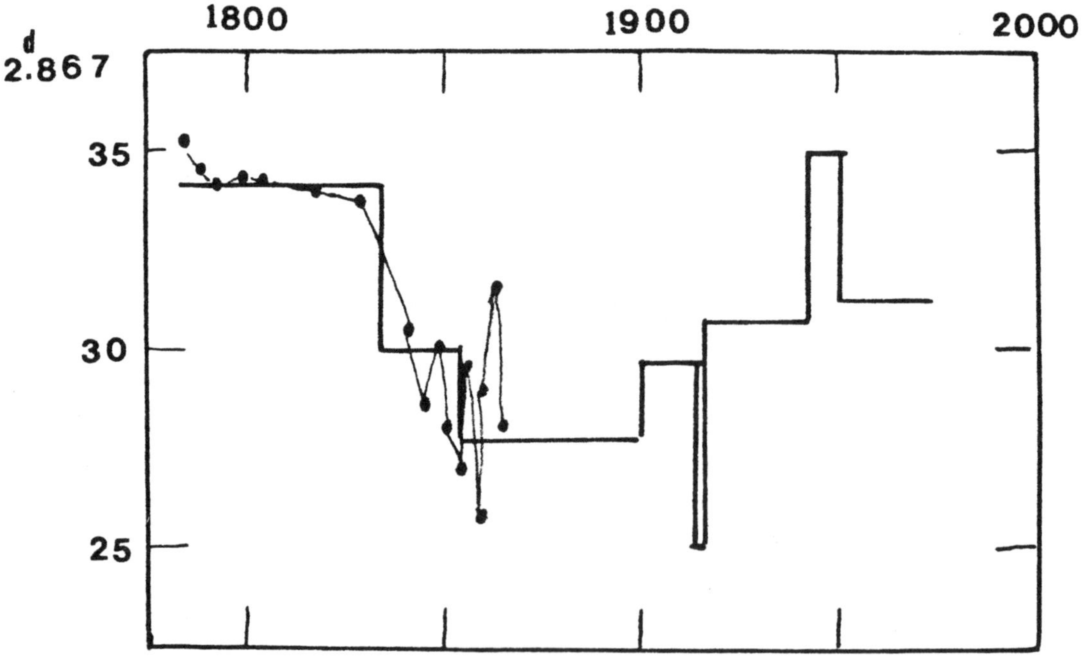

Fig. 7.3. The changing period of Algol. Dots from the Loomis dissertation; horizontal segments from GCVS. The ordinates give the 4th and 5th decimals of the period.

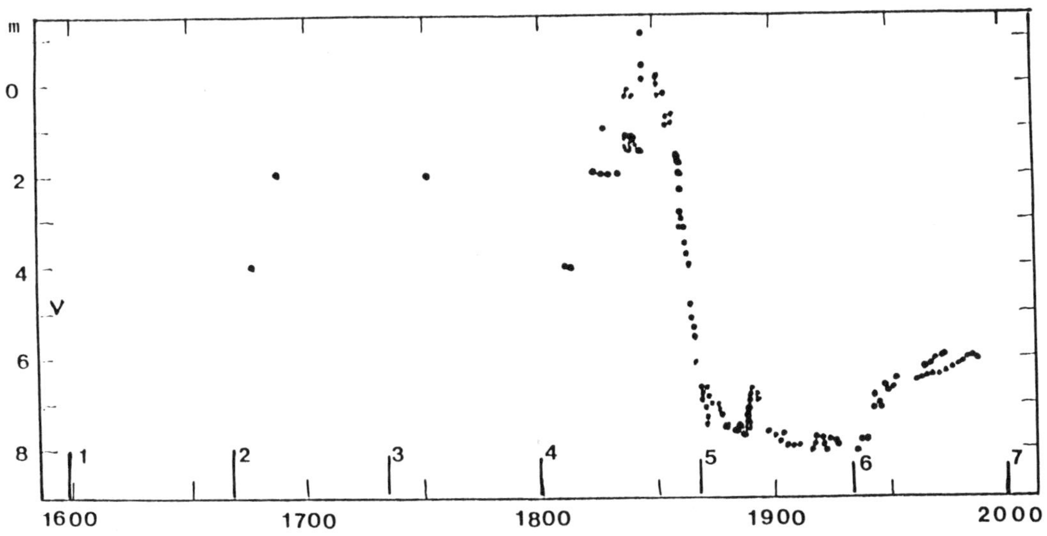

Fig. 7.4. Complete light curve for η Carinae from 1600 to 1984. After 1960 the upper set of observations gives photoelectric, the lower visual magnitudes. Predicted minima numbered 1–7.

Carinae may not be as peculiar or significant as the sudden decline after 1850 to a possibly previously unprecedented low minimum. This nova-like variable seems to remain a perpetual enigma. Observations in the next two decades should prove important for the interpretation of the star's behavior and evolution. Loomis's forgotten period may possibly play a crucial role.

Loomis tabulated all the available observations of T CrB by author. John Birmingham[17] has always been credited with having been the first to discover the outburst of the nova on May 12, 1866. Yet Loomis lists a few earlier observations for that same year by W. Barker—observations that had already been discredited as fraudulent by Hind[18] in August and by Stone[19] in December of 1866. Of the light curve of T CrB the younger Loomis supposed that it is "of the same kind as those of Mira and η Argus, but the period is probably a very long one, perhaps several centuries." On the other hand, he did note that in 1842 (only 26 years before its major outburst) Sir John Herschel[20] had observed a sixth-magnitude star in the same position, while Sir William Herschel[21] reported a double star of the ninth magnitude also in the same position in 1782. These identifications with T CrB are not mentioned in the extensive bibliography in the *Geschichte und Literatur des Lichtwechsels,*[22] or in the GCVS.[11] Has John Herschel's discovery simply been forgotten, or is it discredited because it could not be confirmed in any other source? In 1946, after an interval of 80 years, T CrB flared again; now, of course, the outbursts are no longer expected to be periodic.

In addition to his four special stars, Francis Loomis summarized the available published results for 100 variable stars (out of the 120 then known) for which periods had ostensibly been determined. It is of interest to compare these early determinations with currently accepted results for the same stars. In 89 cases the agreement is as good as can be expected on the basis of the relatively small numbers of epochs represented by the early observations. The periods of most of the Mira-type stars differ by less than 60 days, which is considerably less than the spread of cycle-lengths of Mira itself that Loomis had indicated. Table 7.1 lists the periods he quoted, together with the data from the GCVS for the eleven stars for which the old and the new results are discordant. For R Lib the old period is spurious, related to the new through the annual gaps in the observations. μ Cephei is a complex variable with three interfering periods—730, 904, and 4,400 days. These are uniquely related to the 5- to 6-year period that Loomis quoted from Chambers,[23] the inferred beat period computed from the three modern being 5.6 years.

The 315-year period given for Tycho's nova of 1572 was taken from Huggins and Klinkerfues[24] who pointed out that there had been novae with poorly determined positions in Cassiopeia in the years 945 and 1264. If these were indeed the same as that of 1572, the period of 315 years would imply that Ty-

Table 7.1
Discordant Results

VAR	Loomis Period	GCVS Period	GCVS Type	Remarks
S Mon	3.45d	—	Ia	Blue eruptive irregular variable.
R Sct	71.7	146.5d	RV	Like RV Tau, having alternate deep and shallow minima; early period did not differentiate between deep and shallow and therefore gave half the real period.
T Psc	146.5	252	SR	Semi-regular period.
R Mon	204	—	Ina	Like type Ia but associated with nebulosity.
V Sgr	316	—	Cst	Misnamed U Sgr by Loomis.
R Lib	723	241.85	M	Mira type: long period variable.
				Spurious: 1/241.85 − 1/365.25 = 1/716*.
μ Cep	5–6yrs	730	SRc	Misnamed U Cep by Loomis.
				Red semi-regular variable.
		(904)		An earlier rejected period.
				1/730 − 1/904 = 1/3788.
		4400		1/3788 + 1/4400 = 1/2036 = 1/5.6 yrs.
P Cyg	18yrs	—	S Dor	High luminosity eruptive variables, like the prototype star.
η Car	67	—	S Dor	
R Cep	73	—	Cst	
B Cas	315?	—	SNI:	Tycho's Nova 1572. Misnamed β Cas in Loomis's table but not in his text. Supernova of Type I, at maximum the intrinsically most luminous kind.

*Spurious periods are usually related to each other or to the true period (P_1, P_2, P_3 . . . through the common factor p of the intervals between the observations (one day or one year) by formulae of the form, $1/P_1 \pm 1/p = 1/P_2$.

cho's star would have been bright at the time of the birth of Christ, hence the Christmas star. As Tycho's star is now confirmed as a supernova (as well as for other reasons) this hypothesis is untenable.

Having concluded his own analysis of data and periods, Francis Loomis speculated on possible causes for the variations. He discussed four hypotheses, all of which he found wanting in one respect or another, especially as he was searching for a common cause for all types—Algol, the genuinely periodic stars, and "temporary" stars:

1. Stellar rotation of a thin disc-shaped star, alternately presenting a flatface and edge-on appearance to the observer, could not explain the light curve of Algol, or the "seemingly capricious" irregularities of Mira and η Carinae. "This hypothesis should then be absolutely rejected."
2. An annular eclipse of Algol by a companion adequately explains the major variations of that star, but is not compatible with the variations of Mira or η Carinae.
3. "It has been supposed that a cluster of dark bodies, or a nebulous body of great extent may revolve about the variable star so as at times to intercept a portion of its light." This hypothesis seemed to him to have some merit. However, he concluded, "The remarkable fluctuations in the light of Mira, of η Argus and T Coronae, seem to indicate changes in the condition of the star itself: so that although this third hypothesis may be to some extent true, it does not seem sufficient to explain all the phenomena."
4. "It has been supposed that the variable star may not be uniformly luminous upon every part of its surface, but by rotation upon an axis may occasionally present to the earth a disc partially covered with dark spots, and shining therefore with a dimmer light." Again this would not account for all the observed phenomena, even when it was additionally stipulated that the star-spots themselves were variable.

Already in the 1866 edition of his *Treatise on Astronomy,* the elder Loomis had enunciated these four hypotheses, and concluded that temporary stars were probably also periodic stars but of longer period. In the 1869 edition of the *Elements of Astronomy* Elias Loomis concluded, "If we suppose that the spots are periodically developed upon the surface of the variable star, similar to those which are observed upon our sun, but of vastly greater extent, the phenomena of most of the variable stars may be explained." Later he seemed to vacillate: in the 1871 and 1874 editions of the *Treatise on Astronomy* he favored the hypothesis of the interception of light by a nebulous body; but once more in 1888 (his latest edition) he reverted to the star-spot theory: "It seems most philosophical to conclude that the changes in the periodic stars are due to causes analogous to those which exist in our own system," adding that it should not be surprising if a few stars were to be found for which the changes are far greater than in the sun.

In textbooks it is not always necessary to cite specific sources. Hence it is not clear from the elder Loomis's treatises who actually originated the various hypotheses. (That the elder Loomis must have been familiar with the work of foreign astronomers seems attested by the numerous old English, German, and French journals now shelved in the Yale Department of Astronomy, bearing bookplates signifying his original ownership.) Obviously, father and son must have shared thoughts and discussions. The analogy between some of the light curves and the sunspot cycle was evidently not original with either Loomis. R. Wolf (see Clerke[25]) had been credited with this suggestion as early as 1852, while the same suggestion by Goodricke[26,27] as early as 1783, as an alternative for the binary theory, seemed to have been entirely forgotten.

Concerning his fourth hypothesis, Francis Loomis (like his father[28]) looked into the variations of sunspots, speculating whether disturbing effects of the planet Jupiter could be responsible, Jupiter having nearly the same period of revolution as the sunspot period. He suggested that the difference might be due to magnetic effects, noting how the magnetic pole of the earth changes progressively. If indeed Jupiter were responsible for the sunspots, he assumed that the other planets should likewise, but to a smaller extent, affect the sunspot cycle. He seemed, in fact, to find evidence for secondary periodicities related to Saturn, Venus, and the Earth. He therefore reasoned that unseen planetary companions might similarly be responsible for stellar variations. His reasoning reflects his father's interest in terrestrial magnetism and its relation to sunspots and aurorae. However, the younger Loomis did not estimate the amplitude of the total light variation of the sun during a sunspot cycle (only a few hundredths of a magnitude), which is far smaller than the variations of the mainly Mira-type stars he was trying to explain. Regarding temporary stars like T CrB, Francis Loomis recalled that their outburst had generally been ascribed to the direct collision of some foreign body like a comet. If the star observed by John Herschel in 1842 was really the same as that of 1866, then he strongly favored the hypothesis that T CrB was influenced by a large planet in a highly eccentric orbit approaching close to the surface of the star.

In conclusion, Francis Loomis favored star spots caused by planetary disturbances as the major cause of stellar variability. While this theory is untenable for the relatively high-amplitude stars he studied, it is of historic interest to note that modern photoelectric photometry is detecting for other stars of low amplitude the star spots that the Loomises advocated more than a hundred years before, but now without implicating planets. Pulsation theory, now taken for granted, did not come into vogue until nearly 1920. It had been first suggested by A. Ritter[29] in 1878–79, but seems to have been ignored until Shapley[30] in 1914 revived it in his refutation of the then prevalent hypothesis that Cepheid variables were binary systems with ellipsoidal components, and Eddington[31] in 1918–19 put the pulsation theory on firm ground by his mathematical development. Evidently Elias Loomis was not aware of Ritter's work when he revised the latest editions of his treatises.

Tremendous advances in the understanding of these variables have been accomplished since the time of the Loomises. While older generations of astronomers struggled to interpret their observations they came upon ideas that future generations could not support. Yet often the very hypotheses that ultimately failed to account for the observations in hand, independently recurred later to account for new data on other types of stars. The elder Loomis, in particular, through his teaching and textbooks, whether right or wrong from our present standpoint, did much to stimulate the progress of astronomy. Francis Loomis, after a brilliantly promising early career, was less fortunate. Although he lived to age 76, he spent most of his life as an incurable invalid. In 1871 he had been appointed Professor of Physics and Industrial Mechanics at Cornell University, but within eight months, after losing one lung in a bout with pneumonia, he resigned. He moved to Switzerland, hoping against hope that the climate there would be beneficial to his health. His Göttingen thesis was his only published contribution to astronomy.

While still at New York University, Elias Loomis had written to Yale, offering to work there at his own expense (in view of the financial success of his books) provided Yale would build and equip an observatory.[32] This generous offer Yale could unfortunately not afford to accept. Curiously, after Loomis had returned to Yale, the Sheffield Scientific School in 1866 acquired a 9-inch Alvan Clark telescope, but there is apparently no record that Loomis made any use of the Sheffield observatory. His name is not even mentioned in Chittenden's comprehensive two-volume history of the Sheffield School.[33] Loomis took great interest later in the plans Oliver Winchester proposed for financing the establishment of an observatory for Yale College. At a memorial service for Loomis in 1890 H. A. Newton[34] said,

> To establish an Astronomical Observatory had been through life a cherished object. He entered into and aided heartily the plans of Mr. Winchester, both before and after Mr. Winchester asked his Trustees to transfer his magnificent endowment to the University. Professor Loomis looked forward to a large institution in the future on the observatory site. To endow this public service, after making liberal provision for his two sons, he bequeathed his estate. The income from more than $300,000 will eventually be available to continue the work of his life. With clear judgment of what was most important he limited the use of that income to the payment of salaries of persons whose time should be exclusively devoted to the making of observations for the promotion of the science of astronomy, or to the reduction of astronomical observations, and to defraying the expenses of publication. He knew that if he provided observers, other benefactors would furnish buildings, and instruments, and the costs of supervision and maintenance.

As early as 1838 E. Loomis had expressed his philosophy

> that it is essential to the best interests of society that there should be a certain class of men devoted exclusively to the cultivation of abstract science without any regard to its practical applications; and consequently that such men instead of being a dead weight upon society are to be ranked among the greatest benefactors of their race.[34]

He would have been happy to know that the various testimonials to his memory and that of his sons effectively support this ideal. In 1893 friends bought, for $1,250, an 835-pound Canyon Diablo meteorite and presented it to Yale in his memory. (H. A. Newton said of this meteorite that it disclosed both black and white diamonds.) In 1902 Francis Loomis gave Yale $10,000 to found the Francis E. Loomis Fellowship in Physics; and in 1911 to the Yale School of Medicine $20,000, the income of which was to be used for aiding research. The other son, H. B. Loomis (1853–1939), a lawyer, in 1905 established the Henry Bradford Loomis Fellowship in Chemistry. Arthur Hill, Chairman of the Department of Chemistry, wrote to H. B. Loomis on September 26, 1935, "Your fellowship is the only one of our present offerings which takes adequate care of the expenses of a graduate student, and I do want you to know how grateful we are for what you are doing for the Department of Chemistry."[35]

H. B. Loomis also made gifts to increase the fellowship in physics established by his brother. But the one memorial that would have pleased Elias Loomis most came in 1923 when the unique Loomis Coelostat was dedicated to his memory[36] (see Chapters 15 and 17). A street in New Haven has been named for him (Loomis Place) and a bronze tablet in Woolsey Hall proclaims him to have been "An exact scholar, an astronomer of wide repute, in meteorology a pioneer and a large benefactor to this University." As E. Loomis had been the first to construct weather maps,[37] which eventually became the prototypes of those regularly issued by the U.S. Weather Bureau, his biographer A. P. Stokes remarked that the U. S. Weather Bureau "is more his monument than that of any other man."[38]

REFERENCES

1. H. A. Newton, *Elias Loomis LL.D.*, Memorial Address at Yale, April 11, 1890.
2. W. I. Milham, *Early American Observatories,* Williamstown, Mass., p. 44, 1938.
3. T. Dwight, *The New Englander,* **51**, 310, 1889,
4. E. Loomis, *Amer. Jour. Sci.,* (2), **29**, 137, 298, 447, 1860.
5. H. A. Newton, *Amer. Jour. Sci.,* (2), **30**, 186, 1860.
6. E. Loomis, *Treatise on Astronomy,* N.Y., p. 284, 1871.
7. ———, *Proc. Amer. Assoc. Adv. Sci.,* **15**, 5, 1866.
8. ———, *Monthly Notices Roy. Ast. Soc.,* **29**, 298, 1869.
9. R. Wolf, *Ast. Nachrichten,* **60**, 61, 1863.
10. F. E. Loomis, *Periodic Stars,* Inaugural Dissertation Univ. Göttingen, 1869.
11. P. N. Kholopov, *General Catalogue of Variable Stars,* Moscow, 1985–87.
12. D. Proust, *Bull. Assoc. Française des Obs. d'Étoiles Var.* No. 24, p. 57, 1983.
13. D. Hoffleit, *Jour. Amer. Assoc. Var. Star Obs.,* **13**, 67, 1984.
14. G. Müller and E. Hartwig, *Geschichte und Literatur des Lichtwechsels,* Leipzig, **1**, 304, 1918.
15. A. Winnecke, *Ast. Nachrichten,* **51**, 383, 1859.
16. A. Feinstein and H. G. Marraco, *Ast. and Astrophys.,* **30**, 271, 1974.
17. J. Birmingham, *Monthly Notices Roy. Ast. Soc.,* **26**, 310, 1866.
18. J. R. Hind, *Ast. Nachrichten,* **67**, 257, 1866.
19. E. J. Stone, *Monthly Notices Roy. Ast. Soc.,* **27**, 57, 1866.

20. J. Herschel, *Monthly Notices Roy. Ast. Soc.,* **26**, 299, 1866; **27**, 213, 1867.
21. W. Herschel, *Phil. Trans.,* **75**, 40, 1785.
22. Ref. 14, **2**, 435, 1920.
23. G. F. Chambers, *Ast. Nachrichten,* **63**, 117, 1864.
24. W. Huggins, *Spectral-Analyse,* Trans. into German with additions by W. Klinkerfues, Leipzig, p. 81, 1868.
25. A. Clerke, *History of Astronomy in the Nineteenth Century,* London 4th ed., p. 128, 1902.
26. J. Goodricke, *Phil. Trans.,* **73**, 483, 1783.
27. H. Shapley and H. Howarth, *Source Book in Astronomy,* McGraw-Hill, N.Y., p. 152, 1929.
28. E. Loomis, *Amer. Jour. Sci.,* (2), **50**, 153, 1870.
29. A. Ritter, *Ann. d. Physik u. Chemie,* n. s., **5**, 405, 543, 1878.
30. H. Shapley, *Astrophys. Jour.,* **40**, 448, 1914.
31. A. Eddington, *Monthly Notices Roy. Ast. Soc.,* **79**, 2, 177, 1918–19.
32. Ref. 1, p. 16.
33. R. H. Chittenden, *History of the Sheffield School,* New Haven, 1928.
34. Ref. 1, p. 31.
35. A. Hill to H. B. Loomis, Sept. 26, 1935, *Yale Obs. Archives.*
36. F. Schlesinger, *Pop. Ast.,* **31**, 77, 1923.
37. Ref. 1, p. 21.
38. A. P. Stokes, *Memorials of Eminent Yale Men,* New Haven, **2**, 28, 1914.

8
THE SHEFFIELD OBSERVATORY

In 1847 Yale added a *Scientific Department* to its regular, mainly classical curriculum. This new department was converted into a separate *Scientific School* within Yale in 1854, and in 1861 it was renamed the *Sheffield School* in honor of its chief benefactor (Fig. 8.1). The innovation of science and engineering had indeed been looked upon as inferior to the aims of the University as such. As Chittenden[1] describes Yale's position in 1845, there existed "a deep seated aversion to new ideas, especially so if the latter threatened to diminish in any degree the amount of time available for the dead languages; and science, so far as it was understood, was considered of relatively small importance and to have little educational value." Hence the courses offered in the Scientific School generally differed from, and were kept separate from, those in Yale College. The students in the Scientific School were, however, given the privilege, upon payment of a $15 fee, of attending the lectures of the elder Professor Silliman and Professor Olmsted in the "academic department."[2]

William Augustus Norton (1810–83) was a distinguished graduate of the military academy at West Point, where he became professor of natural sciences and served from 1833 to 1839, after having taught for six years at the University of the City of New York. Then he was enticed to successively more congenial positions, first to the University of Delaware, then to Brown University, and finally to Yale. On July 24, 1852, Norton sent the following proposal to the Yale Corporation: "I propose to open a School of Engineering, sustain it at my own expense, and provide it with the necessary apparatus;—receiving all the emoluments of the same. The School to be attached to the Department of Philosophy and the Arts in the College . . ."[3] By July 27 the Corporation had so voted, appointing Norton the Professor of Civil Engineering, and adding; "None of the existing funds of the College to be applied in support of this professorship." How stimulating he must have been as an instructor is exemplified by the fact that his first class at Yale included his Brown University students who had followed their professor to Yale.[4] In 1854–55 the Chemical and Engineering Schools were united under the title, "The Yale Scientific School." Norton held his post for 31 years. Besides his major duties of teaching engineering and physics, Norton was greatly interested in astronomy (not originally included in the Engineering School curriculum). He had written *An Elementary Treatise on Astronomy* which went through three editions (1839, 1845, 1852) before he came to Yale; *A First Book of Natural Philosophy and Astronomy* (1868), and three editions of *A Treatise on Astronomy, Spherical and Physical* (1867, 1875, 1881). In addition he published numerous research papers, mainly in the *American Journal of Science.* Among these he revealed considerable thought on the physical constitution of the sun and of comets.

Agnes Clerke[5] credits Norton as having been (1845) "The earliest formal advocate of the Expulsion Theory of the solar surroundings," for accounting for the appearance of the corona. Norton was even more concerned with the appearance and origin of cometary tails. Already in 1812 Olbers in Germany had proposed a theory that an electrical repulsive force was exerted by the sun, driving particles off the head of the comet to form the tail. This theory was further investigated by such eminent astronomers as John Herschel, Friedrich Bessel, and Benjamin Peirce. Norton pursued the problem more deeply, and with the appearance of Donati's Comet of 1858, he felt that the observations of the shape of the comet's tail were in good accord with the theory that particles evaporated from the head of the comet by the sun's heat were repulsed by an electrical force from the sun. Not until "the space age" nearly a century later was it discovered that the solar corona is expanding,[6] causing a corpuscular solar wind consisting of ions and electrons. This wind now accounts not only for the fact that comet tails basically point away from the sun, but also for the complex distortions of some of them.[7]

Fig. 8.1. Joseph E. Sheffield (1793–1882). (Courtesy Yale University Archives)

For Olmsted's course in Practical Astronomy, observations were carried out from 1856 with a transit instrument, meridian circle, zenith telescope, and sextant. In 1859, the year of Olmsted's demise, Chester S. Lyman (1814–90)[8–10] was appointed Professor of Industrial Mechanics and Physics, a title later changed to Professor of Physics and Astronomy, and finally, in 1884, simply to Professor of Astronomy in the Sheffield School. Prior to the acquisition of the Sheffield telescope in 1866, Professor Lyman used his own private observatory for training students in observational astronomy.

In 1852 a schedule had already been set up to lead to the degree of Bachelor of Philosophy.[11] That first year six men were awarded the degree[12,13] two of whom later became professors at the Sheffield School (W. H. Brewer, Professor of Agriculture; and G. J. Brush, Professor of Metallurgy and Mineralogy, and Director of the Sheffield Scientific School[14]). On July 20, 1860, on the recommendation of the faculty of the Scientific School, the Yale Corporation voted to establish the degree of Doctor of Philosophy along the general lines of German universities. The *Historical Register* of Yale University[15] notes, "Certainly Yale and Sheffield are entitled to the credit of introducing among American institutions the Degree of Doctor of Philosophy, demanding for it a high standard of attainments, and never bestowing the honor by any irregular promotion."

The first Ph.D. degree in America[16] was awarded by the Sheffield School in 1861, to Arthur W. Wright for his dissertation on the computation of meteor orbits:[17] *Given the Velocity and Direction of Motion of a Meteor Entering the Atmosphere of the Earth, to Determine its Orbit about the Sun, Taking into Account the Attractions of Both of these Bodies.* By 1869 the Scientific School had awarded 155 Bachelor of Philosophy and eight Doctor of Philosophy degrees[18] (two for topics in astronomy, the second to Francis E. Loomis). Until 1919 students at Sheffield matriculated in just three years, in contrast to the usual four in Yale College.

Joseph Earl Sheffield (1793–1882),[19] a merchant and philanthropist, took great interest in the Scientific School. In 1866 he presented it with a 9-inch Alvan Clark refractor of focal length 9 feet 10 inches with an equatorial mount. The tube is reputed to have been one of the last wooden tubes made by the Clarks. The telescope was first installed atop a building on the corner of Grove and Prospect Streets (Fig. 8.2) where, since 1931, Sheffield-Sterling-Strathcona Hall now stands.[20] The original building had been erected as a hotel by James Hillhouse, Treasurer of Yale College, where he hoped to entertain distinguished guests of the College.[21] However, it was never used for this purpose, and from 1814 until 1858 it became Yale's Medical Institution. At that time it was in bad repute because of an unfounded rumor that a tunnel connected the building with the Grove Street Cemetery across the street, enabling students to steal corpses![21] In 1859 Sheffield bought the building for the Scientific School and it became known as Sheffield Hall. He added a clock tower in 1865, topped by a gun turret for the telescope (designed after the turret of the battleship *Monitor*), at a cost of some $60,000 for the building, equipment, and library. In 1870 the Hillhouse Mathematical Library, containing a large collection of books on pure mathematics, astronomy, mechanics, etc., was purchased by Sheffield for the School.[22] His total benefactions to the Sheffield School ultimately amounted to over a million dollars.

Professor Lyman, now Director of the Sheffield Observatory, used the telescope both for teaching and research. He observed sunspots regularly[23] and made some important observations of Venus at its closest approach to the sun.[24] The presence of an atmosphere on Venus had been surmised[25] from observations of the transits of Venus in 1761 and 1769; and in 1842 F. Guthrie[26] in England appears to have been the first to see a complete annular ring of nebulosity about the narrow crescent image, attributing the ring to refraction in a Venusian atmosphere. But according to Agnes Clerke,[25] this phenomenon was

Fig. 8.2. South Sheffield Hall showing the Monitor-type turret housing the Alvan Clark 9-inch refractor donated by Sheffield in 1866. (Courtesy Yale University Archives)

seen "with more satisfactory distinctness by Mr. C. S. Lyman of Yale College, before and after the conjunction of December 11, 1866, and during nearly five hours previous to the transit of 1874 when the yellowish ring of refracted light showed at one point an approach to interruption, possibly through the intervention of a bank of clouds."

Like Olmsted and Loomis, Professor Lyman considered teaching as his major lifework. He was also a gifted instrument designer, using his skill to make equipment for teaching demonstrations. In 1852–53 he had constructed a combination transit-instrument and zenith-telescope for the determination of latitude,[27] built some ten years before the English geodecist-astronomer, George Davidson,[28] invented essentially the same instrument. Lyman's was used for many years in the survey of India.[9] He was also interested in practical horology and devised improvements in escapements and compensation pendulums for observatory clocks.[29]

In 1869 M. D. Collier (Yale 1866) gave the Sheffield School $5,000 for the purchase of scientific instruments in memory of his brother, T. F. Collier, who had attended the School in 1866–68. Lyman was asked to travel to Europe to select and purchase equipment that was not readily available in America. Lyman invited his friend, Professor H. A. Newton, to accompany him. Before leaving, Lyman gave a lecture to friends of the School on the purpose of his trip. It was so inspiring that the friends spontaneously donated more than $4,000 additional funds for the purpose, a really substantial sum for those times.[30] When Lyman retired in 1889 he wrote to the Director of the Sheffield School,[31]

> I wish that I could leave to the School some gift which would adequately express the interest I have felt in it since my first connection with it thirty years ago. All that I have which would be of any use are, 1st, The Transit Instrument of my own make and as a substitute for the Zenith Telescope of my own invention; 2nd, The Astronomical Clock by Howard which is in the same room; 3rd and 4th, Two pieces of apparatus of my own invention; one for the illustration of Ocean Waves, and the other for describing Lissajous' Acoustic Curves. These four articles I very gladly present to the School.

A Chester S. Lyman Lectureship Fund was established in 1910 by a gift of $5,000 from his son, Chester W. Lyman, to maintain a course of lectures on water conservation. This was combined in 1926 with the Chester S. Lyman Memorial Fund for teaching hydraulic engineering.[32]

There is little evidence what use, if any, was made of the Sheffield Telescope between Lyman's death in 1890 and 1920 when Schlesinger became the Director of the Yale Observatory. Schlesinger wanted to use the Sheffield telescope for public open nights and for undergraduate teaching.[33] However, he found that the mounting had deteriorated, the room in which the telescope was housed was too small, and the stairway leading to it was too hazardous to accommodate more than a few visitors at a time. By 1926 in Schlesinger's *Annual Report* we find, "Efforts are being made . . . to move the nine-inch Lyman [*sic*] telescope from its present location in the tower of Sheffield Hall to a place that will be more convenient and accessible to undergraduates." He first contemplated mounting it on top of the tower of the old library; then Thomas Farnam suggested the roof of Bingham Hall. The following year the telescope was dismantled and temporarily removed to the Yale Observatory (originally named the Winchester Observatory) for renovation and storage until it could be moved to the roof of Charles W. Bingham Hall where it would be erected in "a fine revolving dome" that cost about $6,900, and would be ready for use by the undergraduate class in the 1928–29 term.[34] A bronze tablet was mounted in the dome, inscribed as follows:

THE SHEFFIELD TELESCOPE
MADE BY ALVAN CLARK
DONATED IN 1866 TO THE
SHEFFIELD SCEINTIFIC SCHOOL
BY JOSEPH EARL SHEFFIELD
MOUNTED UNTIL 1927 ON
SHEFFIELD HALL AT THE
CORNER OF PROSPECT AND
GROVE STREETS WHEN IT WAS
MOVED TO THIS SITE

In 1931 a small transit instrument was also set up at Bingham for undergraduate instruction. Here, from 1930 to 1942 Dr. Dirk Brouwer taught observational astronomy.

Among the many students who used the Sheffield telescope was Thornton Page[35] (Director of the van Vleck Observatory from 1959 until 1971) who recalls, "I certainly remember the 9-inch on Bingham Hall roof . . . it was the first telescope I ever used, in Brouwer's astronomy class in 1932. He had us time occultations and the 4 or 5 students also used it to look at the Moon, planets, nebulae and clusters." This was the era when E. W. Brown was soliciting the help of amateur groups for observing occultations or carrying out relevant computations in connection with his investigations of the motions of the moon.

Brouwer's *Annual Report* for 1944–45 notes that undergraduate use of Bingham Hall ceased temporarily when Old Campus became off-limits to civilians during World War II. He added, "Presumably the instrument will be used again for the same purpose, but plans should be considered for its removal to a better site." Bingham Hall he considered "very unfavorable and the construction of the dome leaves much to be desired." None of his succeeding annual reports mentions when the telescope was actually dismantled. By 1950 plans had been made to dismantle the Southern Station at Johannesburg, South Africa, and move it to the site of the Mt. Stromlo Observatory in Australia. The transfer was expected to be completed by 1952 when the lens of the 9-inch Sheffield would be sent there to be used as a guide telescope for the 26-inch refractor.[36] When the Australian site was also found to be unsatisfactory for Yale's purposes, the 26-inch was presented to the Australian National University, proprietors of the Mt. Stromlo Observatory. The 9-inch, however, was removed with the intention of using it as a guide-telescope for the unique camera-cluster at the new observatory location at El Leoncito, Argentina.[37] That instrument was never completed. The Yale observers at Mt. Stromlo had unfortunately found the historic Sheffield telescope unsatisfactory as a guide telescope, nicknaming it "the 9-inch Oddie."

The dome atop Bingham Hall is still there. It had remained idle for perhaps a decade when, in 1955, a young instructor, Harlan J. Smith, mounted a toy $15 Spitz planetarium on the pier. Projecting the constellations on the inside of the dome "actually worked surprisingly well for the purpose of showing coordinate systems and bright stars."[38] After Smith left Yale the dome was abandoned for astronomical purposes. Thus ends the history of the second observatory at Yale.

REFERENCES

1. R. H. Chittenden, *History of the Sheffield Scientific School of Yale University,* New Haven, p. 21, 1928.
2. ———, *Idem,* p. 49.
3. ———, *Idem,* p. 57.
4. W. P. Trowbridge, *Biog. Mem. Nat. Acad. Sci.,* **2**, 189, 1886.
5. A. Clerke, *History of Astronomy in the Nineteenth Century,* London, 4th ed., p. 193, 1902.
6. *Cf.,* J. M. Pasachoff, *Contemporary Astronomy,* Philadelphia, p. 153, 1989.
7. J. C. Brandt., *Introduction to the Solar Wind,* San Francisco, 199 pp., 1970.
8. Ref. 1, p. 77.
9. W. J. Youmans, Ed., *Pop. Sci. Monthly,* **32**, 116, 1888.
10. Anon, *Sidereal Messenger,* **9**, 400, 1890.
11. Ref. 1, p. 53.
12. G. W. Pierson, *Yale College, An Educational History, 1871–1921,* p. 704, 1952.
13. Ref. 1, p. 54.
14. W. W. Skiddy, *Biographical Records, Classes 1852–67,* Sheffield Scientific School, 1909.
15. *Historical Register of Yale University, 1701–1937,* New Haven, 1939.
16. J. F. Fulton, *Bull. History of Medicine,* **20**, 104, 1946.
17. *Doctors of Philosophy,* Yale University 1861–1960, New Haven, 1961.
18. Ref. 1, p. 127.
19. *Webster's Biographical Dictionary,* Springfield, Mass., 1943.
20. R. C. Carroll, Ed., *Buildings and Grounds of Yale University,* Yale Printing Service, New Haven p. 31, 1979.
21. R. Holden, *A Pictorial History of Yale,* New Haven (pages not numbered), opp. Figs. 85–86, 1967.
22. Ref. 1, p. 335.
23. C. André and A. Angot, *Les Observatoires en Europe et en Amérique,* Part 3, p. 127, 1877.
24. C. S. Lyman, *Amer. Jour. Sci.,* (2), **43**, 129, 1867; (3), **9**, 47, 1875.
25. Ref. 5, p. 253.
26. F. Guthrie, *Monthly Notices Roy. Ast. Soc.,* **14**, 169, 1854.
27. C. S. Lyman, *Amer. Jour. Sci.,* (2), **30**, 52, 1860.
28. G. Davidson, *Monthly Notices Roy. Ast. Soc.,* **28**, 181, 1868.
29. C. S. Lyman, *Amer. Jour. Sci.,* (2), **12**, 251, 398, 1851.
30. Ref. 1, p. 78.
31. *Idem,* p. 80.
32. *Idem,* p. 535.
33. F. Schlesinger to R. H. Chittenden, Feb. 27 and Mar. 4, 1920, *Yale Obs. Archives.*
34. *Idem,* Nov. 1926, *Yale Obs. Archives.*
35. T. Page, *Private communication,* Mar. 27, 1983.
36. D. Brouwer to R. Woolley, Oct. 29, 1952, *Yale Obs. Archives.*
37. D. Brouwer to C. Jackson, Oct. 25, 1962, *Yale Obs. Archives.*
38. H. J. Smith, *Private communication,* Apr. 20, 1983.

9
HUBERT A. NEWTON: ACE NINETEENTH CENTURY AMERICAN METEORITICIST

While Olmsted and Herrick laid the foundation for better research on meteors, it was Hubert Anson Newton (1830–96)[1–3] who had been the American most credited with the advancement of this science, as Lyman points out, both for his own research and as a guide for others.[4] His outstanding research represents an important period in the history of astronomy at Yale. He was trained as a mathematician, first as a Yale undergraduate and later under the tutelage of the French geometrician, Michel Chasles. In 1853 Newton became a tutor at Yale and two years later, when his superior in the Mathematics Department died, Newton, at the age of only twenty-five, was given a professorship in Yale's Mathematics Department. He remained at his alma mater until his own death in 1896.

Newton's chief interest, however, was not pure mathematics but rather the use of mathematics to analyze astronomical problems, particularly those involving the dynamics of meteors. His first published paper,[5] in 1853, was theoretical, *On the Deviation of Falling Bodies to the South of a Perpendicular.* Considering the effects of the Earth's rotation as well as gravitation, he found that a body dropped from any height would deviate to the south if dropped in the northern hemisphere, and to the north if dropped in the southern hemisphere. This was clearly a forerunner of his later investigations on meteors and meteorites. At this time it was not entirely clear whether meteors were solid bodies, whether they had a cosmic origin, or what the nature of such a cosmic source might be. There were many unanswered questions pertaining to meteoric orbits, the composition of meteorites, and the relationship of comets and asteroids to meteor showers, all of which Newton attempted to solve.

Newton's earliest papers in meteoritics dealt with some spectacular fireballs, those of November 15, 1859, August 2 and 6, 1860, and August 10, 1861.[6–8] From the positions, durations of flight, and path lengths observed from different localities, he attempted to derive heights and velocities. Although he was aware of the tendency of observers to overestimate the time of flight, and made allowance for this, he did not take into account a like tendency to overestimate the path length. Consequently, his velocities for two of the fireballs came out too large and he optimistically concluded that the velocities were hyperbolic, indicating that the fireballs had originated outside the solar system. In the case of the August 6 fireball he commented, "Though an elliptic orbit is possible I can hardly think it probable." Ever since, various investigators have searched for, and often seriously suspected meteors of having hyperbolic velocities, but none has yet been definitely confirmed. Of the three early fireballs that Newton investigated, none yielded any meteorites. He deduced that atmospheric friction was the source of the heat of fireballs, and that only the slower and/or more massive could survive passage to the ground as meteorites.[9] Thus early in Newton's career he believed that meteors, fireballs, and meteorites all represented the same types of bodies differing only in mass and/or velocity.

The August 10, 1861, fireball appeared to come from the Perseid radiant. From the data for this and other Perseids Newton derived an orbit[9] differing widely from modern determinations (Table 9.1). The discordances, especially in period, are reminiscent of Olmsted's determination of a short period for the Leonids. As the Perseids were an annual shower it was even less obvious that their period should be much longer than a year. From the daily rates during the shower and the assumption that the meteoroids were distributed uniformly around their orbit, he guessed that the ring would be about five to ten mil-

Fig. 9.1. Hubert A. Newton (1830–1889). The photograph on the wall is of E. Loomis.

lion miles in thickness (as compared with current estimates of about twenty million), with a membership of 3×10^{14} fragments.

Incidentally, it is H. A. Newton who in 1864 coined the word *meteoroid* to indicate the solid bodies that orbit the sun before they collide with the Earth's atmosphere.[10,11] *Meteors* are the luminous atmospheric effects of such impacts; *meteorites* are the stones or irons that survive passage through the air to the ground.

While Newton believed that the Perseids moved in a ring or disc around the sun, he was careful to point out that this ring was unique to the Perseids, that other shower meteors moved in different rings. However, Newton was not the first to suggest that the August meteoroids might be distributed around their entire orbit, not in just one cloud. This had been proposed by Erman[12] as early as 1839 in a paper possibly forgotten because of Peirce's criticism of Erman's orbital calculations.[13] (Although Peirce was correct in regard to Erman's omission of terms for the Earth's attraction, he himself was proved wrong in his deduction that meteors move almost in the plane of the earth's orbit.[14]) Only about half a dozen years after Newton's investigation, from observations of the Perseids of 1866, Schiaparelli[15] derived an orbit close to the modern determinations and showed that it was in good agreement with the orbit of Comet 1862 III.

The work for which Newton is probably best remembered deals with the Leonid meteors. The eminent French astronomer H. A. E. Faye,[16] in 1867 was

Table 9.1
Newton's and a Modern Orbit for the Perseids

	Newton	Modern
Semi-major axis	0.84AU	22.6AU
Ellipticity	0.28	0.96
Perihelion distance	0.60AU	0.96AU
Inclination	96°	116°
Period	281 days	108 yrs
Relative velocity	43k/s	60k/s

inspired to proclaim "We may find in the works of M. Newton, of the United States, the most advanced expression of the state of science on this subject, and even the germ, I think, of the very remarkable ideas brought forward in these last days by M. Schiaparelli and M. Le Verrier."

Although his predecessors had clearly identified the showers of 1799 and 1833 as belonging to the same Leonid swarm, it was H. A. Newton who traced that shower back to A.D. 902 on the basis of old observations of thirteen showers recorded in Oriental and European records prior to 1833. He then successfully predicted its return in 1866.[17] Olbers in 1837 had already surmised that the shower would not return before 1867, but he had been less specific.[18] Recognition that the older showers represented the same swarm as the Leonids of 1833 had been hampered by the fact that the time of maximum display had shifted from October 13 in A.D. 902 to November 13 in 1833. Newton attributed this change in date to an annual drift of 1′.711 along the ecliptic of the node of the orbit of the meteors. Like Olmsted and Herrick before him, Newton noted that the showers sometimes occurred in consecutive years, after which they became scarce or altogether missing for about a third of a century. Newton envisioned the swarm as moving in a ring around the sun, with a cloud of meteors concentrated in one part of the ring, but that perturbations caused other meteors to be distributed heterogeneously in other parts of the ring. The 33-year period would thus correspond to the passage of the Earth through the densest part of the cloud. The nightly maximum duration of the shower, about 5 hours, indicated that the meteors were spread out over 40 million miles along the orbit, with a stream thickness of over 100,000 miles.

From the difference between observed times and those computed on the basis of a 33.25-year period, Newton inferred that meteors might be seen over a span of five or six years. Using formulae that are now especially familiar to variable and double-star observers for relating true (P) and spurious (P′) periods with (p) the common-factor intervals between observations, namely, $1/P = 1/P' \pm n/p$, n being an integer, Newton found that five periods satisfied the observed times of shower apparition about equally well.[17] The reciprocals of these periods in years were 1/33.25, 1 ± 1/33.25, and 2 ± 1/33.25. The first he rejected because the corresponding orbit would have its perihelion point nearly on the Earth's orbit, considered an unlikely coincidence. Similarly two periods were considered unlikely because the aphelia of the orbits were at the Earth's distance from the sun. The remaining two periods were nearly equal to the period of the Earth; he noted that observations in successive years should progress toward an earlier day of the month in one of these cases, whereas the opposite should be true for the other case. On this score the available observations favored the period of slightly less than a year, namely 354.621 sidereal days. Although Newton appeared to favor this period, he had evidently not yet completely rejected the others, for he pointed out that the progression of the nodes could be predicted for all five periods and compared with the actually observed trend.[17] Unfortunately he seems not to have carried this out, and it remained for the British Professor J. C. Adams[19] to demonstrate, after the 1866 shower, that by Newton's own criteria only the 33.25-year orbit was compatible with the motion of the node. Previously, however, in 1866 Newton had commented[20] that whether or not meteors would occur that year would depend on the perturbing actions of the Earth, Jupiter, and other planets. Explicitly involving Jupiter in this way would imply that he had not ruled out the 33.25-year period as being simply spurious.

Meanwhile Comet Tempel (1866I) had been discovered on December 19, 1865. Three European astronomers independently and almost simultaneously announced their discovery that the November Leonids traveled in the same orbit as this comet: C. F. W. Peters[21] on January 29, G. V. Schiaparelli[22] on February 2, and Th. Oppolzer[23] on February 6, 1867. Apparently H. A. Newton's earliest published reference to this association came in 1868 only as a quotation from Simon Newcomb[24] who stated, "It is now known that the entire stream of November meteoroids follows Comet I of 1866, commonly called Tempel's, in its orbit." By 1879 Newton believed that not just shower meteors but all meteorites as well had once been parts of comets.[25]

From the work of European astronomers,[26] by 1867 it was known that the Andromedid, or Bielid, meteors moved approximately in the orbit of Comet Biela. The comet had first been discovered by Montaigne in 1772, again by Pons in 1805, then by Biela in 1826. Upon its return in 1846 it was observed to have split in two. In 1852 the pair was again observed but widely separated. Thereafter at expected returns it was only dubiously seen if at all. It is assumed to have been lost, more likely completely disintegrated. Meteor showers with a radiant point in the vicinity of γ Andromedae had been observed on December 6, 1798, and late in November in 1830, 1838, 1841, and 1847. From these various observations the German astronomers E. Weiss, H. L. d'Arrest, and J. G. Galle in 1867 independently concluded that the meteors moved in the same orbit as Biela's comet, a conclusion supported by the brilliant display of November 30, 1867. Weiss then predicted a return of the shower in 1872, but somehow this prediction was overlooked in America.

A young Yale student, Arthur T. Hadley, who later (1899–1921) was to become President of Yale, noticed them by pure chance at 7:35 P.M. on November 24, 1872. He informed his uncle, Professor A. T. Twining, who in turn notified Newton. "The appearance so early in November of the meteors supposed to be connected with Biela's comet," wrote Newton, "had not been anticipated and, therefore, came upon us by surprise."[27] Hadley counted 43 in one hour; Newton 250 before midnight on November 24. The main part of the shower, on November 27, was over before nightfall in America but it was well observed in Europe.[28] Newton commented that it had been a true shower according to Herrick's definition, namely that it consisted of over 1,000 meteors per hour.[29] While most of the major meteor showers show hourly rates at maximum under 200 meteors per hour, there have been a few spectacular exceptions (Table 9.2) ranging upward of several thousand an hour.[30]

Table 9.2
Peak Shower Frequencies

Shower	Year	Hourly Rate
Leonids	1833	10,000?
	1866	5,000
Bielids	1872	2–6,000
	1885	75,000
Giacobinids	1933	4–6,000
	1946	2–4,000 visual
	1946	10,000 radio

The Bielid shower of 1872 was the best shower since the Leonids of 1833. Newton collected and analyzed all available reports and compared them with earlier observations by Brandes in Germany on December 6, 1798, and by Herrick on December 6, 1838. From the area of the observed radiant he inferred that the individual meteors did not travel in precisely the same orbits. He felt there could no longer be any doubt but that the meteors were fragments or companions of the comet and concluded that the force responsible for the rupture of Biela's comet must have acted near perihelion. His conclusions discounted the hypothesis of the famous Schiaparelli on the extrasolar-system origin of meteors. Newton commented,[27] "We have no evidence as yet that any radiant of meteors is so small as is apparently required by the supposition of the distinguished Italian astronomer, that the meteors were drawn as a stream into the solar system from stellar spaces." By 1873 Newton was convinced that all shower meteors were related to periodic comets, and that there was no evidence for supposing that any meteors came from interstellar spaces, i.e., have hyperbolic orbits.[31] In regard to meteorites, on the other hand, Schiaparelli[32] in 1871 had stated categorically,

> The hyperbolic velocity, which has been observed for certain meteorites destroys every probability of a lunar origin, makes the assumption of a planetary origin for meteorites almost impossible, and does not fit in well with their cometary origin, unless we will assume errors for the observers out of all probability. It therefore becomes necessary to place the origin of meteorites in the region of the fixed stars.

The 1885 return of the Bielids as seen in Europe was presumably the most intense display ever observed. Newton calculated that the total width of the stream was over 200,000 miles. Surprising at the time was the fact that the radiant point appeared to move with respect to the stars. This Newton interpreted as due to the Earth's attraction which "changes the radiant . . . of the meteors toward the vertical of the observer, one-tenth of the observed zenith distance of the radiant."[33]

Newton noted that the 1885 Bielids probably left the comet not earlier than 1841.5, a time of closest approach of the comet to Jupiter.[34] He also noted that there would have been a similar close approach in

1772, which probably accounts for the shower of 1798.[35] He did, however, remark that the Bielids of 1838 had a different orbit (different radiant and node) from those subsequent to the breakup of the comet in 1846. The disintegration of the comet would thus have progressed in installments at different times. For the disruption of the comet he invoked Jupiter and the gravitational effects of the close approach of the comet to the sun; he finally suggested that the breakup could have come about from excessive heating as a cold body approached the sun.[36,37] These studies led him to investigate how meteors burn in the atmosphere and how melting changes their shape.[38]

That comets seemed to belong to "families" had been recognized at least since 1857.[39] One of the more stimulating of Newton's investigations concerned the capture of comets by the major planets, leading to comet families with the aphelia of the cometary orbits near the orbits of those planets. He showed that comets originally traveling in parabolic orbits could by planetary perturbations be drawn into elliptical orbits.[40] Jupiter would have captured more comets than any other planet.[41–43] Indeed, the Bielids belong to the Jupiter family.

From his theoretical analysis Newton generalized,[44] "Those comets which are changed by the perturbing action of Jupiter, or other planets from parabolic orbits of every possible inclination to the ecliptic into short period ellipses ... will as a rule (but with exceptions) move in orbits of moderate inclination to the ecliptic, and with direct motions."

In practice he did find that most short-period comets do so move while long-period or parabolic comets could be found to move at all possible inclinations. In the case of meteor showers, Lovell's text on meteoric astronomy[45] gives orbits for eighteen meteor showers, most of them associated with known comets. Of these, eleven have direct motion with inclinations under 35°, three have retrograde motions with inclinations 162° or 163°, while four have high inclinations, 56° to 112°.

Although Newton devoted most of his attention to shower meteors and the relationship between comets and meteors, he also determined heights for 342 sporadic meteors[46,47] and studied their frequencies: 58 percent ranged between 60 and 120 km, with an average of 96 km. Only 7 percent were found above 180 km and 4 percent below 30 km, values consistent with modern determinations. These figures may be compared with corresponding results he obtained for the Leonid and Perseid meteors of 1863.[48] For 78 Leonids observed at U.S. Naval Observatory and Haverford College he found the beginning altitudes ranged from 56 to 327 km, with a weighted mean of 155 km; the end points from 32 to 209 km with a mean of 98 km; and a weighted midpoint average of 126 km. However, he believed it unlikely that any meteors actually appeared at heights greater than about 200 km, these determinations being the effects of uncertainty in the observations. (Eleven of the Leonids were computed as appearing at heights over 200 km.) For about forty Perseids observed from various localities in the northeast he found the mean beginning height at 112 km, the end 90 km with a weighted midpoint of 101 km, about 25 km lower than for the Leonids. He concluded,[48] "The most plausible explanation of this remarkable fact is, that the two groups of bodies differ in their chemical and mechanical constitution; the November group being more inflammable than that of August." Now, of course, it is known that the heights of first appearance of the meteors are correlated mainly with velocity of impact with the atmosphere. Although Newton quoted a value of 44 miles per second for the 1967 Leonids,[24] he did not compare this with his earlier determination of about 27 miles per second for a Perseid.[7]

From observed frequencies of sporadic meteors Newton attempted to determine the total numbers of meteoroids likely to encounter the Earth daily. Newton estimated the number of sporadic meteors visible over the entire Earth at 10,460 times the number visible from a single locality.[49,50] Considering one meteor every two minutes as a reasonable average, he estimated that more than 7.5 million could be expected to enter the atmosphere every day.[50,11] (A more realistic modern evaluation amounts to five per minute from a given site or ten times as many over the globe as Newton inferred.[51]) Telescopic meteors, he found, should amount to around 400 million a day.[52] He was, however, little interested in making observations of telescopic meteors, as they yielded too short tracks for accurate radiant determinations.

Newton assumed that most of the naked-eye meteors were like grains of sand, and he estimated that each volume the size of the Earth should contain 13,000 such particles capable of producing naked eye meteors, and at least fortyfold as many telescopic meteors. Since the sporadic appeared not to show definite radiants, he inferred that their velocities must average greater than that of the Earth. The velocities estimated from path lengths and durations of flight indicated mean velocities between 78 and 130k/s, with the smallest velocities over 2.5 times the Earth's orbital velocity. The estimates of velocity, he pointed out, could be too large because the path-lengths were overestimated, the times of flight under-estimated, or the meteors might indeed have hyperbolic orbits. The meteors could not all belong to a single ring but might form (1) a number of rings; or (2) a disk in or near the plane of the ecliptic (like the zodiacal light); or (3) they might simply be distributed at random.

(1) If there were a number of rings, more radiants should be found. However, if the number of meteors belonging to a given ring were small, the discovery of the radiant would not be readily apparent. It remained for Öpik[53] in his analysis of the data from the Arizona Meteor Expedition of 1931–33 to investigate this problem, based upon 26,000 observations of 22,000 meteors, yielding 279 mostly marginally probable radiants. Even he recommended caution in their acceptance.

(2) If the meteors are concentrated in a disk like the zodiacal light, the radiants should lie near the plane of the ecliptic.

(3) If the radiants are distributed at random, the mean observed velocity should show a correlation with the time of night. Newton remarked,[54] "For if the velocity is very small few would be seen in the evening, while if it is very large there should be nearly as many in the evening as in the morning." Quoting observations by R. A. Coulvier-Gravier,[55] Newton found that nearly twice as many sporadic meteors were observed after than before midnight. Earlier Herrick[56] had estimated that three times as many meteors occurred in the morning than in the evening hours. As the mean velocity considerably exceeded the velocity of the Earth, Newton concluded that the sporadic meteors must have orbits like those of the comets. Thus he believed that sporadic meteors, like shower meteors, were also cosmic with similar characteristics and origin,[49,57] and concluded, "These bodies cannot be regarded as the fragments of former worlds. They are rather the materials from which the worlds are forming."[57] This constitutes an endorsement of the Laplace nebular hypothesis for the formation of the solar system, which at the time was considered controversial.[58] Incidentally, Isaac Orr,[59] a Yale graduate of the class of 1818, a minister, independently evolved a nebular theory very nearly the same as that of Laplace.[60]

Two similar nebular theories of the origin of the solar system were current in Newton's time, those of Kant (1755) and Laplace (1796). Although usually loosely considered as essentially the same, one was concerned mainly with a particular nebula out of which the solar system originated, the other with a larger primordial system embracing all of interstellar space. Kant had closely associated the comets with the planets; considering them the outermost bodies condensed out of the nebula he argued,[61]

> The eccentricity in the movements of the planets increases with their distance from the sun; and the more remote planets thus come nearer the definition of comets. It may therefore be supposed that there are other planets beyond Saturn which are still more eccentric in their orbits, and thereby have nearer affinity to comets, so that through a continuous gradation the planets finally become comets.

Newton, in an essay entitled *On the Origin of Comets,*[62] reasoned that the distribution of the orbits of the comets could offer a clue as to which theory was preferable. If the solar system, including the comets, had condensed from a single nebula the comets should have elliptical orbits with aphelia distributed near the plane of the ecliptic, and their inclinations should be small. If the comets were derived from primordial material randomly scattered throughout interstellar space, the orbits of the comets should show no such concentration. Also, if the comets derived from interstellar matter rather than from a solar-system nebula, their orbits should be hyperbolic. As fairly reliable orbits had already been computed for 247 comets, Newton felt the alternatives could be put to a test. As only 33 of the inclinations were under 30 this appeared to favor the Laplace theory.

Observational uncertainties, however, were too great to differentiate accurately between hyperbolic and parabolic or highly eccentric elliptical orbits. Moreover, the original orbits could have been altered considerably either by a resisting medium, if such did in fact exist within the solar system, or by perturbations by close approaches to the planets, especially Jupiter. The perturbations could either accelerate or decelerate the motions, converting hyperbolic to elliptical, or elliptical to hyperbolic. Thus the test could be inconclusive. Although he favored the Laplace theory, he believed that the short-period comets with small inclinations must always have belonged to the solar system. In a subsequent paper[40] Newton further discussed the short-period comets. Assuming that they, too, had originally moved in nearly parabolic orbits, he investigated the effects of close approaches to Jupiter, summarizing his investigations thus:[63] "So far then from the direct motion of the periodic comets being a reason for assigning to them a separate genesis from that of the other comets, that direct motion is just what we ought to expect upon the supposition that the comets have been thrown into their orbits by Jupiter or by other planets."

Somewhat wistfully Newton ended his 1864 National Academy paper, *Shooting Stars,* by speculating: "If astronomy furnishes any measure of their total mass we may therefrom obtain some idea, rude [sic] though it be, of the mean mass of the individuals."[57] Some twenty years later,[64] in a theoretical paper entitled *On the Effect upon the Earth's Velocity Produced by Small Bodies Passing Near the Earth,* Newton concerned himself with the effect of meteoroids on the motions of rotation and revolution of the Earth and the consequent changes in the length of the day and month. This subject had already been studied by Oppolzer,[65] especially pertaining to the Moon. E. C. Herrick[66] had broached the subject as early as 1838. Considering the enormous numbers of meteoroids meeting the Earth head-on over a span of thousands of years, Herrick had asked if this could not have shortened the year. At the time, the answer appeared to be negative. Newton extended the investigation not just to the effects due to the meteors actually entering the Earth's atmosphere, but particularly to the effects by those merely passing nearby. He found that the effect upon the Earth's motion of those entering the atmosphere is at least a hundredfold that of the near misses.

As a collector of meteorites Newton was interested not only in their dynamical and orbital properties, but also in how their surfaces were altered in their flight through the atmosphere, and in their chemical composition. From the breakup of Comet Biela it could be surmised that meteoroids entering the Earth's atmosphere at high velocities would be completely burned up before reaching the ground. Larger masses would probably have irregular shapes which would be modified by melting. At first, unsymmetrical air resistance would result in the diversion of their paths from an initially almost straight line. But later, with the melting of the forward portions, their paths would again become more nearly straight. Newton had several specimens that showed both a side smoothed by melting and a rear side still rough.

A spectacular fireball passing over Ansonia, Connecticut, on January 13, 1893, provided Newton with more information on the behavior of meteoroids during their flight through the atmosphere.[67] The Ansonia fireball was serendipitously photographed by an amateur astronomer, John Lewis, who was attempting to photograph Comet Holmes (discovered by Edwin Holmes in London, Nov. 6, 1892) when the fireball happened to cross the field of his photograph. Lewis himself did not see the bolide, but others did. According to an observer in Massachusetts its brilliancy exceeded that of first-magnitude stars. To have left as strong a trail as it did on Lewis's plate it must have been very much brighter over Connecticut. Fig. 9.2 shows a reproduction of a segment of Lewis's original 33-minute exposure, when the meteor was at an altitude between 16.5 and 12.5 miles, shortly before it disintegrated.

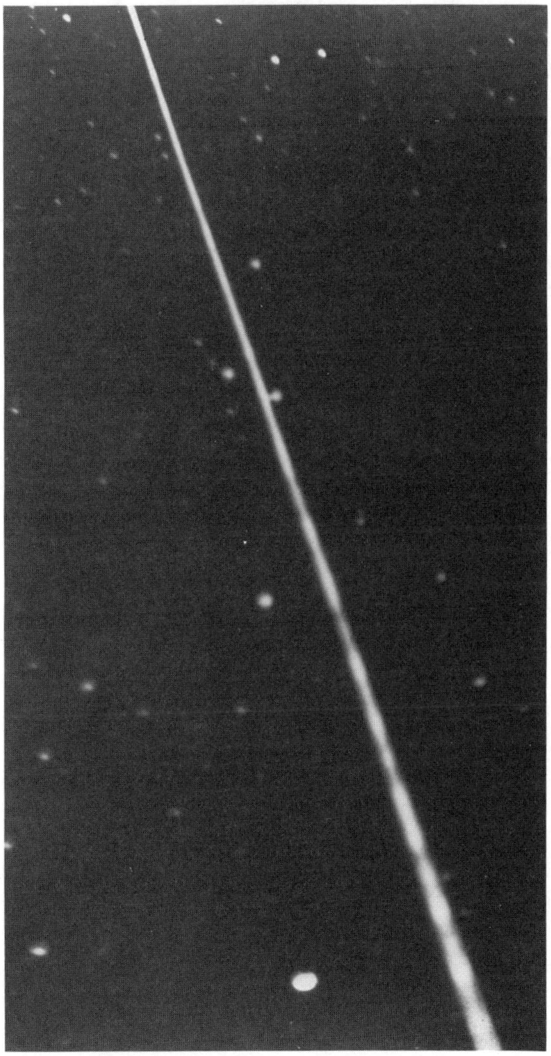

Fig. 9.2. Segment of J. E. Lewis' photograph of the fireball of 13 January 1893.

Lewis gave Newton the plate for analysis. Using a filar micrometer Newton measured the distances of a number of stars from the track of the meteor. Since the positions of the stars were accurately known from the *Bonner Durchmusterung,*[68] Newton was able to obtain good positions of selected points along the trail. From seven such points he computed an orbit. Newton noted the fluctuations in brightness along the trail, at first recurring "only after hundreds of feet, while near the end they are perhaps ten times as frequent." Initially he attributed the fluctuations in brightness to small fragments breaking off the main mass. To this, however, he found objection because he would have expected the trail to show small spurs of light which he could not detect. Instead he favored another hypothesis, namely, "It seems much more reasonable to suppose that the stony mass was in rotation, more rapid at the end than at the beginning, and that unequal amounts of burned material were thrown off according as a well burned or a raw surface was for the instant in front."[69]

Newton thus appears to have been the first to point out the now well-known phenomenon of meteoroid rotation. Not knowing the velocity of the meteoroid, he did not venture to calculate the period of rotation. The fuzziness at the end of the trail he thought might be attributed to small fragments shearing off the main body. Eventually, as indicated from visual observations, beyond the range of the photographed part of the path, the entire meteoroid broke into several large pieces, none of which was recovered. From Danbury, Connecticut, the fireball had been seen nearly overhead and almost stationary (i.e. approaching head-on). An observer there estimated its duration as about 7 seconds. However, this interval would probably have included the duration of a persistent train, not just the duration of flight; hence this timing could not help to determine the velocity.

Newton's theory of rotation did not remain unchallenged. S. Tolver Preston, a physicist at Hamburg, Germany, wrote a letter to *Nature* (a copy of which the editor transmitted to Newton) reverting to Newton's rejected hypothesis and arguing that the explanation of rotation "seems to me rather forced . . . considering that dense air acts like a target on a meteor owing to its immense speed. Therefore the flashes of light or sparks not unnaturally would augment until the bolide probably destroyed itself at last by collision against the dense air."[70] The whole picture must evidently be a combination of both rotation and fracture.

The possibility that the photographed fireball of 1893 might have deposited meteorites in the vicinity of Danbury reminds one of Newton's wide interest in historic falls, including in particular one portayed in a painting by Raphael. E. S. Holden, Director of the Lick Observatory, in 1890 called attention to Raphael's *Madonna di Foligno,* in which a fireball is depicted under a rainbow.[71] The painting had been commissioned in 1511 by Sigismondo dei Conti da Foligno, secretary to Pope Julius II. Sigismondo is shown as a kneeling figure in the painting. Holden's article inspired Newton to call attention to a copy of this painting by Terry in the Yale Art Gallery, and to elucidate on the significance of the fireball, which some had previously thought to represent a bomb.[72] Actually, Newton had already called attention to the Raphael painting in a public lecture in New Haven on March 29, 1889, entitled "The Worship of Meteorites," but this was published only posthumously in 1897.[73] In that article Newton displayed a remarkable knowledge of classical literature, describing numerous falls of meteorites that the ancients had thought were either actual deities or else had been sent by God foreboding good or evil. In regard to the Raphael painting Newton pointed out that a shower of meteorites had fallen at Adda, near Crema, southeast of Milan, on the night of September 4, 1511, and that several accounts of the detonating fireball were extant. Newton believed this was the only detonating meteor to have fallen in Italy for several years preceding 1512 of which any account was preserved; hence this must have been the event being portrayed.

The Crema fall had occurred at a time when there was much superstition about the significance of such displays. Only nineteen years previously the famous Ensisheim, Alsace, fall of 1492, a 280-lb. mass, had been interpreted by Maximilian as a sign from heaven that he would defeat his enemies, the Turks. By analogy the portrayal of the Crema stonefall in a religious painting would appear appropriate at a time when the Papal forces were at war with France. The fireball with the rainbow would thus symbolize the promise of divine reconciliation and assistance. Unfortunately, there are no records indicating that any specimens of the Crema meteorites have been preserved.

In regard to the Ensisheim fall, Newton remarked that it represented the earliest record of a meteorite for which a trajectory could be computed. He found that it had moved in a direction nearly the same as the direction of revolution of the Earth and in an orbit only slightly inclined to the ecliptic.[74]

Newton's success in interpreting the light variations of the Ansonia fireball may later have encouraged others to use photography as a means for studying meteors. As early as August 9 and 10 of that same year William Lewis Elkin, who was to become Newton's successor at the Yale Observatory, mounted a 6-inch Voigtländer portrait lens of 32-inch focal length on the same mounting with the 8-inch Reed equatorial telescope,[75] pointing it close to the Perseid radiant. Three trails were photographed in a total of four hours, two of them Perseids, one almost six degrees long. John Lewis, ten miles away, also tried photographing meteors with a stationary 4-inch lens. He happened to obtain a trail of the same bright meteor as one photographed at Yale, displaced some six degrees from the Yale trail.[76,77] Newton reduced the observations, finding the beginning and end heights to be 68.0 and 29.3 miles, respectively. Attempts at photography were again made on December 10, 1893, when three Geminid meteors were photographed at Yale, but Lewis captured only one weak, doubtful trail. The three Geminid trails, extended backward, intersected in an unbelievably small radiant area, only one minute of arc in diameter. Newton concluded that "The experiment shows that with a reasonable amount of labor we can get photographs of meteor tracks. The accuracy of results obtained from the photographs is unspeakably greater than those gotten by any other known method."[77]

REFERENCES

1. J. Willard Gibbs, *Biographical Mem. Nat. Acad. Sci.*, **4**, 99, 1902; *Amer. Jour. Sci.*, (4), **3**, 101, 1897.
2. A. W. Phillips, *Bull. Amer. Math. Soc.*, (2), **3**, 169, 1897.
3. O. Callandreau, *Monthly Notices Roy. Ast. Soc.*, **57**, 228, 1897.
4. C. S. Lyman, *The New Englander*, **27**, 146, 1868.
5. H. A. Newton, *Ast. Jour.*, **3**, 126, 133, 1853.
6. ———, *Amer. Jour. Sci.*, (2), **30**, 186, 1860.
7. ———, *Ibid.*, (2), **32**, 448, 1861.
8. ———, *Ibid.*, (2), **33**, 338, 1862.
9. ———, *Idem*, p. 344.
10. ———, *Ibid.*, (2), **39**, 198, 1865.
11. ———, *Mem. Nat. Acad. Sci.*, **1**, 300, 1866.
12. A. Erman, *Ast. Nach.*, **17**, 3, 1840.
13. B. Peirce, *Trans. Amer. Phil. Soc.*, n.s., **8**, 83, 1843.
14. C. P. Olivier, *Meteors*, Baltimore, p. 48, 1925.
15. G. V. Schiaparelli, *Ast. Nach.*, **68**, 331, 1867.
16. H. A. E. Faye, *Comptes Rendus*, **64**, 551, 1867; quoted in translation in Ref. 1a, p. 110.
17. H. A. Newton, *Amer. Jour. Sci.*, (2), **38**, 53, 1864.
18. H. W. M. Olbers, *Schumacher's Astronomisches Jahrbuch*, p. 278, 1837.
19. J. C. Adams, *Monthly Notices, Roy. Ast. Soc.*, **27**, 247, 1867.
20. H. A. Newton, *Amer. Jour. Sci.*, (2), **43**, 78, 1866.
21. C. F. W. Peters, *Ast. Nach.*, **68**, 287, 1867.
22. Ref. 15, p. 332.
23. T. Oppolzer, *Idem*, p. 333.
24. H. A. Newton, *Amer. Jour. Sci.*, (2), **45**, 237, 1868.
25. ———, *Nature*, **19**, 315, 340, 342, 1879.
26. Ref. 14, p. 65.
27. H. A. Newton, *Amer. Jour. Sci.*, (3), **5**, 53, 1873.
28. ———, *Idem*, p. 150.
29. ———, *Idem*, p. 56.
30. A. C. B. Lovell, *Meteor Astronomy*, Oxford, pp. 330, 332, 339, 350, 1954.
31. Ref. 27, p. 62.
32. J. V. Schiaparelli, *Sternschnuppen*, Stettin, p. 228, 1871. Trans. by Olivier, Ref. 14, p. 255.
33. H. A. Newton, *Amer. Jour. Sci.*, (3), **31**, 418, 1886.
34. ———, *Idem*, p. 409.
35. ———, *Idem*, p. 88.
36. ———, *Proc. Amer. Phil. Soc.*, **32**, 29, 1893.
37. ———, *Amer. Jour. Sci.*, (3), **47**, 152, 1894.
38. ———, *Encyclopaedia Britannica*, 9th ed., **16**, 112, 1883.
39. M. Hoek, *Ast. Nach.*, **45**, 49, 1857.
40. H. A. Newton, *Report*, 49th Meeting, *Brit. Assoc. Adv. Sci.*, p. 272, 1879.
41. ———, *Mem. Nat. Acad. Sci.*, **6**, 7, 1891.
42. ———, *Amer. Jour. Sci.*, (3), **42**, 183, 482, 1891.
43. ———, *Astron. and Astrophys.*, **11**, 12, 1892.
44. ———, *Idem*, p. 16.
45. Ref. 30, Chap. 13–18, 1954.
46. Ref. 17, p. 135.
47. Ref. 11, p. 292.
48. H. A. Newton, *Amer. Jour. Sci.*, (2), **40**, 250, 1865.
49. Ref. 11, p. 299.
50. Ref. 10, p. 294.
51. F. Watson, *Between the Planets*, Cambridge, Mass., p. 92, 1952.
52. H. A. Newton, *Amer. Jour. Sci.*, (2), **36**, 145, 1863.
53. E. Öpik, *Cir. Harvard Obs.*, No. 388, 1934.
54. Ref. 11, p. 308.
55. R. A. Coulvier-Gravier, *Recherches sur les Météores*, Paris, p. 220, 1859.
56. Ref. 10, p. 206.
57. Ref. 11, p. 312.
58. R. L. Numbers, *Creation by Natural Law: Laplace's Nebular Hypothesis in American Thought*, Seattle, 184pp., 1977.
59. ———, *Jour. Hist. Astron.*, **3**, 49–51, 1972.
60. I. Orr, *Amer. Jour. Sci.*, **6**, 128, 1823.

61. W. Hastie, *Kant's Cosmogony* (an English translation of I. Kant's *Allgemeine Naturgeschichte und Theorie des Himmels,* 1775), p. 66, Glasgow, 1900. This and its Appendix B are copied verbatim, including pagination, in I. Kant, *Universal Natural History,* but with a new Introduction by M. K. Munitz, Ann Arbor Paperbacks, University of Michigan, p. 66, 1969.
62. H. A. Newton, *Amer. Jour. Sci.,* (3), **16**, 165, 1878.
63. Ref. 40, p. 274.
64. H. A. Newton, *Amer. Jour. Sci.,* (3), **30**, 409, 1885.
65. T. v. Oppolzer, *Ast. Nach.,* **108**, 67, 1884.
66. E. C. Herrick, *Amer. Jour. Sci.,* **33**, 362, 1838.
67. H. A. Newton, *Amer. Jour. Sci.,* (3), **46**, 161, 1893.
68. F. Argelander, *Bonner Durchmusterung,* Bonn, **1–4**, 1859–62.
69. Ref. 67, p. 172.
70. S. Tolver Preston, unpublished letter to *Nature,* H. A. Newton file, *Yale Obs. Archives.*
71. E. S. Holden, *Pub. Ast. Soc. Pacific,* **2**, 19, 1890; *The Sidereal Messenger,* **9**, 120, 1890.
72. H. A. Newton, *Ibid.,* **3**, 91, 1891; *Amer. Jour. Sci.,* (3), **41**, 235, 1891.
73. ———, *Amer. Jour. Sci.,* (4), **3**, 4, 1897.
74. ———, *Comptes Rendus,* Paris, **113**, 234, 1891.
75. *See* Chap. 11.
76. W. L. Elkin, *Ast. Jour.,* **13**, 132, 1893.
77. Ref. 37, p. 154.

10
H. A. NEWTON: COLLECTOR OF METEORITES

During his lifetime Newton collected meteorites from about 100 different localities. The Yale archives contain voluminous correspondence relative to his purchases and exchanges of meteorites with other collectors, both in the States and abroad. He contributed to Yale's purchase of some Homestead and Estherville meteorites, and donated to Yale a 55-lb iron from Hammond, St. Croix County, Wisconsin, and a fragment from a 1.75-lb meteorite that had been found in one of the Hopewell Mounds in Ohio, but was later identified as having originated at Brenham, Kansas. His daughters ultimately donated Newton's entire collection to Yale.[1]

Farrington's classical treatise on meteorites[2] lists the 19 largest meteorite showers that had been recorded up to 1915, of which all but one fell before 1890. Two of these fell in Iowa: Homestead in Iowa County, on February 12, 1875; Estherville in Emmet County on May 10, 1879; and a third (not included among the 19) in Forest City in Winnebago County on May 2, 1890. Newton acquired, or helped Yale acquire, specimens of all three.

The Homestead, or Iowa County, shower Farrington described as "typical of the more violent form." Newton estimated that nearly 200 stones weighing about 800 lbs fell in an area 4 by 7 miles.[3] Prior's catalogue[4] gives Leonard's[5] earlier determinations as over 100 stones of aggregate weight 500 lbs, the largest weighing 74. Dana's Catalogue[6] of the meteorites at Yale in 1886 indicates there were 20 stones of which the largest weighed 12 kg or about 26 lbs. Farrington's catalogue of 1909 indicates that Yale possessed a total of 35 kg or 77 lbs from this fall.[7] Newton believed, in 1879, that these were once parts of a comet.[8]

Gustavus Hinrichs[9] described this fall as "The Great Iowa Meteor," giving details of the flight of the fireball and the discovery of the meteorites. The village, Homestead, became headquarters for "meteor brokers" gaining huge profits from the sale of the meteorites. One landowner was paid $2.50 for a lot weighing 44 lbs, which the "brokers" then sold at $2 a pound to an astronomer for $88, making a profit of $85.50!

Estherville in Emmet County was even more impressive. It is estimated that between one and two thousand pieces of this stony-iron meteorite fell in an elliptical area about 8 miles long by 1 mile wide. The two largest specimens weighed 437 and 170 lbs, the aggregate probably over 700. Of this shower Yale acquired some 650 samples. The largest, 92.5 lbs, had been found nine months after the fall in a 5-ft-deep pit. Estherville is a mesosiderite (its constituents are almost equally divided between stony minerals and nickel—iron inclusions). Nininger[10] many years later examined both the mostly marble-sized specimens at Yale and the larger meteorites from the same fall at other institutions. He found that when the meteoroid fragmented in the atmosphere there was sufficient velocity remaining for melting of the forward surfaces to occur, so that some of the iron fragments as well as a few stony remnants did show signs of orientation in their melting patterns. Nininger writes,[11]

> A careful examination of 650 of these in the Yale Collection revealed that 183 showed recognizable degrees of orientation. Many of them were perfect examples. That these small irons may have been free individuals in space is out of the question because their shapes and dimensions were such that they could not possibly have survived the atmospheric flight that was observed in connection with this fall. The large masses of 431 and 150 lbs and smaller which contained hundreds of similar metallic nodules evidence no orientation.

The Forest City, Winnebago County, fall is a typical chondrite (a stony meteorite in which chondrules, or small rounded grains of minerals are present). It is of interest not only for its scientific value but also for having instigated court cases that finally led to adoption of a law that meteorites belong to the owner of the property upon which they fall, not to the finder. Soon after the first meteorite had been found, an attorney approached Newton on the question of ownership. A farmer, whose wife had witnessed the fireball, found the stone buried more than three feet deep in hard prairie soil and dug it up.[12] Geology Professor Horace V. Winchell of Minneapolis heard about the discovery and promptly bought the meteorite for $105 from the finder on behalf of the University of Minnesota. Then it turned out that the land upon which the meteorite had fallen was a neighbor's. The land-owner sued for the return of what he considered his property. It was his attorney who contacted Newton.[13] Although he stated that *The elementary principles of real estate law are in favor of landowners,* the lawyer desired precedents in regard to finders of meteorites. Newton made a search of legal records both for the United States and foreign countries, finding that France, in particular, favored the land-owner. The Municipal Court judge also decided in favor of the owner, but the defendant countersued on the theory that finders are keepers of objects that had no previous owner before falling from the sky, and the case dragged on. On December 8, 1891, Municipal Court Judge of Minneapolis, Charles B. Elliott, sent Newton the *Brief of the Case of the Meteorite.* Ultimately (October 1892) the Supreme Court of Iowa ruled in favor of the landowner, and Professor Winchell finally succeeded in acquiring the meteorite for the University of Minnesota.

Meanwhile another larger meteorite of the same fall, weighing 80 lbs was found about three-fourths of a mile from the first. This was acquired by Professor Oliver Huntington for the Harvard meteorite collection.[14] Although N. H. Winchell (father of Horace) was credited[12] with having noted that, in addition to seven sizable meteorites, some 5,000 small fragments had fallen, Prior's[4] catalogue (quoted by later editions of the British Museum catalogues[15]) noted that only some 500 small stones were collected with an aggregate weight of about 270 lbs. By 1952 Nininger[16] had increased this number to about 1,800 pieces distributed over an ellipse of 1 by 2 miles. Newton acquired between two and three hundred small specimens. Then in 1891, Henry F. and Alice K. English donated funds to Yale for the purchase of 900 to 1,000 more for the Peabody Museum, and Yale President Timothy Dwight (grandson of a former president by that name) happily commented,[17] "It is doubtful whether any other museum can show an equally numerous collection from a single meteorite shower." Torrey and Barbour point out[12,18] that there had been reports of a larger, 104-pound mass from Kossuth County, up for sale in Forest City, but this proved to be a fraud.

The Hammond Township, St. Croix County, Wisconsin, iron octahedrite meteorite of approximately 55 lbs was among the largest acquired by Newton for Yale. (Octahedrites are nickel—iron meteorites with an octahedral crystallization pattern.) The Hammond meteorite had been plowed up in a cornfield in 1884. D. Fisher[19] described the condition of the meteorite to be such that it must have fallen "not many months before its discovery."

At Brenham Township, Kiowa County, Kansas, about twenty meteorites with an aggregate weight of 2,000 lbs were discovered in 1882; the largest weighed 466 lbs. The largest Yale specimen, weighing 101.5 lbs, was donated by E. M. Reed (who gave Yale its 8-inch refractor) and P. N. Welch, Yale class of 1862.[1] Later finds included four larger masses, 1,000, 740, 474, and 466 lbs, and smaller ones bringing the aggregate to about 4.5 tons, the largest of any known pallasite fall as described by Nininger.[20] The term *pallasite* is reserved for stony-iron meteorites in which olivine is a prominent constituent, a continuous network of iron enclosing the silicates. The Brenham objects were known to ancient Indian tribes who made practical use of them as metals and hammered many into artifacts. The meteorites were collected later as curiosities by the first white settlers and cowboys.

The Brenham meteorites were first recognized as such on the Kimberly farm in Kansas, notably by Mrs. Kimberly who realized their resemblance to an Estherville stony—iron meteorite she had been shown in her childhood.[21-23] Many years later, in 1929, Harvey Nininger was to discover on the same property that a depression then believed to be a buffalo wallow was in reality a meteorite crater (the Haviland Crater[22,24]). Nininger conducted an excavation in 1933, the first such thorough investigation. The fully oxidized meteorites Nininger found in the crater he named *meteorodes.*

Yale's Hopewell Mounds sample, called *Anderson,* is a fragment of a meteorite weighing 1.75 lbs, that had been found at an altar at the Turner Mounds in 1882. Newton surmised it must have been used by the Hopewell Indians (who lived[25] about A.D. 900 to 1300) "in some ceremony of a religious, possibly one of a funereal character."[26] The term *Hopewell Mounds* has been applied not to only one mound but to some 5,396 mounds of similar characteristics in and near Ohio.[27,28] Some of these had been excavated by Professor F. W. Putnam of Harvard in the 1880s. Moorehead[29] in 1922 noted that the numbers of meteoric irons in the Hopewell mounds outnumbered all that had been found in Ohio in recent years, and he inferred, "Either the fall of meteorites was more frequent, or the Hopewell people had access to some good-sized meteoric iron deposit the location of which is at present unknown." But by 1890 G. F. Kunz,[30] of the American Museum in New York, had already discovered their probable source.

It turned out that the meteorites found in the Little Miami Valley in Ohio had the same mineralogical peculiarity as those of Brenham—namely, unusual proportions of olivine. Hence Kunz believed that the meteoric irons found in Ohio must have been transported by early Indians from Kansas to the Ohio Valley, "since no other olivine meteorite had up to that time been found in North America." A modern reevaluation of the Brenham and Hopewell Mounds meteorites effectively substantiates this conclusion.[31] These meteorites are classified as stony-iron pallasites in Prior's Catalogue. The one meteorite called "Hopewell Mounds, Ross County, Ohio" in Prior's catalogue, however, is classified as a medium octahedrite iron, not as a pallasite. This 5-oz specimen, at the Field Museum in Chicago, was found at a mound only 75 miles from the Turner Mounds. Farrington[32] compared its composition with the Brenham and the Turner Mound meteorites and found the Hopewell differed from the others by percentages greater than previously found between any members of the same fall. Hence it was assumed that it could not have been a part of the Brenham fall. However, later investigators found that octahedrites and pallasites could co-exist. For example, Krinov remarks,[33]

> There also exist interesting intermediate types of meteorites, for example, the meteorite Brenham, which consists of several individual specimens. Some of the specimens are iron (with octahedral structure), while others belong to the pallasites. Most remarkable of this group, however, is one specimen. A part of this meteorite appears to be entirely iron and has octahedral structure; its other part is pallasitic.

Newton's collection contained meteorites of widely varying composition and internal structure. Widmanstätten figures had already been recognized in iron meteorites since 1808. Stony meteorites contain nickel—iron inclusions of varying proportions, some microscopic, others more pronounced. In 1893 Newton[34] published his analysis, "Lines of Structure in the Winnebago Co. and in Other Meteorites." In all of the seven meteorites he sampled, some Widmanstätten structure was detectable. (Widmanstätten figures are defined as the pattern of crystallization that appears when the cut and polished face of an octahedrite is etched with acid.[35]) Newton concluded,

> These markings are such as we might expect if the forces which determine the crystallization of the nickel iron of the iron meteorites also dominated the structure of the rock-like formations of the stony meteorites and the distribution therein of the iron particles. The relation of quartz crystals to the structure of graphic granite is naturally suggested by these meteorite markings.[34]

The earliest chemical analysis of a meteorite in America was by Benjamin Silliman[36] on the Weston fall of 1807. In 1875 Yale physicist and Professor of Chemistry Arthur W. Wright published a paper, *Spectroscopic Examination of Gases from Meteoric Iron.*[37] His purpose was to learn whether the unknown gaseous elements assumed to be present in the solar corona and chromosphere might be detected in the gases occluded in meteorites. Ordinary chemical analysis he considered too insensitive for detecting the small amounts likely to be present, and he devised new techniques. He analyzed samples well below the fusion crust of the Red River (Texas) and the Tazewell (Tenn.) iron meteorites but found no new substances. The varying character of the oxygen and nitrogen lines in the presence of hydrogen, and their near coincidence with two of the lines in the spectrum of the corona, far from indicating the presence of otherwise unknown elements, proved them "due simply to hydrogen and the gases of the air, namely, oxygen and nitrogen."

In a second paper[38] Wright reported on a similar examination of a sample of the Homestead stony meteorite. He noted a marked difference between the iron and stony meteorites, hydrogen being the most prominent gas in irons, oxides of carbon in the stones. The brightest carbon bands were "precisely the ones most conspicuous in the spectra of some of the comets, and this fact is a remarkable confirmation of the received theory as to the meteoric character of these bodies." Previous work by Newton and others had indicated that meteorites were probably of the same nature as shower meteoroids, which were related to comets. Hence Wright felt that examination of the gases of freshly fallen meteorites should furnish important information about the tails of comets. He wrote,[39] "There is a high degree of probability that, if a large comet should hereafter approach near enough to the sun to have its nucleus intensely heated, the hydrogen lines will be found in its spectrum, in addition to the bands heretofore observed." Not until the advent of the space age and ultraviolet spectroscopy was this prediction on the observability of atomic hydrogen in cometary spectra verified[40] in 1970. Wright deplored the fact that Comet Donati (1858) departed "just early enough to escape observation with the spectroscope." In two further papers he gave results of an examination of several other meteorites, both iron and stony, confirming his previous findings.[41,42]

In his retirement address as President of the American Association for the Advancement of Science in 1886 Newton[43] summarized the history and then current knowledge and theory on meteorites and meteors. He emphasized, "This Association exists, it is true, for the advancement of science, but science may be advanced by rejecting bad hypotheses as well as by framing good ones." He enumerated a sample of seven that he considered open to revision and moderate changes, propositions having "practically universal acceptance among scientific men." These dealt with the paths, velocities, and relations of several meteor showers to periodic comets, the differences between meteorites and terrestrial rocks, and the lack of detection of any trace of organic life in meteorites. He then discussed the concept that the only real difference between ordinary meteors, fireballs, and meteorites is a matter of size. To the objection that no meteor showers have appeared to yield meteorites, he determined how many such meteorites might be expected, reaching the estimate of two or three annually. The failure to acquire these meteorites he considered "slender ground upon which to rest a denial of the common nature of objects that are so similar in appearance and behavior as the large and small meteors." As the shower meteors were found to be related to comets, he found it reasonable to expect a relation between the meteorites and other astronomical bodies or systems of which the comets form a part.

Newton rejected then current hypotheses that meteoroids are the fuel of the sun; that they are responsible for accelerations in the moon's motion; their producing geological changes by adding to the earth's strata; their being responsible for the light of the solar corona; that meteorites are products of volcanic eruptions; or that they came from other planets either by volcanic activity or the rupture of a planet. He was left with his favorite assumption that meteoroids "are in fact fragments of comets, or it may be in some cases minute comets themselves." He advocated the usual hypothesis on the origin of the comets, namely that they condensed from nebulous matter either in the outer parts of the solar system or from matter entirely independent of the solar system and scattered throughout space.

Believing that meteorites had been broken off comets, in 1878 Newton contributed some thoughts on how meteoroids had first become solid, since meteorites more nearly resemble masses to be found in the deep interior of the Earth than in its crustal layers. He commented:[44]

> It seems very improbable that iron masses whose like in the earth is found only in the igneous trap rocks, especially in Greenland traps, should have become consolidated in the cold of space in small parcels. The internal structure of the comet fragments are records of an interesting early history. To decipher the legends belongs rather to the mineralogist and the physicist than to the astronomer. My effort has been to find *where* they were written.

Other scientists had assumed that the structure of the meteorites could have come about only under conditions of high temperature and high pressure. Later it was found that great heat at the time of first consolidation was not really essential. Newton quoted Daubrée as stating that the fine-grained structure of meteorites resembles those of hoarfrost or snow which are due to the immediate passage of vapor into the solid state. But, asks Newton, is heat necessary for the making of meteorites? Some crystallizations do take place in the cold, being direct changes from the gaseous to solid forms. He reasoned: "Wherever crystallization can take place at all, if there are present silicon and magnesium, and iron and nickel, with a limited supply of oxygen, there silicates ought to be expected in abundance, and the iron and nickel in their metallic form. . . . The smallness of comets should not then be an objection to considering the meteoric stones and irons as pieces of comets."[45]

These masses as they travel from the cold of outer space to the hot proximity of the sun undergo further changes, as the dispersal of meteoroids around the Leonid ring, and the splitting of Comet Biela testify. The conditions encountered by the comets are "so strange to our experience, and so impossible to reproduce by experiment, that our physics can not as yet explain it."[44] But, under his assumption that meteorites had been components of comets, he expressed as follows his optimism that solutions would eventually be found:

Of those strange bodies, the comets, we shall have far greater means of study than of any other bodies in the heavens. The comets alone give us specimens to handle and analyze. Comets may be studied, like the planets, by the use of the telescope, the polariscope and the spectroscope. The utmost refinements of physical astronomy may be applied to both. But the cometary worlds will be also compelled, through these meteorite fragments with their included gases and peculiar minerals, to give up some additional secrets of their own life and of the physics of space to the blowpipe, the microscope, the test-tube and the crucible.[46]

Modest spectroscopic equipment was available for use with the 8-inch Reed refractor. However, there is no indication that Newton himself made use of it; and spectroscopic observations by Sherman[47] failed to yield satisfactory results.

The great love Newton felt for comets and meteors was uniquely expressed by the way he decorated the wallpaper in his home. A biographer of J. Willard Gibbs, Newton's neighbor, relates,[48] "The walls of this house [Newton's] were different from all other walls. . . . here on the steady gray background of the wall, stylized in pale gold, curved the tracks, twined in all-over patterns of the most graceful curves imaginable—fireballs and comets twisting in variety in every room." This description is reminiscent of Trouvelot's painting of meteors he observed during the Leonid shower in 1868 (Fig. 10.1).[49]

Agnes Clerke, in her *History of Astronomy During the Nineteenth Century,*[50] averred that Professor of Mathematics H. A. Newton by his "diversion from purely abstract studies stimulated their effective exercise, and constituted him one of the founders of meteoric astronomy." J. Willard Gibbs, Yale's outstanding Professor of Mathematical Physics from 1871 until 1903, who had been Newton's student (Yale class of 1856), friend, neighbor, and colleague, eulogized Newton's achievements in the *Biographical Memoirs of the National Academy of Sciences* and compiled his bibliography.[51]

Although Gibbs is hardly remembered as an astronomer, under the inspiration of Newton he did make some relevant contributions in this field.[52] As a student he joined Newton, staying up late into the night observing meteors. While Newton was fighting for the adoption of the metric system in America, Gibbs delivered a lecture at the Connecticut Academy of Arts and Sciences urging its acceptance. His paper *On the Determination of Elliptic Orbits from Three Complete Observations*[53] was described by his biographer in 1909 as possessing distinct advantages over the older methods of the experts, Gauss and Oppolzer.[54] This evaluation, on the other hand, is not shared by modern celestial mechanicians. As Marsden points out,[55]

Gauss solved the orbit-determination problem completely, and Gibbs' approximations are both unnecessary and inexact. Gauss had trouble with his geometry, it is true, but Gibbs really did not do much better, in spite of his work on vectors. Twentieth-century contributors since, as Merton and Cunningham, did very much better in this respect. Oppolzer also made some important contributions that will have longevity over those of Gibbs.

In 1884 Gibbs published a paper *On the Fundamental Formula of Statistical Mechanics, with Applications to Astronomy and Thermodynamics.*[56] This was followed in 1902 by his book, *Elementary Principle in Statistical Mechanics,* which is reputed to have served as an inspiration to Edwin Hubble.[57] Jaffe[58] notes Gibbs's "monumental contributions in thermodynamics," ranking him along with Benjamin Franklin as one of the two most influential scientists in America in the eighteenth and nineteenth centuries. Gibbs's contributions in thermodynamics, vector analysis, and statistical mechanics certainly have bearing on the future development of theoretical astronomy and astrophysics.

As a teacher Gibbs was not especially successful. Jaffe[59] writes, "He is said to have remarked in 1902, after thirty years of teaching, that not many of his graduate students were really getting anything valuable out of his courses. Among this small group were Lee de Forest, inventor of the radio tube and pioneer in television and the sound film."

While a student at Mt. Hermon, a college preparatory school, de Forest claimed astronomy as a favorite subject.[60] The fifth satellite of Jupiter, Amalthea, had recently been discovered by Barnard (1892). Young de Forest found that it and the Galilean satellites obeyed the hypothesis known as "Bode's Law." He therefore believed it was really a law of nature and precociously predicted that Neptune should have another satellite besides Triton! When Neptune's second satellite, Nereid, was discovered in 1949 its distance would, according to Bode's formula, have corresponded to the fourth satellite out from Neptune, not the second. Bode's Law, an empirical relation, had first been discovered by J. D. Titius[61] in 1776 but was later named for Bode who more successfully publicized it when he demonstrated that the then newly discovered asteroid, Ceres, was at a position corresponding to the gap

Fig. 10.1. Trouvelot painting of meteors during the Leonid shower of 1868 (not simultaneous meteors, but representing all types actually seen). (Courtesy Mount Holyoke College Library)

between Mars and Jupiter.[62,63] The relation predicted that the relative distances of the successive planets from the sun should be as 4, 4+3, 4+6, 4+12, 4+24, etc. = 4, 7, 10, 16, 28 ... Thus the second satellite out from Neptune would have been expected at less than twice Triton's distance but was actually found at four times that distance. In 1989, *Voyager II* recorded six additional Neptunian satellites and four rings,[64] all well within the orbit of Triton, and none at relative distances compatible with Bode's Law.

De Forest's experiences as a Yale student were often disappointing. One of his chief teachers was Physics Professor Charles S. Hastings. Instead of encouraging the young student's inventions (some of which years later were independently discovered and patented by others), Hastings severely criticized them. In Graduate School, on the other hand, de Forest came under the influence of Gibbs, who recognized the genius of this student. After a short stint with the Yale Artillery Company at the time of a Cuban crisis, de Forest returned to his studies under Gibbs. In his autobiography de Forest wrote, "and again I was to listen for two hours a week to his solemn discussions on the Orbits of Comets and Asteroids." Perhaps it was as a consequence of these lectures that he took an interest in meteors. Possibly in conjunction with Elkin's photographic work on meteors, visual observations were being solicited. Having had difficulty finding a summer job, de Forest commented, "Finally, the Yale Observatory could use several keen-eyed young observers during summer nights, to observe and take note of meteors. Many a lonely tryst I kept after observatory work in the predawn beside Lake Whitney or in East Rock Park, then to steal home to attempt a prose poem of description." No acknowledgment of these observations is found in the annual reports of the Observatory.

Newton, for his investigations of meteors, especially the Leonids of 1888, was the first ever to be awarded the J. Lawrence Smith gold medal of the National Academy of Sciences,[65] which had been established in 1885 by Mrs. Smith as a memorial to her husband for the purpose of promoting one of his favorite interests, the study of meteoric bodies.[66] In his acceptance speech Newton remarked, "To discover some new truth in nature, even though it concerns the small things in the world, gives one of the purest pleasures in human experience."[67] In his appreciative biographical note on Newton, Yale mathematician A. W. Phillips wrote,[68] "One prominent characteristic of his investigation was always its exhaustive character. For whatever Professor Newton did, it was not worth the while of anyone else to cover the same field." Newton himself appeared to minimize his own accomplishments when he concluded his article in the *Encyclopedia Britannica,*[69] signed only as was the custom, "(H.A.N.)," with the statement, "Meteoric science is a structure built stone by stone by many builders. In this article no attempt has been made to assign to each builder the credit for his contribution." Had such credit been assigned, the modest Newton might have appeared conceited!

REFERENCES

1. B. Narendra, *Discovery,* **13**, 10, 1978.
2. O. C. Farrington, *Meteorites,* Chicago, p. 46, 1915.
3. H. A. Newton, *Nature,* **19**, 315, 1879.
4. G. T. Prior, *Catalogue of Meteorites,* London, 1923.
5. N. R. Leonard, *Amer. Jour. Sci.,* (3), **10**, 357, 1875.
6. E. S. Dana, "Catalogue of the Collection of Meteorites in the Peabody Museum of Yale College," *Amer. Jour. Sci.,* (3), **32**, 246, 1886.
7. O. C. Farrington, *Mem. Nat. Acad. Sci.,* **13**, 239, 1909.
8. Ref. 3, p. 342.
9. G. D. Hinrichs, *Pop. Sci. Monthly,* **7**, 588, 1875.
10. H. H. Nininger, *Out of the Sky,* Denver, p. 173, 1952.
11. ———, *Idem,* p. 174.
12. J. Torrey and E. H. Barbour, *American Geologist,* **8**, 67, 1891; *Mem. Nat. Acad. Sci.,* **13**, 195, 1909.
13. W. E. Bradford to H. A. Newton, June 12, 1890, *Yale Obs. Archives.*
14. O. Huntington to H. A. Newton, November 25, 1891, *Yale Obs. Archives.*
15. A. L. Graham, A. W. R. Bevan, and R. Hutchison, *Catalogue of Meteorites,* Tucson, 4th ed., 1985.
16. Ref.10, p. 189.
17. T. Dwight, *Ann. Report of the President of Yale* for the year ending Dec. 31, 1891.
18. J. Torrey, Jr., and E. H. Barbour, *Amer. Jour. Sci.,* (3), **39**, 521, 1890.
19. D. Fisher, *Amer. Jour. Sci.,* (3), **34**, 381, 1887.
20. Ref. 10, p. 139.
21. G. F. Kunz, *Amer. Jour. Sci.,* (3), **40**, 312, 1890.
22. Ref. 10, p. 200.
23. H. H. Nininger, *Find a Falling Star,* N.Y., Chap. 6, 1972.
24. ———, *Idem.,* p. 86.
25. D' A. McNickle, *Encyclopaedia Britannica,* **12**, 206, 1958.
26. H. A. Newton, *Amer. Jour. Sci.,* (4), **3**, 1, 1897.
27. W. K. Moorehead, *The Hopewell Mound Group of Ohio,* Anthropological Series, Field Museum of Natural History, **6**, No.5, 1922.
28. H. C. Shetrone, *The Mound Builders,* N.Y., London, 508 pp., 1941.

29. Ref. 27, p.128.
30. Ref. 21, p. 316.
31. J. T. Wasson and S. P. Sedwick, *Nature,* **222**, 22, 1969.
32. O. C. Farrington, *Pub. Field Museum of Chicago, Geol. Ser.,* **1**, 310, 1902.
33. E. L. Krinov, *Principles of Meteoritics,* trans. into English by I. Vidziunas, Oxford, p. 204, 1960.
34. H. A. Newton, *Amer. Jour. Sci.,* (3), **45**, 61, 1893.
35. Ref. 23, p. 250.
36. B. Silliman, *Mem. Conn. Acad. Arts and Sci.,* **3**, 213, 1815.
37. A. W. Wright, *Amer. Jour. Sci.,* (3), **9**, 294, 1875.
38. ———, *Idem.,* p. 459.
39. ———, *Ibid.,* (3), **10**, 44, 1875.
40. T. E. Houck and A. D. Code, *Bull. Amer. Ast. Soc.,* **2**, 321, 1970.
41. A. W. Wright, *Amer. Jour. Sci.,* (3), **11**, 253, 1876.
42. ———, *Ibid.,* (3), **12**, 165, 1876.
43. H. A. Newton, *Proc. Amer. Assoc. Adv. Sci.,* **35**, 1, 1886; *Popular Science Monthly,* **29**, 733, 1886.
44. ———, *Amer. Jour. Sci.,* (3), **16**, 179, 1878.
45. ———, *Proc. Amer. Assoc. Adv. Sci.,* **35**, 15, 1886.
46. ———, *Idem,* p. 18.
47. *See* Chap.11.
48. M. Rukeyser, *Willard Gibbs,* N.Y., p. 292, 1922.
49. E. L. Trouvelot, *Astronomical Drawings,* N.Y., pp. 115, 166, 1881.
50. A. Clerke, *A Popular History of Astronomy During the Nineteenth Century,* London, 4th Ed., p. 330, 1902.
51. J. W. Gibbs, *Biog. Mem. Nat. Acad. Sci.,* **4**, 99, 1902.
52. B. Jaffe, *Men of Science in America,* N.Y., pp. 320, 328, 1944.
53. J. W. Gibbs, *Mem. Nat. Acad. Sci.,* **4**, 79, 1889.
54. C. S. Hastings, *Biographical Mem. Nat. Acad, Sci.,* **6**, 375, 1909.
55. B. G. Marsden, Private communication, Aug. 24, 1990.
56. J. W. Gibbs, *Proc. Amer. Assoc. Adv. Sci.,* **33**, 57, 1884.
57. Ref. 52, p. 493.
58. ———, *Idem,* pp. xxxi, xxxvii.
59. ———, *Idem,* p. 310.
60. L. de Forest, *Father of Radio,* Chicago, p. 65, 1950.
61. I. Asimov, *Biographical Encyclopedia of Science and Technology,* 2nd ed., N.Y., item 301, 1982.
62. ———, *Idem,* item 344.
63. J. E. Bode, *Von dem neuen, zwischen Mars und Jupiter entdeckten achten haupt Planeten des Sonnensystems,* Berlin 1802; trans. into English by H. Shapley and H. Howarth, *Source Book in Astronomy,* N.Y., p. 180, 1929.
64. Anon., *Sky and Telescope,* **78**, 351, 1989.
65. I. Remsen, *The Semi-Centennial Anniversary Meeting of the National Academy of Sciences, 1863–1913,* p. 9, 1913.
66. F. W. True, Ed., *A History of the National Academy of Sciences, 1863–1913,* Washington, pp. 58, 66, 371, 1913.
67. Ref. 53, p. 118.
68. A. W. Phillips, *Bull. Amer. Math. Soc.,* (2), **3**, 172, 1897.
69. H. A. Newton, *Encyclopedia Britannica,* 9th ed., **16**, 112, 1883.

11
THE WINCHESTER OBSERVATORY

Yale's prestige, resulting from the early rediscovery of Halley's comet in 1835 with the largest refracting telescope then in America, was soon followed by years of frustration.[1] While other institutions were acquiring better equipment and building observatories, Yale seemed incapable of acquiring a proper observatory for its famous telescope, making it practically useless for research. After prolonged struggles, Denison Olmsted, toward the close of his long career, sadly succumbed to the conclusion that "to advance astronomical science is not what college observatories are wanted for . . . What we want of an observatory is as a part of the means of education, as auxiliary to instruction in astronomy."[2] With the grossly inadequate housing for the famous old Dollond telescope, not even this modest goal could really be achieved. The Sheffield Observatory, established in 1866 with its 9-inch refractor, was a great improvement, at least from the teaching standpoint. But by that time no fewer than twelve other American observatories boasted larger refractors,[3] all more favorably situated than this one, high atop a potentially swaying building.

As early as 1858 Mrs. Cornelia Hillhouse and her daughters (Mary and Isaphene), in memory of her father-in-law James Hillhouse, donated to Yale six acres of prime land at the crest of Prospect Street as a site for an observatory.[4,5] James Hillhouse had been Treasurer of Yale for fifty years, 1782 to 1832. His son, James A. Hillhouse, Cornelia's husband, was a poet who wrote a moving 310-verse poem in 1838, *Sachem's Wood,*[6] memorializing the history of the family property where Gibbs Hall and Kline Tower now stand:

> I've seen the world, from side to side,
> Walked in the ways of human pride,
> Mused in the palaces of kings,
> And know what wealth to grandeur brings;
> The spot for *me,* of all the earth,
> Is this, the dear one of my birth.

It may seem puzzling that the Hillhouse land gift was not taken into account when Sheffield gave his telescope in 1866. The land, however, was given to Yale College, whereas Sheffield wanted his school to remain independent of the College, even though both were parts of the University.

The benefactor whose philanthropy finally made possible the establishment of an observatory for the College was an industrialist. Oliver Winchester (frontispiece) was head of the Winchester Firearms factory in New Haven. In 1871–72 he donated 36 acres of land at a cost of $100,000, adjacent to the six previously given by the Hillhouse family. Winchester speculated that land values were increasing and that within ten years the sale of the land should yield some $500,000, sufficient for the erection of a first-rate observatory. Winchester appointed six gentlemen to a Board of Trustees with himself as chairman, and including astronomy professors from both the College (Loomis and Newton) and the Sheffield School (Lyman). In a letter to the President and Fellows of Yale College dated July 10, 1871, Winchester stipulated,

> I shall direct that they and their successors shall annually and whenever requested, report to the Corporation of the College the condition of this trust and the state of the Observatory, whenever it shall have been established; and I shall further direct that the observers nominated by said Trustees shall not enter upon their office without the approval of the Corporation of Yale College; it being understood that the appointment shall not turn upon political or ecclesiastical considerations.[7]

Moreover, Winchester optimistically went to Paris to acquire a 29-inch optical disk from Feil and Company, one of the few concerns then capable of making such large ones. Jacob Campbell, one of the Trustees, donated four 10-inch disks.[8] The 29-inch and two of the 10-inch disks Winchester deposited with the Alvan Clark Company in Cambridge, Massachusetts, who were eventually expected to build the telescope. He wanted Yale to have the largest telescope in America, larger than the 26-inch equatorial acquired by the U.S.Naval Observatory in 1873. Oddly, while Winchester wanted to surpass the Naval Observatory refractor, which was an achromatic with an objective consisting of both a crown and a flint lens, Winchester apparently ordered only the flint component from Feil. That company seems to have had more difficulty in casting crown disks. After the original crown for the 36-inch Lick refractor was broken in packing for transit in 1882, the company had cast another 19 disks before being successful in 1885.[9] Then also the first 33-inch crown photographic corrector lens was found defective and cracked while Clark was attempting to repair it.[10]

The large Yale 29-inch refractor never materialized. In 1887 the report of the Board of Trustees indicated that they were considering selling the disk to Lick Observatory for not less than $3,200, but there was no follow-up.[11] Then correspondence in 1921 between Schlesinger and Carl Lundin,[12] successor to the Clarks, indicates that the lens was still in Cambridge and that Schlesinger had offered to sell it to the Company for a fair price, estimated at about $1,000 minus insurance costs for 50 years of storage. Lundin replied bitterly,

> Our Company has never been favored with any orders from Yale Observatory although we know that during the past fifty years orders from your observatory have been placed elsewhere. We would be perfectly willing to forgo any charge covering the storage and insurance cost of this disc provided we were definitely assured of receiving some substantial orders from you.

Nothing further seems to have come of this until after Lundin left the Alvan Clark Company in 1929. The new owner, C. H. Sawyer, reported that the Clarks had declared the disk unsuitable for lens purposes,[13] and he returned it to Yale at no cost other than shipping expenses. In 1933 it was sent to the Fels Planetarium at the Franklin Institute in Philadelphia for exhibit purposes. Later it was returned to Yale, but no practical use has ever been made of it.

On Winchester's munificent endowment H. A. Newton was to comment that it differed "from most of the benefactions of Yale College in that its direct object is to endow scientific research rather than to furnish educational facilities."[8] Certainly the upshot was a greater diversity of research projects after the erection of the Observatory. While the bulk of the publications of Newton's predecessors were textbooks, the publications now emanating from the observatory primarily concerned research results. Although Newton (unlike Olmsted and Loomis) did not publish a single textbook, he nevertheless was highly regarded by his students. In regard to Newton's research, Gibbs recalled,

> In fact, we might almost call the studies which we have been considering the recreations of a busy life of one whose serious occupation has been that of an instructor. If from all those who have come under his instruction we should seek to learn their personal recollections of Professor Newton, we should probably find that the most universal impression he made on his classes was that of his enthusiastic love of the subject he was teaching.[14]

In 1873 architect R. G. Russell drew up elegant plans for a building (Fig. 11.1). These great expectations were soon to be followed by hard times as the financial panic of 1873–74 frustrated most of the plans. Finally, in 1880, the land was transferred from the Trustees of the Winchester Observatory to the Yale Corporation. The Trustees were then appointed as a Board of Managers, with Professor Lyman as President and H. A. Newton as Secretary. Newton was also appointed Director, without pay, of the still nonexistent observatory. Land was sold, and a modest part of Russell's proposed building was erected by 1882. (Compare the architect's plans, Fig. 11.1, with a photograph of the building, Fig. 11.2, soon after it was erected, and Fig. 20.1 as it appeared in about 1930. Winchester, who died in 1881, unfortunately did not live long enough to see even this degree of completion of his dream come true. Lyman, Loomis and Newton remained on the Board of the Winchester Observatory from its inception until their decease. Newton's last report as Secretary of the Board was dated June 23, 1894; he died on August 12, 1896. Lyman and Loomis both died in 1889. The minutes of the Board for 1889–90 paid them tribute:

> They were selected by Mr. Winchester to be Trustees of the institution at its first inception. He relied upon them for advice and assistance in giving wise direction to his generous designs for the advance of the Science of Astronomy. During eighteen years they have attended the meetings of the Trustees and the Board, and by their counsel and labors have done what they could to build up the Observatory, and make it efficient.

These remarks, as worded by Newton in honor of his colleagues, apply to an even greater extent to Newton himself.

Upon the application of representatives of the Winchester family, the name of the Winchester Observatory was officially changed on October 28,

Fig. 11.1. R. G. Russell 1873 plans for the Winchester Observatory, frustrated by the financial panic of 1873–4.

1881, to *The Observatory in Yale College,* but the original title continued to be used informally for many years thereafter.

Leonard Waldo (Harvard Sc.D., 1879) had been appointed as an Observatory Assistant in November 1879. Already in 1880, at his suggestion,[15] the Board of Managers had ordered from Repsold of Hamburg a 6-inch heliometer at an expected cost of about $7,200, and a 15-foot dome from Grubb of Dublin. The Alvan Clark Company of Cambridge had originally been offered the contract for the heliometer lens but refused because they did not want to run the risk of mishap in bisecting the lens. The annual report for 1880–81 explained,[16]

> The heliometer is an instrument hitherto unknown in the United States, and will enable us to make exact measurements rather than optical discoveries. Of large equatorials the country has a goodly number, either already in place or in process of construction. Of transit circles there are also several of the highest class. But a large and important class of problems in astronomy are open to investigation by the use of the heliometer which are not easily attacked with the ordinary equipments of our observatories.

In March of 1880 Professor Newton and Dr. Waldo were appointed as a committee to raise funds for the heliometer and a proper dome, the total cost being estimated at about $10,000. They traveled to Cincinnati, where Waldo addressed the Yale alumni of that city in his most appealing style:[17]

> There is no science whose cultivation is more germane to the highest dignity of the University than Astronomy. . . . One of our greatest astronomers . . . said on hearing what we proposed to do, "Well, Harvard is radical and Yale is conservative; but when Yale makes up her mind to do a thing, it is pretty sure to be done." . . . Just so sure as there is inherent in the heart of every man who has been trained within her walls a love of learning for its own sake, an appreciation of the mighty power which science contains for making men better . . . so sure is it that there will be found givers who will support this effort to make the University strong where she is now weak, to make her glorious where she is now rapidly sinking into oblivion.

Newton was especially hopeful that this new equipment would be on hand to prove its value during the upcoming transit of Venus on December 6, 1882. Soon, thanks to a then still anonymous donor (E. M. Reed, Superintendent and Vice President of the New York, New Haven and Hartford R.R.), Newton was also able to order an 8-inch refractor from Grubb, and both instruments did arrive on time for the transit of Venus. The heliometer (invented by J.

Fig. 11.2. The Winchester Observatory *ca* 1880.

Dollond in 1754, and best remembered as the type of instrument with which F. W. Bessel in 1838 published the first reliable parallax of a star) has an objective split in half along a diameter. By sliding the two halves relative to each other and rotating the entire assemblage, the images of two stars can be superposed and the separations of the stars accurately measured.[18,19] Larger separations can be measured with the heliometer than is usually the case with the more familiar filar micrometer generally used in double-star measurements.

While the observatory was being built and the instruments installed and tested, the shortage of operating funds was apparent. In 1880 Leonard Waldo initiated a plan, carried over from his Harvard experiences and then prevalent in some European and other American observatories, of establishing a public time service. While he had been employed at Harvard (1875–79) he had been in charge of its time service and had written a pamphlet advocating and explaining a standard time system.[20] In another article[21] he stressed his belief that railroads ought to be the most influential means in securing uniformity; and for establishing standards, active cooperation with the telegraph companies was essential.

Already between 1856 and 1862 a free public time service for Boston and vicinity had been initiated at Harvard under the Bonds (the first two directors of Harvard College Observatory), but Winlock (the third Director) in 1871 needed an assistant but found himself in a financial pinch. He therefore seized upon the precedent of Langley at the Allegheny Observatory who in 1869 started selling time to the Pennsylvania and affiliated railroads that entered Pittsburgh.[22] Later, in 1878, Carleton College in Minnesota also established this custom and by 1881 served all the railroads terminating in Minneapolis—St. Paul, seven railroads serving six states.[23] The Yale Horological Bureau became operative in 1880, and shortly thereafter (following the example of the Kew Observatory in England) also established a Thermometric Bureau which would rate thermometers both for meteorological purposes and for hospitals.[24] Waldo had previously advocated a similar scheme at Harvard but, E. C. Pickering reported,[25] "A plan for establishing a bureau for the testing of fine watches and thermometers was considered, and abandoned on the ground that such work would become commercial rather than scientific, and therefore not within the scope of the observatory." Yale was evidently the first American observatory to raise funds by checking thermometers.[5,26]

Railroads at that time had been handicapped by the fact that almost all stations along their routes ran on independent time systems. By 1870 over fifty different times were being used by the railroads, as many as five in the Pittsburgh terminals alone.[27,28] Waldo planned to check clocks and watches primarily for the railroads in Connecticut, but also for watch and clock makers and for any scientific purposes. He succeeded in getting the state of Connecticut to subsidize the undertaking at $2,000 a year, the first state to do so. The standard time to be adopted in Connecticut was to be the local time at the City Hall of New York City,[29] as that was the time already being employed by four-fifths of the manufacturing and railroad interests of the state.[30] This was before standard time zones were established in the United States. As early as 1850, Greenwich Observatory time had already been established as standard time throughout England, Wales, and Scotland. By 1879 Sanford Fleming had proposed a standard time system for Canada, but it was impractical for civil purposes.[31] In the United States at a convention in St. Louis in 1872 the railroads proposed a system, devised by C. F. Dowd,[32] that the States be divided into four time zones with times in successive zones differing by exactly one hour.[33] Eleven years passed before this plan was finally approved by the railroads in October 1883. Then, in August 1884, Congress authorized President C. A. Arthur to convene an international conference to consider a worldwide system of standard time zones.[27] This was held in Washington in October 1884 and a system was adopted. However, setting the boundaries at exact intervals of 15° in longitude could inconveniently bisect cities or put adjacent towns in different zones. The actual boundaries to alleviate these problems were not fixed until 1918, some even later.[27] Waldo's early efforts toward standardizing time in Connecticut were an important step in the right direction.

When the Horological and the Thermometric Bureaus were first established it had been with the clear understanding that they should be entirely self-supporting, not dependent in any way upon Yale resources. In this they were at first eminently successful. During the first year of operation, 1,957 thermometers were examined and issued certificates of accuracy, and in the succeeding four years between 4,552 and 6,390 were tested annually. This was a thriving business since most of the thermometers rated were clinical. Some of special interest were those to be used in Arctic explorations by the U.S. Signal Service.[5] In 1883 Waldo looked upon his success with the Thermometric Bureau as a seed adventure for establishing a meteorological observatory on Yale's astronomical observatory premises, but this did not materialize.[34]

On November 20, 1883, the Prudential Committee of the University appointed Mason Young and F. J. Kingsbury (Yale graduates and successful businessmen) a committee "to confer with the Director of the Observatory and Dr. Waldo respecting the status of the Horological Bureau and to report a plan for its future relation to the Observatory.[35] Further, the Yale Corporation voted on February 19, 1884, that these two gentlemen" be also directed to inquire into and report upon the expediency of additional regulations for the organization and administration of the Observatory and its several parts." These committeemen were not astronomers but as businessmen they admired Waldo's business acumen and were strongly influenced by Waldo's wishes in regard to the two Bureaus. They submitted their report directly for approval to the Corporation without first conferring with the Director of the Observatory as they had been instructed to do. Only after the Prudential Committee had approved the report did Newton receive a letter from Young on April 9, 1884, stating, "The Committee of the Corporation of the Observatory has prepared its Report—as a matter of courtesy I send you the enclosed copy of the Statute and resolutions which it will recommend for adoption."[35]

The recommendations included the abolition of the Board of Managers of the Observatory to be replaced by a faculty committee consisting of the President of the University, the Director of the Observatory, and the officer in charge of the Verification Bureaus (Horological and Thermometric); that the Bureaus be made independent of the Director of the Observatory; that three-fourths of the office space of the Observatory be given over permanently to the Bureaus; that Mr. Winchester's donations to the Observatory be applied to the expenditures of the Bureaus.

The reactions of Newton and the Board of Managers were prompt. On April 15 they sent the President and each of the Fellows of the Corporation a strong and detailed letter of protest.[36] In particular they pointed out that Winchester's bequest stipulated that it be used for astronomical and physical research, whereas the Bureaus were a commercial and utilitarian undertaking; and to put the head of the Bureaus on a par with the Director would make it more difficult to find a Director for the Observatory. Newton's position was elaborated:

> Two years ago the present Director very reluctantly consented to take the position he now holds, and has performed a large amount of labor, much of it being of the most disagreeable kind, without pay, hoping that the time might soon come when we could afford to have a salaried Director. We have been struggling to secure the services of a Salaried Director who should give his whole time to the Observatory, but we have not been able to secure the requisite amount of money. Even if the money were procured, neither the eminent Astronomer whom we had in view, nor any other who would be satisfactory to us, would consent to become the Director of only a part of the Observatory, [of] which there was another part using the name of the Observatory which was practically independent of him.[37]

A few days after this letter was transmitted to the Corporation, Newton, on April 19, 1884, sent in his resignation as Director of the Observatory, and Loomis also resigned from his membership on the Board of Managers. The actions of the Prudential Committee and the resignations of the two esteemed professors were widely publicized in the public press.[38] Newton was highly respected, both by the public and in the University and his resignation, leaving the Observatory without a director, was especially deplored by the press. Yale alumni who had elected Mason Young to the Corporation were puzzled and outraged by his major involvement in this crisis. He was consequently not reelected.

Soon the Records of the Corporation show that it "Resolved, that, for the present, the President of the College shall be added to the Board of Managers of the Observatory, and that, (with this exception) no change shall be made in the organization of the Observatory until the President [Noah Porter] has become familiar with the working of the same and has satisfied himself that other changes were desirable." Evidently Porter and the Corporation were satisfied with the status quo. They then accepted Newton's resignation as Director, but he remained the Secretary of the Board of Managers through June 1894, continuing to perform all the duties of a director. Although the Board of Managers made efforts to find a new Director, and at one time considered Asaph Hall (Sr.) of the Naval Observatory, no new Director was appointed until shortly before the death of Newton in 1896—largely for the usual financial reasons. The Board as a whole directed the Observatory.

In October 1884, Waldo relinquished the Thermometric Bureau when the Board appointed O. T. Sherman Astronomer in Charge of the bureau.[39] But Waldo was instructed to make himself available for advice when needed. The Secretary of the Observatory, Robert Brown, Jr., was appointed to issue certificates of verification of thermometers, in accordance with Mr. Sherman's observations, and to look after the correspondence and business matters of the Thermometric Bureau. Orray Taft Sherman (1856–1921, Yale B.A. 1877, Phi Beta Kappa) had been associated with the Thermometric Bureau since its inception in 1881. Until his resignation in November 1886, well over 30,000 thermometers had been tested, mostly by himself—an average, after the first year, of nearly 5,700 a year or about 20 a day.[40]

The Horological Bureau, of which Waldo continued to be in charge for several more years, met with mixed success. Waldo was to be paid up to $2,000 a year but only from any excess of income over expenditures. In 1884 the Bureau showed a deficit of over $1,500. At first the railroads were evidently happy with their improved time services. Then in March of 1885, ostensibly for reasons of economy alone, the State Appropriations Committee, without first consulting with any of the officers of the Observatory, the railroad commissioners, or any of the other recipients of the time service, terminated its contract, thereby threatening serious additional deficits at the Bureau and the Observatory.[41] However, four railroads and seven of the watch companies being served raised a subscription to meet the deficits the following year.[42] Watch companies in the initial year of the Horological Bureau submitted 219 watches for verification. Then the numbers declined, amounting to between 41 and 87 in the following five years, and by 1887 to only a negligible number.[43] Finally at a meeting of the Board of Managers on May 3, 1887, it was resolved that "the public do not desire the maintenance of a Horological Bureau for testing the rate of watches, and we regard it as inexpedient to continue the experiment any longer."

The Board consequently recommended both the discontinuance of the Bureau and the severance of Dr. Waldo's connection with the Observatory. But as Waldo was given the opportunity to resign before these resolutions were presented to the Corporation, they were not put into effect.[44] Instead, appreciating the previous support of several watch companies in helping to meet the financial deficits, the Bureau stood ready to reactivate the watch-checking service should a future demand arise. At the following meeting, on June 7, Asaph Hall, Jr. was appointed to take charge of the Time Service, a position he held until August 1889 when he returned to a post he had previously held in Washington. Thereafter a succession of other assistants continued the work. In 1916 the Astronomer then in charge, Mason F. Smith, reported that the Time Service had been maintained in its usual efficient condition.[45] The Time Service remained active, providing time to both the New Haven Railroad and the Southern New England Telephone Company until World War I, 1918, when the Observatory was temporarily closed.

The activities of the Thermometric Bureau were not included in the annual reports of the Director of the Observatory. At first they were prepared in separate reports by Waldo, then for 1884–86 by O. T. Sherman, and finally they were included in the *Annual Reports* of the Secretary of the Observatory, Robert Brown. His last available report, for 1898–99, indicated that the work of the Thermometric Bureau had continued as usual at least until that time. *The Catalogue of Yale University* for 1899–1900 specifically states,[46]

> Besides the ordinary astronomical work the Observatory maintains two public services. Continuous time signals are transmitted from the distributing clock at the Observatory to the railroads and other parties. The Observatory offers facilities also to persons interested in accurate thermometry for the comparison of thermometers with standard instruments.

These statements, substantiated in numerous reports, appear to contradict others[1] that the Thermometric Bureau ceased in 1887, the year that Waldo left Yale.

When Waldo left the University he must have been considerably in debt to the Observatory, for the record of the meeting of the Board of Managers on November 3, 1888, states, "The President read a letter from Dr. Waldo offering [a building] built by him on the observatory grounds as a donation to the University and requesting that a paper certifying that his pecuniary obligations to the University are satisfied be given him. Voted that a statement of his account with the University be prepared."[47]

In 1885 Waldo had requested permission to erect a laboratory building at the rear of his house which belonged to the Observatory. Permission had been granted on condition that he erect it entirely at his own expense, and remove it at any time, should he be asked to do so.[48] At the Board meeting on November 10, 1888, the matter was settled: "A report of Dr. Waldo's indebtedness having been made it was voted that the building which he erected upon the Observatory land be accepted as a settlement of all his indebtedness to the Observatory."

Waldo was a very able and industrious young man (only 35 when he left Yale) whose pushy self-importance became his greatest obstacle to real success. His future occupations were mainly in engineering, not astronomy. But during his tenure at Yale Waldo had participated in much valuable work in addition to that of the two Bureaus. In 1882, before the Winchester Observatory was erected, he and Professor W. A. Rogers of Harvard had exchanged telegraph signals for the purpose of determining the longitude difference between the meridian circle at Harvard and the transit instrument at that time situated in the northeast corner of North Sheffield Hall, about a mile south of the Winchester Observatory.[24,49] These observations were reduced by J. F. Kershner[50] who in 1885 obtained a Ph.D. degree from Yale for his thesis on the longitude difference between Yale and Harvard. The value they found, $0^h\ 7^m\ 10.353^s \pm 0.0091^s$, differs by only $+0.86^s$ from the difference in the longitudes of the two observatories as given in the 1882 and 1885 *Nautical Almanacs,* and by -0.82^s from the 1955 values (before the Yale Observatory was moved to a new location in 1956). Waldo also participated in the erection and testing of the heliometer when it arrived, and with it made measurements of the diameters of the sun, moon, and the planets Venus, Jupiter, Saturn and Uranus. What had been expected to become his most important contributions were his heliometer measurements of the transit of Venus, elaborated below.

Before the observatory was erected in 1882, complete with two 15-foot domes, its equipment had been inconveniently housed in numerous widely scattered quarters. Now, besides the heliometer and the 8-inch Reed telescope, the equipment included a 4-inch Troughton and Simms transit instrument donated by Dr. W. Hillhouse, nephew of the former Yale Treasurer, three astronomical clocks, one by Bond, one by Howard and the third made and presented by Dr. W. Hillhouse, two chronographs, and thermometric equipment.[51] The heliometer lived up to expectations. While it was still at the factory in Hamburg, Waldo and the eminent German astronomer

Arthur von Auwers were there to inspect it. They agreed, "It is a very beautiful instrument—perhaps the most perfect the art of man has so far constructed."[52] In October of 1882 the *Sidereal Messenger* noted "The new heliometer of Yale college observatory which is already mounted, is said to be one of the most perfect instruments of its kind in the world."[53]

Sir Edmund Halley in 1716 is reputed to have been the first to suggest that the solar parallax could be determined from transits of Venus observed from widely separated positions on the earth.[54] No such transits occurred in his time.[55] The parallaxes derived from observations of the transits of 1761, 1769 and 1874 yielded highly discordant values (8.49″ to 10.10″ for 1761, 8.43″ to 8.85″ for 1769, and 8.754″ to 8.884″ for 1874).[56] Hence the transit of 1882 (the last until 2004), for which more sophisticated equipment had become available, was eagerly anticipated. Auwers in Germany,[57] as previously in 1874, had organized four expeditions, two in the northern, two in the southern hemisphere, each equipped with a three-inch Fraunhofer heliometer.

In the United States, Congress appointed a commission to take charge of all affairs connected with the transit and authorized the appropriation of $75,000 for necessary expenses and an additional $10,000 for repairs of instruments.[58] This Commission organized eight expeditions, four in the northern and four in the southern hemisphere. In England and Germany, experiments with photography at the 1874 transit had been disappointing,[56] so that astronomers in these countries considered the photographic method for determining the solar parallax a failure. Harkness cryptically commented: "The Americans, all the same, put most of their trust in photography. Exaggerated ideas prevailed that the sun's distance could be settled by transits of Venus alone, but this is not so. However, the real reason for spending so much on transits of Venus is that they are so rare that it would be unpardonable to neglect any opportunity of observing them."[59]

Elsewhere[60] Harkness remarked, "the astronomers of the twentieth century will not hold us guiltless if we neglect in any respect the transit of 1882," even though he added: "it will not settle the value of the solar parallax, but it will contribute to that result." Most of the programs of the American Commission dealt with photographic procedures. Harkness believed "The photographic method employed by the United States Transit of Venus parties in 1874 is among the most accurate known, and should not be neglected."[59] Aside from the expeditions organized by the Commission, many observers at American colleges participated in visual observations of the contacts.[61]

At Yale, several types of observations were undertaken.[62] Waldo and Kershner collaborated in the observations at the heliometer while R. W. Willson, mostly at his own expense, used the 8-inch Reed refractor to make photographs of the progress of the transit, of which 86 were ultimately measured. He devised an ingenious method for securing a horizontal reference line by simultaneously photographing a mercury cistern just in front of the photographic plate. The 9-inch Sheffield telescope was used by O. T. Sherman and mathematics Professor W. Beebe who inserted inclined lines at the focus and made 80 observations of the transit of the sun and Venus. Others, to the extent that telescopes could be furnished, observed and timed the contacts, especially Professor Lyman who in 1866 had reported that when the crescent Venus came within 108′ of the limb of the sun, refraction by the atmosphere of Venus made the crescent extend into an annular ring.[63] He again observed a complete 360°-ring at the transit of 1874.[64] This effect was observed at the Allegheny Observatory in Pittsburgh at the time of the 1882 transit.[61,65]

Unfortunately, no positive results seem to have materialized from any of the careful Yale observations. Willson's photographs were turned over to the Transit Commission in Washington for measurement.[66] However, they were not incorporated into the solutions derived from other photographs, Harkness stating in 1888,[67] "As the New Haven photoheliograph was of an entirely different kind from those used at other stations, the New Haven equations have been temporarily laid aside for further discussion." In 1887 Harkness reported that the plates had indeed been measured but the reductions were not yet completed. In 1889 he gave the same story, adding, moreover, that contact observations by the official parties and by nearly a hundred volunteers still remained to be reduced, for which he was desperately in need of assistance. No records are available to indicate that anything further was done, and Willson, in the meantime, had left Yale in 1884. Sherman's records met a similar fate. In the *Annual Report* for 1883–84[66] he stated that his reductions awaited only final revision before being ready for the press; but there is no record that they were ever published and he, too, left Yale in 1886. The heliometer observations, so critical for justifying the reputation of that instrument, fared no better: by June 1884 the observations had been reduced as far as the equations of condition. It is not clear whether or not Harkness received either these observations or their partial reductions. At the 1882 transit, only Yale and the four groups associated with Auwers used heliometers, and Harkness[67] commented, "Until the resulting values of the solar parallax are published the accuracy of their work can not be satisfactorily estimated." By that time all the Yale participants in the transit observations had left Yale. (Auwers published his own results in 1891.[57]) Later work by Elkin on stellar parallaxes as well as solar parallax based on observations of asteroids by Elkin in New Haven and Gill at the Cape, were finally to justify the original expectations regarding the excellence of the Yale heliometer.

Notwithstanding his heavy load of testing thermometers (in his capacity as head of the Thermometric Bureau) and other types of instruments, Sherman spent considerable time observing at the 8-inch Reed telescope. At the request of Elkin[68] he made measurements of the diameter of Venus, using a filar micrometer and various powers of eyepieces, and he used the lens both at full aperture and reduced to 6 and 4 inches. The results were of dubious value as the measurements differed according to the circumstances, the 8-inch aperture giving a diameter larger by ″.27 than with a 4-inch aperture.

In the *Report* of the Board of Managers for 1886–1887 we find that "Previous to his departure Mr. Sherman had been engaged in spectroscopic studies with the Reed equatorial. Some interesting results of his work have been published by him in the scientific journals." No reason was given for Sherman's resignation. He had devoted considerable time to observations with a direct vision spectroscope of the sun, planets,[69] comets,[40,70] Nova Andromedae,[71,72] and particularly stars,[40,73,74] publishing several papers in the *American Journal of Science* and the *Astronomische Nachrichten*. Strangely, none of Sherman's publications (either while he was at Yale or later) was cited in Poggendorff's comprehensive bibliography.[75] Perhaps the compilers considered him a crank and intentionally rejected his papers after Maunder in Great Britain[76,77] and Vogel in Germany,[78] in articles that are cited in Poggendorff, severely criticized both Sherman's methods of observing and nearly all of his conclusions. Sherman's self-defense[79] proved of no avail.[77] His observations of two bright lines in the supernova, S And 1885, quite fortuitously were approximately correct. Also his determinations of wave lengths for Comet C 1886[80] compared fairly well with those of "hydrocarbon" as given by Hasselberg.[81] The three brightest bands had originally been found by Huggins[82] in the Comets of 1868 to be identical with those of the spark spectrum of "olefiant gas," a hydrocarbon molecule. The bands were later identified as the Swan bands of C2, named in honor of the British physicist William Swan (1818–94), a pioneer in the laboratory observation of the spectra of hydrocarbons. Sherman had, however, observed Comet C of 1886 with a Rowland grating with 14,347 lines per inch, not with the Nicol prism (made by the French optician Jules Duboscq, 1817–86) which he had used in his other measurements. The grating had been obtained some time between October 24, 1885, and April 17, 1886.[83]

Sherman had reported many emission features in numerous stars and several planets, none of which could be confirmed by astronomers having more sophisticated spectroscopic equipment. Perhaps their harsh criticisms had something to do with his leaving Yale. (Both Maunder and Vogel somewhat tempered their criticism by calling attention to the fact that Secchi had announced emission features in the spectrum of Sirius, likewise due to instrumental effects.) In Sherman's obituary[84] it is stated that he spent the year after leaving Yale at Johns Hopkins studying the laboratory spectrum of hydrogen. There he possibly came under the guidance of H. A. Rowland (1848–1901).[85] In 1892 Sherman obtained an LL.B. degree at the Harvard Law School. Then from 1893 until his death in 1921 he was committed to the Insane Asylum in Worcester, Mass., confined, it was said, because "His mind became affected as the result of too close application to study."

One wonders why this obviously gifted young man did not get the guidance and supervision he needed in his spectroscopic work. But Sherman was Yale's pioneer in the study of celestial spectroscopy, a field in which neither he nor anyone else at the Observatory had had any prior experience. H. A. Newton evidently gave encouragement by cutting Sherman's working hours at the Thermometric Bureau to 36 hours a week to enable him to carry out observations at the Reed telescope[86] and approving the purchase of spectroscopic equipment. At the meeting of the Board of Managers of the Observatory on June 17, 1884, "it was voted that he [Sherman] confer with the Secretary of the Board [Newton] respecting the kind of observations which he had best undertake, and make reports to the Secretary; the whole matter being left to the Secretary and Mr. Sherman."[87] However, Newton seems to have lacked the proper experience in this field either to guide Sherman's research or to evaluate his results prior to their publication.

REFERENCES

1. D. Musto, *Ventures* (Magazine of the Yale Graduate School), **8**, 7, 1968.
2. D. Olmsted, letter to E. Loomis, 1856, quoted by D. Musto, *loc. cit.* p. 11.
3. C. André and A. Angot, *L'Astronomie Pratique et les Observatoires en Europe et en Amérique, Part III, États-Unis d'Amérique,* 167 pp., Paris, 1877.
4. Trustees of the Winchester Observatory, *Records,* ms copy at Yale Dept. Astronomy; memoranda inserted between pp. 62 and 63, 1858.
5. W. E. Decrow, *Yale and "The City of Elms,"* Boston, p. 61, 1882.
6. J. A. Hillhouse, *Sachem's Wood,* New Haven, 30 pp., 1838.
7. Ref. 4, p.7
8. H. A. Newton in W. L. Kingsley, *Yale College, A Sketch of Its History,* New Haven, p. 157, 1879.
9. H. C. King, *The History of the Telescope,* London, p. 309, 1955.
10. D. J. Warner, *Alvan Clark and Sons, Artists in Optics,* Washington, p. 75, 1968.
11. J. Lankford, *Jour. History Astronomy,* **14**, 79, 1983.
12. Schlesinger-Lundin Correspondence, 1921, *Yale Obs. Archives.*
13. C. H. Sawyer to Schlesinger, Nov. 7, 1929, *Yale Obs. Archives.*
14. J. W. Gibbs, *Biog. Mem. Nat. Acad. Sci.,* **4**, 118, 1902.
15. L. Waldo, *Science,* **1**, 91, 1883.
16. The Board of Managers of the Winchester Observatory, H. A. Newton, Sec., *Ann. Report* for 1880–81, p. 2, 1881.
17. L. Waldo, *Address to the Yale Club of Cincinnati,* 1880.
18. For descriptions of heliometers *see* L. Ambronn, *Handb. d. Instrumentenkunde,* Berlin, Chap. 11, p. 552, 1899.
19. W. L. Elkin, *Trans. Yale Univ. Obs.,* **1**, 6, 1887.
20. L. Waldo, *Standard Public Time,* Harvard College Observatory, 1877.
21. ———, *North American Review,* **131**, 528, 1880.
22. B. Z. Jones and L. G. Boyd, *The Harvard College Observatory,* Cambridge, Mass., p. 159, 1971.
23. W. W. Payne, Ed., *The Sidereal Messenger,* **1**, No. 1, 16, 1882.
24. L. Waldo, *First Annual Report of the Astronomer in Charge of the Horological and Thermometric Bureaus of the Winchester Observatory of Yale College,* 38 pp., 1880–81.
25. E. C. Pickering, *Science,* **19**, 87, 1892.
26. H. A. Newton, *The Sidereal Messenger,* **1**, 132, 1882.
27. Van H. English, *Encyclopedia Americana,* **26**, 758, 1985.
28. A. S. Eddington, *Encyclopaedia Britannica,* **22**, 224, 1958.
29. W. W. Payne, Ed., *The Sidereal Messenger,* **1**, No. 1, 22, 1882.
30. S. Withington, *Standardization of Time in Connecticut,* ms of talk to the Railway and Locomotive Hist. Soc., 1938. Schlesinger file *Yale Obs. Archives.*
31. R. N. Mayall, *Pop. Ast.,* **50**, 204, 1942.
32. C. F. Dowd, *Pop. Ast.,* **14**, 638, 1906.
33. W. W. Payne, Ed., *The Sidereal Messenger,* **2**, 278, 1883.
34. Ref. 4, p. 74.
35. *Idem,* p. 66.
36. *Idem,* p. 67.
37. *Idem,* p. 70.
38. Clippings from New Haven newspapers: *The Palladium,* May 22; *Journal and Courrier,* May 23 and June 18, 1884; a few others for which names of paper and dates not indicated. In pocket of volume, *Trustees of the Winchester Obs., Yale Obs. Archives.*
39. Ref. 4, p. 83.
40. O. T. Sherman, *Ann. Report* of June 20, 1886.
41. L. Waldo, *5th Ann. Report* for 1884–85.
42. ———, *6th Ann. Report* for 1885–86.
43. *Board of Managers Report* for 1886–87.
44. Ref. 4, p. 96.
45. M. F. Smith, *Ann. Report* for 1915–16, p. 339, 1916.
46. *Catalogue of Yale University,* p. 354, 1899–1900.
47. Ref. 4, p. 100.
48. *Idem,* p. 89.
49. H. A. Newton, *Ann. Report* for 1881–82, p. 5, 1882.
50. ———, *Sidereal Messenger,* **1**, 132, 1882.
51. Ref. 18, p. v.
52. L. Waldo. letter to H. A. Newton dated London, Jan. 17, 1882. *Yale Obs. Archives.*
53. Editorial Notes, *Sidereal Messenger,* **1**, 165, 1882.
54. E. Halley, *Phil. Trans,* **29**, 454, 1716.
55. H. Woolf, *The Transits of Venus, A Study of Eighteenth-Century Science,* Princeton, 1959.
56. W. Harkness, *Washington Observations for 1885,* Appendix III, 52, 1891.
57. G. F. J. A. Auwers, *Ast. Nachrichten,* **128**, 329, 1891; Abstract, *Ast. and Astrophysics,* **11**, 95, 1892.
58. E. Frisby, *Sidereal Messenger,* **1**, 182, 1882.
59. W. Harkness, *Sidereal Messenger,* **1**, 146, 1882.
60. ———, *Amer. Jour. Sci.,* **22**, 375, 1881.
61. Editorial Notes, *Sidereal Messenger,* **1**, 260, 298, 1883.
62. *Ann. Report* for 1882–83.
63. C. S. Lyman, *Amer. Jour. Sci.,* (2), **43**, 129, 1867.
64. ———, *Ibid.,* (3), **9**, 47, 1875.
65. C. S. Hastings, *Sidereal Messenger,* **1**, 260, 1883.
66. O. T. Sherman, *Ann. Report for 1883–84.*
67. W. Harkness, *Washington Observations for 1888,* A20, 1892.

68. O. T. Sherman, *Ann. Report,* p. 20, June 1, 1885.
69. ———, *Ast. Nachrichten,* **113**, 347, 1886. (The author only 20 days after submitting this note wished it withdrawn but it was already in press. See ref. 67.)
70. ———, *Amer. Jour. Sci.,* (3), **27**, 76, 1884.
71. ———, *Amer. Jour. Sci.,* (3), **30**, 378, 1885.
72. ———, *Ast. Nachrichten,* **113**, 45, 1886.
73. ———, *Amer. Jour. Sci.,* **30**, 475, 1885.
74. ———, *Ibid.,* **33**, 126, 1886.
75. J. C. Poggendorff, *Biographisch-Literarisches Handwörterbuch,* Leipzig, **4**, 1904.
76. E. W. Maunder, *Monthly Notices Roy. Ast. Soc.,* **46**, 282, 1886.
77. ———, *Ibid.,* **47**, 63, 1887.
78. H. C. Vogel, *Ast. Nachrichten,* **113**, 389, 1886.
79. The Secretaries of the R.A.S., *Monthly Notices Roy. Ast. Soc.,* **47**, 14, 1887.
80. O. T. Sherman, *Amer. Jour. Sci.,* (3), **32**, 157, 1886; *Ast. Nachrichten,* **113**, 311, 1886.
81. B. Hasselberg, *Mem. Acad. Sci.,* St. Petersburg, (7), **28**, No.2, 21, 1880.
82. H. Shapley and H. Howarth, *Source Book in Astronomy,* N.Y., p. 296, 1929.
83. H. A. Newton, Sec., *Meeting of Board of Managers,* Oct. 24, 1885 and Apr. 17, 1886. In ms vol, *Records of the Winchester Observatory,* p. 90.
84. Anon., *University Obituary Record of Yale Graduates for the Year ending July 1, 1921,* New Haven, p. 102, 1921.
85. O. T. Sherman, *Ast. Jour.,* **7**, 65, esp. p. 71, 1887.
86. H. A. Newton, Sec., *Meeting of Board of Managers,* May 16, 1884, in ms vol., *Trustees of the Winchester Observatory, Records,* p. 77.
87. ———, *Ibid.,* p. 78, June 7, 1884.

12
ELKIN'S HELIOMETER INVESTIGATIONS

Although the Yale heliometer (Fig. 12.1) had been ordered in 1880 and erected on time for the transit of Venus in 1882, the early annual reports of the Observatory before 1884 made no mention of stellar parallaxes. The only work reported, other than that on the transit of Venus, was on some observations by Waldo for the determination of the diameters of the sun, moon, and the planets Venus, Jupiter, Saturn, and Uranus.[1] Then, in his report of June 19, 1884, to the Board of Managers of the College Observatory, H. A. Newton wrote:

> We had a Heliometer of unsurpassed excellence, and it was due to the College, to the public, and to science, that the best possible use be made of it. This could be done only by an observer who had no other demands upon his time. The opportunity occurred in the later months of 1883, of securing the services of Dr. W. L. Elkin. He had given special study to the Heliometer at Strasburg, and for two years had had very valuable experience in the use of it in connection with and under the guidance of Dr. Gill, the Astronomer Royal at the Cape of Good Hope. To improve this opportunity a subscription was made by ten officers and friends of the Observatory of one hundred dollars each per year for three years, and Dr. Elkin was appointed by you for that period Astronomer in charge of the Heliometer. He entered upon his duties in January.

In his thesis at Strassburg,[2] completed in 1880, Elkin (Fig. 12.2) made use of all available observations of α Centauri for the redetermination of its parallax, taking into account the orbital and proper motions, atmospheric refraction, and air temperature at the times of observation. He made a thorough analysis of all conceivable sources of error. While this work was in progress, Sir David Gill, recently appointed Director of the Observatory at the Cape, visited Strassburg and was impressed by the young scholar's work. Gill invited Elkin to spend some time at the Cape as guest investigator. There the new 4-inch heliometer was installed and together they tested the instrument and made series of parallax observations on nine stars, three by both observers (α Centauri, Sirius and ε Indi), three by Gill alone (Lacaille 9352, o^2 Eridani and β Centauri), and three by Elkin alone (ζ Tauri, ε Eridani and Canopus).[3] They commented, however, that the important question was not so much what the precise parallax of a given star is, but (1) What the average parallaxes of stars of the 1st, 2nd, 3rd, etc. magnitudes are relative to fainter stars; and (2) "What connection does there subsist between the parallax of a star and the amount and direction of its proper motion—or can it be proved that there is no such connection or relation?"

From such data they felt that the law of absorption of light in space could be determined (a subject largely neglected even as late as 1930) and that such data would be of value in the determination of the solar motion. Moreover, they noted that heliometer observations of asteroids relative to surrounding comparison stars could yield a solar parallax free from systematic errors. These early considerations[4] largely determined the course of Elkin's future researches with the Yale heliometer.

Elkin's first use of the Yale heliometer was a remeasurement of the relative positions of the stars in the Pleiades that had been observed by Bessel[5] with the famous Königsberg heliometer almost half a century before. Soon he increased his program by making more detailed measurements and including all *Bonner Durchmusterung* stars near the Pleiades to magnitude 9.2 (69 stars). The results of this extensive investigation were published (with funds contributed by Professor Loomis) in 1887,[6] and later (1889) compared with Gould's[7] reductions of Rutherfurd's wet-plate photographs.[8,9] The Rutherfurd plates taken in 1866–67 yielded positions only slightly less accurate than the heliometer measures. A few plates on the Pleiades taken with the Lick 36-inch and measured by Elkin in 1889, gave relatively poor results for the bright, overexposed images, but excellent results for stars fainter than 7th magnitude. Elkin was already impressed not only with the potential accuracy of the modern photographic methods for relative positional work, but even more so by the fact that they were far less time-consuming.

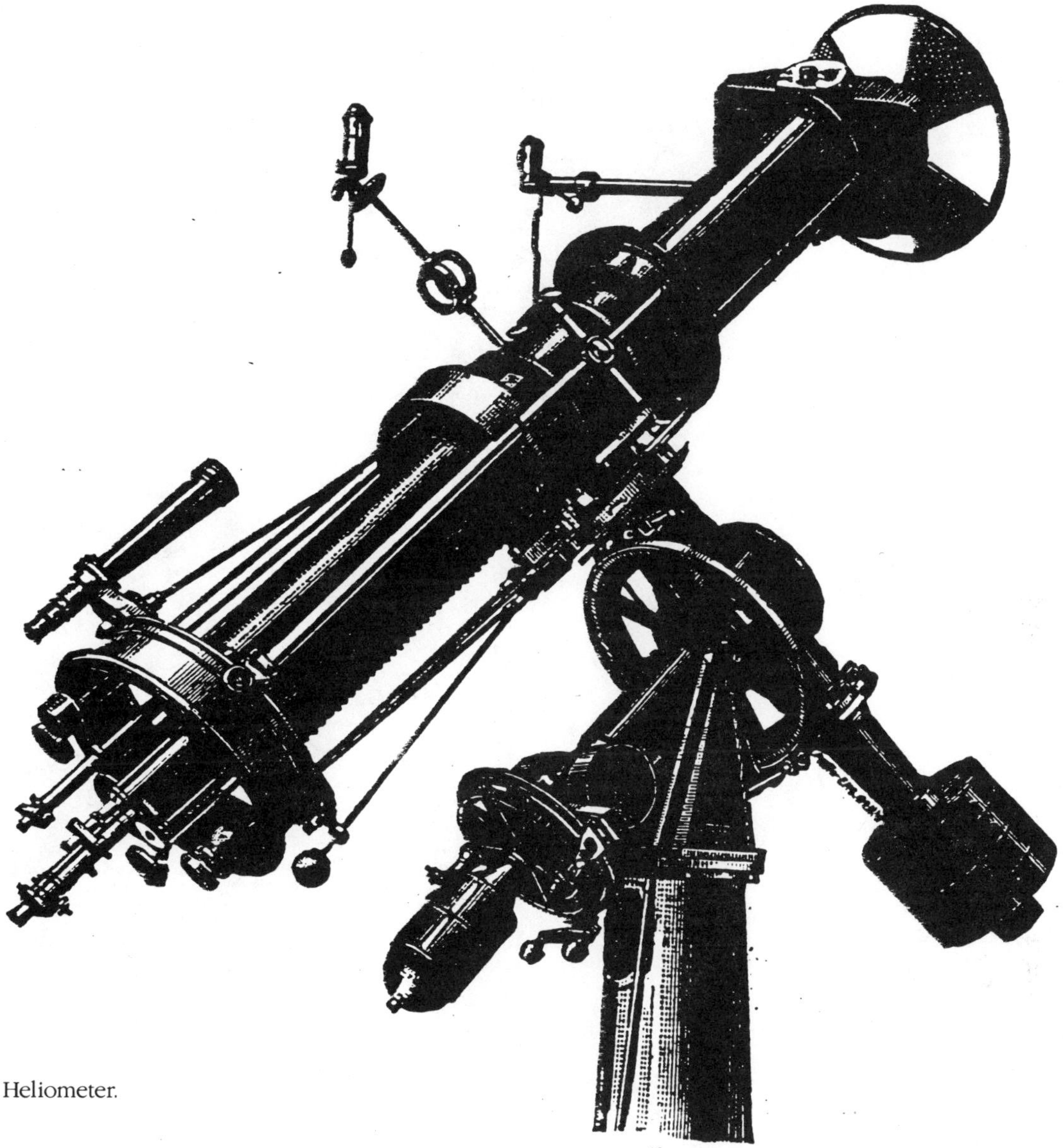

Fig. 12.1. The Yale Heliometer.

He also compared his results with similar visual data obtained with filar micrometers in Paris and Oxford, and reached the conclusion that the heliometer triangulation was the more accurate.

One of the purposes of Elkin's measurements of the relative positions in the Pleiades was for the determination of the internal motions in the cluster (proper motions). In his 1887 report he noted that his comparison with Bessel's results was sufficient "to show the character of the relative motions." However, he deemed the lapse of another half-century necessary for definitive results. Six stars provisionally showed drifts indicating noncluster membership. Later, after he discovered a source of small systematic errors in his reductions, he not only made a completely new reduction but in 1900 also encouraged his assistant, Mason F. Smith, to carry out a new triangulation of the Pleiades after an interval of 16 years. These results were published in 1904.[10,11] Now 9 of the 69 stars were found to have motions inconsistent with cluster membership. However, Elkin preferred not to pursue the matter of the proper motions of the cluster stars further as there were still "a number of photographic impressions under measurement which will materially add to the available material."

Fig. 12.2. W. Lewis Elkin (1855–1933).

In 1891, a similar triangulation of stars in the cluster Coma Berenices was carried out by F. L. Chase (Yale Ph.D. 1891, for a thesis on astrometry), but for this cluster less satisfactory earlier material was available for obtaining proper motions.[12,13]

Another triangulation of considerable interest was carried out by Elkin between 1888 and 1893.[14,15] Professor E. C. Pickering at Harvard had requested that the heliometer be used "to determine relative places of a few stars to serve as fundamentals for a catalogue of the stars within 1° of the Pole by photographic methods." Elkin extended the program to 24 stars brighter than 9th magnitude within 100′ of the Pole. These positions were needed by Pickering who was determining the magnitudes of many stars in the north polar region,[16] the early beginnings of the *North Polar Sequence.*[17]

A few other projects with the heliometer also occupied Elkin's attention during his first year at Yale. He measured the position of the Moon relative to surrounding stars, the principal object being to determine the parallactic inequality in the Moon's motion (a gravitational effect due to the changing distance between the Earth and the sun), which in the past had been "attended with some difficulty."[18] The approaching inferior conjunction of Venus offered a good opportunity to redetermine the diameter of that planet. The following year, measurements were continued on these projects and on measurements of the diameters of Titan and the outer ring of Saturn. In September 1885 Asaph Hall, Jr. began heliometer observations of Titan in order to obtain the mass of Saturn, a project ultimately financed by grants from the Bache Fund of the National Academy of Sciences. This work had been suggested by Elkin because of large discrepancies between the earlier determinations by Bessel and by the elder Hall at Washington.[19] The new value was in better agreement with Bessel's, as well as with later modern determinations.

During the July 1891 to January 1892 opposition of Jupiter, on 114 nights 570 relative positions of its satellites were measured with the heliometer for the purpose of determining their orbits and the mass of Jupiter.[20] Elkin remarked: "For this purpose it seems to me that in one important respect the heliometric method is not likely to be surpassed by other methods, and that is in the accuracy with which the scale-value and the orientation can be ascertained." David Gill at the Cape made similar observations, and Margaretta Palmer, an assistant at Yale, undertook the elaborate reductions,[21] deriving 1,128 equations of condition in 13 unknowns. This colossal work was never completed, partly because ill health delayed her progress but primarily because of the size and complexity of the equations to be solved. Would that she had had the advantage of our modern day computers!

Between 1888 and 1894 Elkin and Gill collaborated on an extensive series of heliometer measurements of three asteroids for the purpose of obtaining a better value for the solar parallax (the angle defined by the ratio of the mean radius of the earth to the mean distance of the sun). Up to this time, from about 1850, the various apparently reliable determinations ranged from 8.758″ to 8.95″. Gill had pointed out that heliometer observations of asteroids obtained simultaneously from stations on opposite sides of the Earth when the minor planet was favorably situated, could yield results of high accuracy. Iris in 1888 offered such an opportunity. Not only the Yale and Cape heliometers, but those at Oxford and Leipzig were also utilized. The apparent success of this adventure in its early stages encouraged extensive international cooperation. The following year further opportunities arose for observations on the minor planets Victoria and Sappho. With the heliometer the distances and position angles between the minor planet and each of two comparison stars were measured. In addition, meridian-circle observations were required for the determination of exact positions for the comparison stars. Heliometer observations were supplied by the Cape, Yale, Leipzig, Göttingen, and Bamberg observatories. Meridian circle observations were supplied by some 22 observatories, extending from Pulkova westward to Cincinnati, including Melbourne and Cordoba as well as the Cape Observatory to the south. A. von Auwers in Berlin was in charge of supplying ephemerides for the asteroids. In 1894 Elkin published a preliminary value of the parallax, 8.825″ ± .008″, based solely on simultaneous pairs or groups of observations of Iris.[22] The final results were published in two huge volumes of the *Annals* of the Cape Observatory in 1897,[23,24] with Gill as the principal author. The parallaxes derived from each of the asteroids separately were 8.812″ ± .009″ (Iris), 8.8013″ ± .0061″ (Victoria) and 8.7981″ ± .0114″ (Sappho), with the weighted mean 8.802″ ± 0.005″.

No sooner had this excellent value been determined when the approach of Eros seemed to present an opportunity for an even better solution, as Eros would come within 0.27 AU of the Earth in contrast to 0.83 or 0.84 for Iris, Victoria, and Sappho.[25,26] Elkin[27] expected to participate as before with the Cape and German observatories, but this hope was abandoned because of the faintness of Eros and its unfavorable location for heliometer observations.[28] In 1968 J. Lieske, in his Yale Ph.D. dissertation,[29] summarized the various solar parallax determinations, based on observations of Eros carried out between 1893 and 1965. The values spanned from 8.7875″ to 8.807″, embracing the Cape—Yale preferred value. Lieske's rediscussion of all the observations, including those by radar, yielded the value of the parallax as 8.″79402 ± .″00012. The visual and radio determinations were consistent.[30] The value currently adopted by the International Astronomical Union[31] is 8.″794148, mainly due to radar.

Meanwhile, in his report of June 1885, after having spent considerable effort on determining the instrumental errors of the heliometer, Elkin was ready to broach his most important plans: "It is now proposed to devote the instrument to systematic investigations in stellar parallax, and a plan of work has been laid out and commenced which, if carried to completion, will furnish, it is hoped, a reliable value of the relative parallax of the stars of the first and eighth magnitudes." According to early plans devised by himself and Gill,[3,32] Elkin began by examining the ten brightest northern stars, each 16 to 20 times relative to two single or double comparison stars; and for the large proper motion star, Arcturus, he made measurements relative to six pairs of comparison stars. As he did not yet expect high accuracy for each individual parallax star, he hoped to obtain a reliable mean parallax for the ten stars. Provisional results were reported in 1888,[14] the parallaxes ranging from −″.042 for Deneb to +″.266 for Sirius, with probable errors in most cases ±″.047, but smaller for Arcturus and Aldebaran which had, respectively, been observed six and four times more frequently. In view of the large dispersion among the individual values he felt no confidence in the significance of the mean parallax of +″.089 ± ″.015. Gill in South Africa, having obtained a new, larger, 7-inch heliometer in 1887, in 1900 published new results for the twelve southern stars of first magnitude,[33] as well as of some high proper-motion stars. The mean parallax of the southern bright stars was found to be +″.118, including the very large parallax of α Centauri, or +″.060 excluding α Centauri. Later, in his report of June 1890, Elkin planned further observations on the bright northern stars, commenting: "It appeared to me that the advantages of the heliometer for these bright stars were such as to make it well worth while to pursue the investigation and strengthen the results already obtained without the fear of their being immediately supplanted by those of the photographic method of observation."

It was to be another two decades before the problem of photographic parallax determinations for the bright stars was satisfactorily solved, mainly by his successor, Frank Schlesinger. Elkin's final paper on the ten brightest northern stars was published in 1902.[34] This had been delayed[35] in order to make measurements on very red stars to ascertain if they were subject to systematic errors. The final results for the ten bright stars were based on 923 observations spanning 1885–94. They yielded smaller probable errors, from ±″.014 to ±″.024. The parallaxes themselves ranged from −″.012 to +″.334, but now no mean value was mentioned. Table 12.1 compares Elkin's results with the most recent modern determinations.[36] About half of the differences are larger (in absolute values) than the probable errors of the

Table 12.1
First Ten Bright-Star Parallaxes

Name	Elkin Par.		Modern		Diff.
α Tau	″.109 ±	.014	″.053 ±	″.005	+.056
α Aur.	.079	21	.082	4	−.003
α Ori	.024	24	.012	5	+.012
α CMi	.334	15	.290	5	+.044
β Gem	.056	23	.104	5	−.048
α Leo	.024	20	.045	7	−.021
α Boo	.026	17	.092	6	−.066
α Lyr	.082	16	.132	6	−.050
α Aql	.232	19	.204	4	+.028
α Cyg	−.012	23	−.006	8	−.006

two series and a small systematic difference of −0.005″ would indicate.

After completion of the observations on most of the above projects, Elkin considered the feasibility of examining the stars of large proper motion for parallax, "as the criterion of propinquity given by brilliancy alone seems too precarious to warrant much dependence being placed thereupon."[37] Some 100 northern hemisphere stars were already known to have proper motions in excess of 0.50″ per year, some of which might prove to be nearby. However, he commented,[38] "But even should this not be the case, the purely negative result that there are probably no more stars in the Northern Hemisphere within a certain distance from us than those we already know of would have some considerable interest." In general he proposed that only a limited number of observations be carried out for each star in order to test for proximity; then the favorable cases would be pursued more extensively to reduce the probable errors.

By 1900[27] only two stars had been found to have parallaxes over +″.200. By 1906 all the planned observations had been completed, and the results for 163 stars were in press the following year,[39] under the joint authorship of Elkin and his assistants Dr. F. L. Chase and M. F. Smith. For this Elkin and Chase were awarded the Lalande Prize for 1908 by the Paris Academy of Sciences, and Smith given honorable mention.[40] Observations, however, continued on other stars of large proper motion, second magnitude stars, and high proper-motion stars for which preliminary Yale parallaxes were either large or negative.[41] This supplementary program was completed by 1910, comprising 35 stars,[42] and was soon fol-

Table 12.2
Parallaxes Determined by Heliometer

Observer	Stars	Observations	p.e. unit wt.
Elkin	36	1102	± 0.175
Chase	160	2876	0.135
Smith	73	1614	0.095

lowed by results on 41 southern stars of large proper motion.[43] Finally, in 1912[44] the combined *Catalogue of Yale Parallax Results* containing 244 stars appeared, two years after Dr. Elkin's retirement (for reasons of prolonged poor health).

It is of considerable interest to compare this first Yale catalogue of trigonometric parallaxes both with other determinations then available, and particularly with modern results. Frank Schlesinger[45] declared, "These results constitute by far the most important single contribution to our knowledge of stellar distances up to that time." Schlesinger found they were nearly free from systematic error and that the published probable errors were close to their true values, an unusual situation even among the more modern determinations at that time (1923).

When this catalogue was first published, the 244 stars it contained represented two-thirds of all of the 365 for which trigonometric parallaxes had been determined up to that time.[46] It included approximately 5,600 individual observations, distributed as in Table 12.2. The majority of the stars had been measured fewer than sixteen times. For these the probable error of the derived parallaxes ranged from ±″.020 to ±″.096. They were mainly stars of small parallax that had not seemed propitious for further work. Those observed thirty or more times by Chase or Smith had final probable errors less than or equal to ″.016. Elkin reduced his errors by making observations of the bright stars upward from 48 to 171 times, his final probable errors being under ″.025. That Elkin's errors should be larger than the others' can be attributed to a number of factors, including frail health. As Schlesinger[47] explained,

> During the first fifteen years or so of his connection with the observatory, Elkin made an astonishing number of observations for a man whose health was never robust, and with an instrument that is notoriously difficult to manipulate. A modern heliometer has eight handles that must be operated with the arms held constantly above the head. In the late nineties, however, his health no longer permitted him to observe so steadily.

His assistants were not only younger and stronger, but they benefited from Elkin's previous experience. The probable errors of unit weight did not, for any of the observers, show any correlation with the magnitudes of the stars, nor, as determined later, with spectral type, declination, proper motion, or the sizes of the corresponding modern parallaxes.

One hundred of the Yale parallax stars had already had parallaxes determined by other methods than by heliometer: 8 by filar micrometer, 83 with the transit instrument, one by chronograph with equatorial, 39 by photography, and 3 by spectrograph.[44] (If an orbit has been computed for a visual binary and the relative radial velocities of both components have been determined by means of a spectrograph, then the parallax can be determined.[48]) In addition, 13 parallaxes for these stars had been determined with other heliometers. The various comparisons indicated generally good systematic agreement though larger dispersion than their individual probable errors would suggest. In his first *General Catalogue of Stellar Parallaxes,* Schlesinger[49] did indicate that although the values determined by Elkin did not require systematic corrections, the parallaxes determined by Smith did require a systematic correction of −″.010, and that the published probable errors of those determined by Chase and Smith were by and large too small.

A comparison of the 1912 catalogue with the most recent available results for the same stars in van Altena's current compilation of trigonometric parallaxes[36] confirms the overall accuracy of the early determinations, as illustrated in the histograms in Fig. 12.3. Here the probable error of a single difference (photographic—heliometer) is ±″.025 for Chase, ±″.029 for Elkin, and ±″.028 for Smith, for all the heliometer determinations, but ±″.019 for those depending on more than forty observations. This compares well with the probable error of ±″.015 of the difference between any two modern photographic determinations given in the *Supplement to the General Catalogue of Trigonometric Parallaxes.*[50]

There seems no better way to summarize Elkin's contributions to stellar parallaxes than to quote from an evaluation presented by S. C. Chandler (volunteer associate of Harvard Observatory) at a meeting of the Boston Scientific Society in 1896. He remarked[51] that Elkin's work is

> as unsuitable for popular presentation as is that of the genius of this age, Dr. George W. Hill, in celestial mechanics. The work of each is of the very highest type. Both names

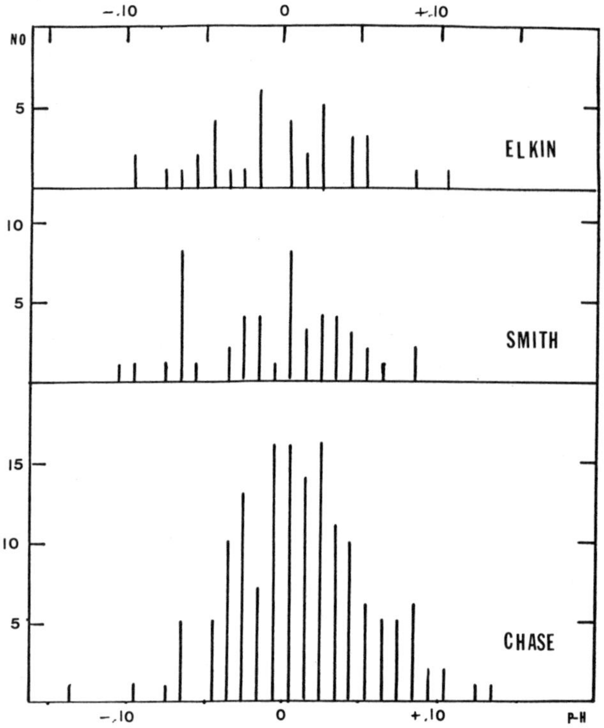

Fig. 12.3. Histogram of modern photographic *minus* Yale Heliometer determinations of stellar parallax.

belong to that small and most select class which any discriminative critic must include in an account of the contributions of the highest class made by Americans to the astronomy of the nineteenth century. Yet the works of both partake, each in its way, of that purely technical character which makes a descriptive analysis of them, for the purpose of securing for them an adequate and intelligent popular appreciation, a difficult task.

Further, Elkin's different accomplishments "constitute a record of faithful, zealous, and vigorous prosecution of astronomy in its very highest plane, and the presence of such a man as Dr. Elkin in our country gives standing and credit to American astronomy."

REFERENCES

1. H. A. Newton. *Ann. Report to Board of Managers* for 1882–83.
2. W. L. Elkin, *Die Parallaxe von α Centauri,* Strassburg Ph.D. Thesis, Karlsruhe, 44 pp., 1880.
3. D. Gill and W. L. Elkin, *Mem. Roy. Ast. Soc.,* London, **48**, 1, 1884.
4. ———, *Idem,* p. 191.
5. F. W. Bessel, *Astronomische Untersuchungen,* **1**, 209, 1841.
6. W. L. Elkin, *Trans. Yale Univ. Obs.,* **1**, 1, 1887.
7. B. A. Gould, *Mem. Nat. Acad. Sci.,* **4**, 173, 1870.
8. W. L. Elkin, *Ast. Jour.,* **9**, 33, 1889.
9. B. A. Gould, *Idem,* p. 36.
10. Ref. 6, p. 331, 1904.
11. M. F. Smith, *Idem,* p. 359.
12. W. L. Elkin, *Report to the Board of Managers,* June 9, 1891.
13. F. L. Chase, Ref. 6, p. 209, 1896.
14. W. L. Elkin, *Report to Board of Managers,* June 7, 1888.
15. ———, Ref. 6, p. 149, 1893.
16. E. C. Pickering, *Ann. Harvard Obs.,* **18**, 121, 1889.
17. H. S. Leavitt, *Ibid.,* **71**, 47, 1917.
18. W. L. Elkin, *Report to Board of Managers,* June 13, 1884.
19. A. Hall, Jr., Ref. 6, p. 111, 1890.
20. W. L. Elkin, *Report to Board of Managers.,* June 1891–93.
21. ———, *Ibid.,* June 15, 1895.
22. ———, *Ibid.,* June 14, 1894.
23. D. Gill with the co-operation of A. Auwers and W. L. Elkin, *Ann. Cape Obs.* **6** and **7**, 1897.
24. W. L. Elkin, *Idem,* **6**, Part 4, 169 pp., 1897.
25. A. Clerke, *History of Astronomy during the Nineteenth Century,* 4th ed., London, p. 238, 1902.
26. A. Pannekoek, *A History of Astronomy,* N. Y., p. 346, 1961.
27. W. L. Elkin, *Report to Board of Managers.,* June 11, 1900.
28. ———, *Ibid.,* May 31, 1901.
29. J. Lieske, *A Dynamical Determination of the Solar Parallax from the Motion of (433) Eros,* Ph.D. Dissertation, Yale University, New Haven, 1968; summary in *Ast. Jour.,* **73**, 628, 1968.
30. D. O. Muhleman, *Monthly Notices Roy. Ast. Soc.,* **144**, 151, 1969.
31. *I. A. U. Trans.,* **16 B**, 5, 1976.
32. W. L. Elkin, *Report to Board of Managers,* June 14, 1886.
33. D. Gill, *Ann. Cape Obs.,* **8**, Part 2, xvi + 173 pp, 1900.
34. Ref. 6, p. 255, 1902.
35. W. L. Elkin, *Report. to Board of Managers,* June 18, 1898.
36. W. van Altena, *General Catalogue of Trigonometric Parallaxes,* 4th ed., in preparation, 1991.
37. W. L. Elkin, *Report to Board of Managers,* May 20, 1902.
38. ———, *Ibid.,* June 8, 1892.
39. F. L. Chase, M. F. Smith, and W. L. Elkin, *Trans. Yale Obs.,* **2**, 1, 1907.
40. *Sci. Amer. Supp.,* **67**, 59, 1909.
41. W. L. Elkin, *Report* for 1906–07.
42. F. L. Chase, M. F. Smith, and W. L. Elkin, *Trans. Yale. Univ. Obs.,* **2**, 209, 1910.

43. F. L. Chase and M. F. Smith, *Ibid.,* **2**, 327, 1912.
44. W. L. Elkin, F. L. Chase, and M. F. Smith, *Idem.,* pp. 385, 397.
45. F. Schlesinger, *Biog. Mem. Nat. Acad. Sci.,* **18**, 181, 1937.
46. K. Lundmark, *Handbuch d. Astrophysik,* **5**, 431, 1932.
47. Ref. 45, p. 183.
48. *Cf.* R. G. Aitken, *The Binary Stars,* N.Y., p. 224, 1935.
49. F. Schlesinger, *General Catalogue of Stellar Parallaxes,* Yale, 57 pp., 1924.
50. L. Jenkins, *Supplement to the General Catalogue of Trigonometric Parallaxes,* Yale, 47 pp, 1963.
51. S. C. Chandler, *Pop. Ast.,* **3**, 392, 1896.

13
METEOR PHOTOGRAPHY

The first systematic attempts at meteor photography in America were carried out at Yale by Dr. Elkin. Elsewhere most of the earliest examples of photographs of meteor trails were obtained by pure chance on photographs exposed for other purposes. The very earliest of which there appears to be a record was, however, obtained by L. Weiner at Prague on November 27, 1885, when meteor photography was indeed the intent of the exposure.[1] Moreover, in addition to a camera at Prague, another had been set up at Dresden, some 75 miles distant, in the unfulfilled hope that the same meteor might be photographed from both stations for height determination. The next earliest successful attempts at meteor photography in Europe were by the Russian, J. Sykora, during the Perseid shower of 1901 when he obtained seven trails, five of them Perseids, yielding an accurate radiant.[2–4] In 1910 he observed a bright Perseid photographed from three stations, Tashkent, Iskander, and Tshimgan, yielding a path from altitude 112 to 81 km.[5] By 1924 he had obtained more than thirty trails.[6] The first Harvard meteor photograph was accidental (Sep. 12, 1887),[7] but systematic meteor photography at Harvard was not carried out until the Leonids of 1897–99.[8–12]

Probably the most spectacular of the early photographs was the one obtained by amateur John Lewis of Ansonia, Connecticut (already described in Chap. 9), which encouraged H. A. Newton and Elkin to experiment with meteor photography. For the 1893 Perseids, Elkin attached a 6-inch Voigtländer portrait lens of 32 inches focal length on the same mounting with the 8-inch Reed equatorial telescope, pointing it close to the Perseid radiant.[13,14] Three trails were photographed in a total of four hours, two of them Perseids, one almost six degrees long. John Lewis, ten miles away, also tried photographing meteors, using a stationary 4-inch lens. He happened to catch the same bright meteor, displaced some six degrees from one of the Yale trails. Newton reduced the observations, finding the beginning and end heights to be 68.0 and 29.3 miles, respectively. He commented that a Perseid upon entering the earth's atmosphere has a velocity of 37.2 miles per second; hence the duration of the photographic path was less than a second. Attempts were again made on December 10, 1893, when three Geminid meteors were photographed at Yale; but Lewis captured only one weak, doubtful trail. The three Geminid trails, extended backward, intersected in an unbelievably small radiant area, only one minute of arc in diameter. Newton concluded: "The experiment shows that with a reasonable amount of labor we can get photographs of meteor tracks. The accuracy of the results obtained from the photographs is unspeakably greater than those gotten by any other known method."[13]

These early successes became the basis of Elkin's pioneering investigations of meteors. He inferred that with a battery of only six or seven wide-angle portrait lenses of five- or six-inch aperture, a sufficient number of trails would be secured to locate the radiant point of the Perseids with certainty.[14] Portrait lenses generally consist of four components with an effective focal length 4 to 6 times the lens diameter.[15–17] Such lenses had been designed by Petzval in 1841 and were first built in modified form by Voigtländer in 1846,[18] and later further modified by others.[19] Of the twelve cameras that Elkin was eventually to acquire three were 6-inch lenses made by the Voigtländer Company.

In 1894 Newton and C. S. Hastings, physics professor and director of Yale's Sheffield Scientific School, made an application to the National Academy of Sciences for a grant to build equipment for photographing meteors.[20] An award was made from the J. Lawrence Smith Fund, and plans (Fig. 13.1) were drawn for a battery of six cameras on a single mounting consisting of a 12-foot long tubular polar axis, with a clock drive at the lower end.[21,22] The cameras were to be pointed at slightly overlapping regions of the sky. Called a meteorograph, the mounting of this

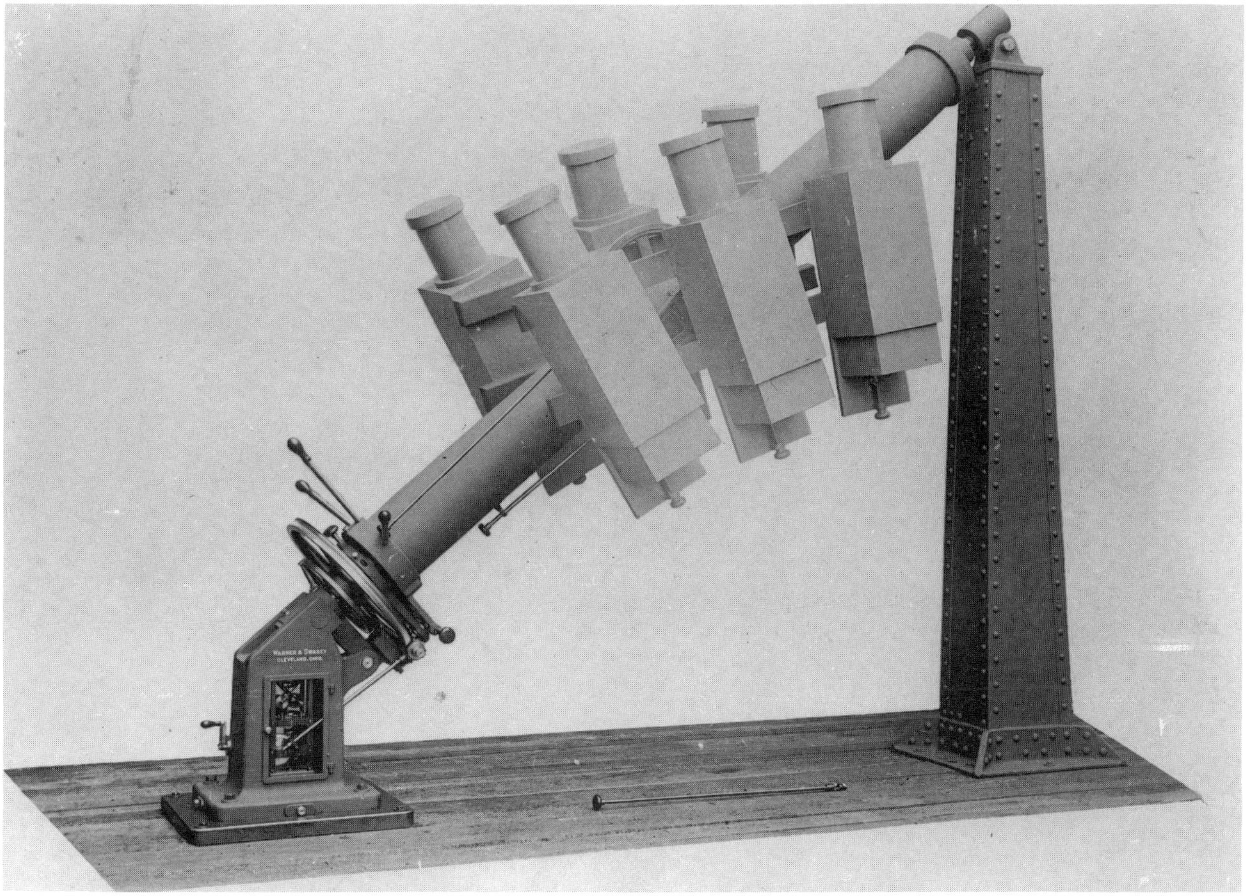

Fig. 13.1. Model for meteorograph.

cluster of cameras was constructed by the Warner and Swasey Company and erected at the Winchester Observatory at Yale in 1894 (Fig. 13.2), at first with only four cameras, two of five-inch, two of six-inch effective aperture. Delivered in July, they were ready for observations of the Perseids in August. The area covered by the four cameras amounted to about 400 square degrees. In a total of 11 hours of exposure on the nights of August 9–11 only two meteors were photographed. It was presumed that none fainter than second magnitude could be captured.[14]

Although Elkin's efforts of the next few years were relatively unproductive, he retained confidence and kept trying. In 1896 (the year that Newton died) he acquired new lenses and had the full quota of six for the meteorograph. In August, with five of the cameras, he secured eight trails, five of which were Perseids.[23] The following year he could report a total of ten trails from various showers.[24] Then, in view of the expected return of the Leonids in 1899, Elkin prepared for a second meteorograph to be erected a few miles distant, in Hamden, Connecticut. From the apparent displacement of the trails relative to the stars as photographed from the two stations, he expected to determine the heights of the meteors.[25] The National Academy again provided a grant from its J. Lawrence Smith fund.[26] (The two grants, the first for $2,000 in 1894, the second for $1,000 in 1898, were the first ever to be awarded from this fund.) The second instrument was of simpler design, with only four cameras and no clock drive (Fig. 13.3). In the 1898–99 season, 28 trails were photographed, 17 at Yale and 11 at Hamden, of which seven were common to both stations. As an interesting by-product, Dr. F. L. Chase (1865–1933, Yale Ph.D. 1891), working at the Hamden station on November 21, 1898, happened to discover a faint new comet on one of his plates,[27] by coincidence almost

Fig. 13.2. The meteorograph at Yale Observatory. Reed Telescope Dome in background.

in the direction of the Leonid radiant. For this Chase was awarded a Donohoe Medal of the Astronomical Society of the Pacific.[28]

At Harvard College Observatory, in contrast to Yale, very little work on meteors was carried out prior to the 1920s except during the anticipated return of the Leonids of 1899. Besides organizing extensive campaigns for visual observations,[8,29] E. C. Pickering planned photographic observations[10–12,30] similar to Elkin's. In 1898 he used twelve cameras in Cambridge directed to the area of the radiant point, and two at Tufts College in Medford about two miles distant. He obtained 31 trails of 8 meteors, on November 14, 1898, four of which were photographed from both stations. (That same night Elkin secured seven trails.) The following year Pickering obtained 34 trails of 11 meteors and ascertained that none was fainter than second magnitude. Pickering seems to have made no attempt to derive heights for the meteors photographed from both stations.

It may seem incredible that Elkin should have assumed that a separation of the stations by only two miles would be sufficient for accurate altitude determinations. Already in 1868 Loomis had pointed out that with a base line of only three or four miles a shooting star would be seen in very nearly the same direction by both observers. Loomis had previously advocated a base line of not less than 40 miles for

Fig. 13.3. Second meteorograph, at Hamden.

visual observations.[31] Elkin, accustomed to the high accuracies required and attainable for stellar trigonometric parallaxes, in which he was a leading expert at that time, presumably assumed that similar accuracy could be attained for photographic meteors. Whereas the largest known stellar parallax was only 0″.75 (determined to an accuracy of ±″.010), the parallax angle, or change in direction of a meteor at an altitude of 50 miles as observed from two stations only two miles apart, is about 2°. Unfortunately, in the case of a meteor varying smoothly in brightness along its path, it is not always possible to match corresponding points on the two photographs. Hence, even increasing the base line to five miles did not yield the desired accuracy. This was particularly unfortunate in his subsequent pioneering efforts (described below) to determine total meteor velocities as distinguished from the projected velocities observed from each station.

As early as 1860 J. Homer Lane (1819–80, Yale 1846), at that time self-employed in Washington, had written an article, *On a Mode of Employing Instantaneous Photography as a Means for the Accurate Determination of the Path and Velocity of a Shooting Star, with a View to the Determination of Its Orbit.*[32] (Lane is best known for *Lane's Law*[33] which states paradoxically that as the sun cools by radiation it shrinks and therefore becomes hotter!) For meteor photography he advocated a battery of cameras at two stations, with some device for marking a time scale along the trail. For timing he proposed putting a thin objective prism in front of the object glass of the camera, the prism made to rotate at an accurately measured rate of, for example, 25 revolutions per second. The prism could be replaced by an eccentric lens, or the object glass itself made to revolve on a slightly eccentric axis. The latter suggestion is much like the rocking mirror E. Öpik was to use in the early 1930s for visual velocity determinations.[34,35] The apparent path across a field of view would be a looped or sinuous curve, each cycle length corresponding to the period of rotation of the lens.

Lane made his suggestions some twenty years before any application of dry-plate photography to astronomy.[36] While he realized the disadvantages of using wet plates, he was keenly aware of the potentialities of both his proposed apparatus and of photographic materials. With the collodion plates first proposed by Archer in 1851[37,38] the photographer had to coat a plate with wet emulsion, immediately expose the plate, and then develop it before the emulsion had a chance to dry. The earliest nonastronomical experiments with dry plates had been carried out in France by Gaudin in 1854.[39] The first application to astronomy, at the transit of Venus in 1874, yielded plates too poor to serve their purpose.[40] The first scientifically successful astronomical dry-plate photograph was a spectrum of Vega obtained by Huggins in 1876.[41]

Wet plates for meteor photography would have been altogether too cumbersome. As the exposure times would have had to be very short, Lane suggested having an observer watch the area to be photographed: when a bright meteor appeared the observer would touch a button to activate a spring to open the shutter and, immediately after the meteor disappeared, close the shutter again. In this way Lane felt that plates would not be wasted should no meteors cross the field of view. One wonders, however, if the observer's reactions would be quick enough to capture an observed meteor. Lane commented, "The expense and trouble of this process will certainly be great, but will not be disproportioned to the importance of the object in view," namely the determination of the orbits of individual meteoroids. Lane realized that the time was not yet ripe for his suggestions to be put into practice. Nevertheless, he published them at the time because he had faith that the rapid development of photography would ultimately make his scheme feasible. A simpler scheme, with a rotating shutter made of cardboard with five sectors occulting the lens ten times a second, was tried in 1885 by W. Zenker[42] in Berlin, but unsuccessfully, in part because of the weather. In December of 1898, G. F. Fitzgerald in Dublin made a similar suggestion.[43] Elkin was aware of these three papers and probably used them as a guide in his own thinking.[44] He used a bicycle wheel with opaque screens inserted between alternate intervals between spokes. The wheels were large enough to occult the lenses of all of the cameras mounted on each meteorograph. The wheels were spun by motors giving an effectively constant spin rate, and a device was added for monitoring that rate. By now both meteorographs were provided with six cameras. Elkin's shutter had twelve sectors occulting the lenses at a rate of six to ten times per second. In 1899, five meteors were photographed in common at both the Yale and Hamden stations.[45] One of the meteors was an Andromedid, which yielded a velocity and an orbit consistent with those of Biela's comet. The other four gave surprisingly low velocities, which would have indicated orbital periods of less than two years. However, the accuracy was low, as only one trail showed more than two breaks, and air resistance was not taken into account. Elkin commented that a higher rate of occultation would have been better.[46]

By the end of May 1901, Elkin had moved the second station from Hamden to Whitneyville, some five miles from the Yale station.[47] At the same time the Whitneyville meteorograph was also provided with a clock drive. In all, through 1910 Elkin acquired some 130 trails from among several thousand plates exposed, but duplicates at both stations, suitable for height and total velocity determination, were available for only 27 meteors. Meanwhile he had meticulously devised mathematical procedures for the determination of radiants, heights, and velocities, and carried out the plate measurements to the nearest second in time and arc in right ascension and declination, respectively.[48] In his last report to the Board of Managers of the Observatory, for 1909–10, Elkin wrote, "All the material so far has been measured and reduced and I am now undertaking a complete revision of the whole work." But evidently the results did not live up to his expectations. In 1910 he resigned his position at Yale, ostensibly because of failing health, but perhaps even more significantly for reasons of frustration. Schlesinger[49] commented that when Elkin "withdrew from astronomy in 1910, he was convinced that his life was near its close." Elkin's two major concerns had been the determination of meteor velocities (which he evidently considered unsuccessful), and the derivation of trigonometric stellar parallaxes with the Yale heliometer. He did not publish the extensive results of his meteor investigations. As a leading expert on stellar parallaxes (having determined both more and apparently more accurate parallaxes than any other observer up to that time) Elkin realized that celestial photography was now making his heliometer determinations obsolete. He lived for another 22 years, devoting his declining energies mainly to music. In supplying Schlesinger with material for his biographical account of Elkin,[49] Mary King Murphy wrote of "Cousin Will,"[50] "His improvisations at the piano—in your Observatory house—used to fill me with wonder. I've heard that he used to put the morning newspaper on the music rack and let his fingers run over the keys and that sympathetic listeners upstairs could tell what kind of news he was reading."

As late as 1925, Edwin B. Frost, Director of the Yerkes Observatory and Editor of the prestigious *Astrophysical Journal,* invited Elkin to review C. P. Olivier's *Meteors,*[51] stating, "I hope your health is such that this would not be a burden," adding,[52] "America used to have a high place in the study of meteors and therefore I think that this book deserves a review by a competent authority like yourself." The expression, "used to have a high reputation" must have been painful to a sensitive man who had devoted so much effort to the advancement of the field. Evidently Elkin did not supply a review; the one in the *Astrophysical Journal* was by George C. Comstock,[53] retired Director of the Washburn Observatory.

After Elkin's retirement, work on meteors at Yale was discontinued and during World War I the Observatory was closed until such time as a new Director could be appointed. When Frank Schlesinger became Director in 1920, he tried unsuccessfully to persuade Elkin to publish. Then it was proposed in 1922 that C. P. Olivier (founder in 1911 of the American Meteor Society[54]) at the University of Virginia should edit and complete Elkin's work.[55] Elkin agreed and turned his records over to Olivier in late January 1922. But when Olivier had finished the task, Elkin was dissatisfied and withdrew permission for the work to be published.[56–58] After Elkin's death in 1933, and with the permission of his widow, Schlesinger again encouraged Olivier to re-examine the article and make such changes as he thought would have satisfied Elkin.[59] Thus the research was finally published by Olivier in the *Astronomical Journal* in May 1937.[48]

As early as 1922 Schlesinger, a member of the J. Lawrence Smith Medal Committee of the National

Academy of Sciences, had proposed Olivier as a candidate for the medal.[60] Olivier, the most active American researcher on meteors at the time, had recently published 349 parabolic orbits of meteor streams[61] (a sequel to several earlier such lists) based on the visual observations of 22,000 meteors observed by members of the American Meteor Society. However, in 1922 the second medal had already been awarded to G. P. Merrill of the U.S. National Museum.[62] According to the deed of the donor (widow of J. Lawrence Smith), the medal could not be awarded more frequently than once every two years, and then for an investigation either accomplished or published since the time of the preceding award.[63] Therefore Whitman Cross, Chairman of the J. Lawrence Smith Fund Committee, wrote Schlesinger, "I presume he merits it, beyond question, and the only difficulty I see can be overcome in all probability."[64] Cross suggested that a publication or some other work done by Olivier since April 1922 could be made the basis for a medal, even if "the major credit may belong to earlier work." Schlesinger was asked to renew the suggestion when further work by Olivier became known; but no subsequent nomination materialized—not upon the publication of Olivier's comprehensive treatise, *Meteors*[51] nor upon his publication of Elkin's results.[48]

Olivier was appropriately generous in his appraisal of Elkin's work. Elkin had been the first to secure exact photographic radiants for meteor showers; he had calculated real heights and orbits for doubly photographed meteors and had determined the retardation produced by the Earth's atmosphere at different heights. Although the base line used was too short to attain the desired accuracy, Elkin had clearly shown what both the observational and mathematical procedures should be. A successful campaign could now readily be carried out with better cameras and faster emulsions than had been available to Elkin, in a more favorable climate, and with a base line in excess of twelve miles.

Elkin's work was the last major Yale contribution to the study of meteors. After Schlesinger became Director of the Yale Observatory he did contemplate reactivation of the project, benefiting from both Elkin's achievements and his disappointments. In 1921–23 Schlesinger and Whitman Cross corresponded regarding the ownership of the apparatus which had been at least partially funded by the J. Lawrence Smith Fund.[65] Schlesinger commented on the desirability of putting the equipment again to use, adding, "I have made two or three tentative proposals to other astronomers to do this work, and have offered to place this equipment at their disposal, but so far have not met with any very enthusiastic response."[66] He did not indicate whether the persons he contacted were within or outside Yale; but he did contemplate that after a few years, when other, now higher priority projects at Yale on stellar parallaxes and proper motions would be running smoothly, Yale itself might resume meteor photography.[66] He wrote to Cross on October 15, 1923, "I have been trying for several years to interest certain persons in this work laying particular stress on the fact that a suitable mounting is available and is now lying in storage here. But the time for agitating this matter more actively is probably several years in the future."[67]

Still later, in 1927, we find Harlow Shapley at Harvard inquiring whether the Elkin meteor apparatus would still be available if Harvard wished to borrow it for work by W. J. Fisher.[68] Although the answer was in the affirmative, it is not clear whether Harvard did in fact acquire either the mounting or any of the cameras. Three of Elkin's cameras had been loaned to Olivier, then at the University of Pennsylvania, for which that University paid $300 in 1945 when it could not locate them. In 1954 the Yale cameras were found, in poor condition, and were therefore not returned to Yale.[69]

In a letter to Shapley dated March 31, 1936, Schlesinger wrote,[70] "You will be interested to hear that we are considering the re-establishment of the meteor stations. It would be interesting and I believe valuable to see what modern cameras and modern plates would give. We already have a number of cameras on hand that would surely be more efficient than anything Elkin had at his disposal." Evidently nothing came of these hopes. Meanwhile Fisher at Harvard (who died in 1934) had already begun such a pilot program and turned it over to graduate student Peter Millman.[71] Although there is no evidence that Yale did anything further with meteor photography, Elkin's efforts, even though disappointing, were of utmost importance in setting the stage for future exploitation by others. The later achievements of F. L. Whipple at Harvard are more than a fulfillment of Elkin's greatest expectations.[72]

The whereabouts of Elkin's plates has been a mystery. Several astronomers had expressed great interest in examining this unique collection; Olivier would have liked to re-examine some of the plates for checking Elkin's results. Willard Fisher wrote in 1933 to enquire if Miss Hoffleit could examine the plates for her work on the photometry of meteor trails.[73,74] Elkin had previously said that the plates were all at the Observatory. Schlesinger and an assistant then examined all of the several thousand Elkin plates that were at the Observatory but could not find a single meteor trail;[75] the plates they examined were probably only the rejected ones. M. F. Smith, who had actually measured the plates with the trails, was also unable to locate them.[76] Rumor had it that Elkin himself had destroyed them. Many years later, in 1957, when the Yale University Observatory had moved to its new location at 135 Prospect Street, a pile of several dozen miscellaneous unidentified plates was found among which there were two plates dated December 12, 1899, one with a single, the other with two very weak trails which proved to be Geminids. These, showing three, three, and four segments, respectively, are evidently Elkin plates, the only two known to have survived (numbers 55 and 56 in Olivier's catalogue[48]).

Except for a suggested student involvement program between Yale and Vassar College for observing the meteors of 1938, there is no further evidence of interest in meteors or meteorites at Yale University Observatory during Schlesinger's regime. In 1937, with the blessing of Schlesinger, the amateur Astronomical Society of New Haven was organized and held its meetings monthly at the Observatory. The members participated in 1937 in observing shower meteors, contributing over 350 observations to Vassar for the purpose of determining meteor heights for any that might have been observed from two or more locations, a project published by Helen Wright in 1939.[77]

Much later, in 1954, Paul Hodge, as an undergraduate working under the guidance of Professor Rupert Wildt, began a study on meteoritic dust.[78] Hodge collected atmospheric dust from four locations: in New Haven, a station ten miles away, in the clear atmosphere of an island 35 miles north of Seattle, Washington, and on the slopes of Mt. Ranier. The diameters of the particles studied ranged from 3 to 20 microns. The maximum rate of fall was indicated to occur on about July 12. On the basis of these early indications a year-long program, supported by the National Science Foundation (July 1955—July 1956), was initiated to collect daily from three widely separated locations: the Smithsonian Astrophysical Observatory's Table Mountain Station in the California Mojave Desert; the Geophysical Institute field station at College, Central Alaska; and at the Canadian Meteorological Service at Resolute, Northwest Territory, Canada. The directors at these stations cooperated in making the collections.[79] From an analysis of these and older data compiled from the literature, Hodge and Wildt concluded that the annual accretion of meteoritic dust particles with a diameter greater than three microns amounted to 5×10^8kg for the entire Earth.[80,81] Upon graduation from Yale, Hodge carried his interest in meteoritics to Harvard. Later investigations of meteorites at Yale came under the jurisdiction of the Department of Geology, not Astronomy.

REFERENCES

1. L. Weinek, *Ast. Nachrichten,* **113**, 374, 1886.
2. J. Sykora, *Mem. Spett. Ital.,* **31**, 81, 1902.
3. ———, *Nature,* **66**, 309, 1902.
4. ———, *Jour. Brit. Ast. Assoc.,* **13**, 43, 1902.
5. ———, *Ast. Nachrichten,* **186**, 111, 1910.
6. ———, *Bull. Soc. Ast. France,* **38**, 64, 1924.
7. D. Hoffleit, *The Telescope,* **8**, 6, 1941.
8. W. H. Pickering, *Ann. Harvard Obs.,* **41**, 133, 1902.
9. E. C. Pickering, *Cir. Harvard Obs.,* No. 33, 1898.
10. ———, *Ibid.,* No. 40, 1899.
11. ———, *Ibid.,* No. 45, 1899.
12. ———, *Pop. Ast.,* **7**, 412, 1899.
13. H. A. Newton, *Amer. Jour. Sci.,* (3), **47**, 154, 1894.
14. W. L. Elkin, *Ast. Jour.,* **13**, 132, 1894.
15. W. B. Brummitt, *Encyclopaedia Britannica,* Chicago, **17**, p. 974, 1966.
16. E. J. Wall and F. J. Mortimer, *The Dictionary of Photography,* London, 10th ed., p. 430. (No publication date.)
17. C. B. Neblette, *Photography, Its Principle and Practice,* N. Y., **17**, 110, 1927.
18. P. W. F. Voigtländer, *Ast. Nachrichten,* **24**, 395, 1846.
19. F. E. Ross, *Jour. Optical Soc. Amer.,* **5**, 123, 1921.
20. O. C. Marsh, *Report, Nat. Acad. Sci.,* p. 13, 1894.
21. W. L. Elkin, *Pop. Ast.,* **2**, 17, 1894.
22. ———, *Ast. and Astrophys.,* **13**, 626, 1894.
23. ———, *Report for 1896–97 of the Board of Managers of the Yale Obs.,* p. 4, 1897.
24. ———, *Ibid.,* for 1897–98, p. 4, 1898.
25. ———, *Astrophys. Jour.,* **9**, 20, 1899.
26. Anon., *Reports, Nat. Acad. Sci.,* p. 10, 1898.
27. W. L. Elkin, *Ast. Jour.,* **19**, 151, 1898.
28. J. E. Keeler, C. Burckhalter, and W. M. Pierson, *Pub. Ast. Soc. Pacific.,* **11**, 37, 1899.
29. E. C. Pickering, *Cir. Harvard Obs.* No. 31, 1898.
30. E. C. Pickering, *Cir. Harvard Obs.* No. 35, 1898.
31. E. Loomis, *Harper's New Monthly Mag.,* **37**, 34, 1868.
32. J. H. Lane, *Amer. Jour. Sci.,* (2), **30**, 42, 1860.
33. ———, *Ibid.,* (2), **50**, 57, 1870.
34. H. Shapley, E. J. Öpik, and S. L. Boothroyd, *Proc. Nat. Acad. Sci.,* **18**, 16, 1932.

35. E. J. Öpik, *Cir. Harvard Obs.,* No. 389, 1934.
36. D. Hoffleit, *Some Firsts in Astronomical Photography,* Cambridge, Mass., p. 32, 1950.
37. C. E. K. Mees, *Photography,* N. Y., p. 11, 1937.
38. B. W. Bt., "Photography," *Encyclopaedia Britannica,* N. Y., **17**, 930, 946, 1966..
39. W. deW. Abney, *Encyclopaedia Britannica,* N. Y., 9th ed., **18**, 821, 1889.
40. A. Clerke, *A Popular History of Astronomy during the Nineteenth Century,* 4th ed., London, p. 236, 1902.
41. W. Huggins, *Observatory,* **1**, 4, 1877.
42. W. Zenker, *Ast. Nachrichten.,* **113**, 228, 1885.
43. G. F. Fitzgerald, *Astrophys. Jour.,* **9**, 50, 1899.
44. W. L. Elkin, *Ast. Jour.,* **20**, 102, 1900.
45. ———, *Astrophys. Jour.,* **12**, 4, 1900.
46. ———, *Monthly Notices Roy. Ast. Soc.,* **61**, 248, 1901.
47. ———, *Yale Reports* for 1900–04, p. 5.
48. C. P. Olivier, *Ast. Jour.,* **46**, 41, 1937.
49. F. Schlesinger, *Biographical Memoirs, Nat. Acad. Sci.,* **18**, 175, 1936.
50. M. K. Murphy to F. Schlesinger, Oct. 26 and Nov. 15, 1935, *Yale Obs. Archives.*
51. C. P. Olivier, *Meteors,* Baltimore, xvii + 276 pp., 1925.
52. E. B. Frost to W. L. Elkin, Apr. 22, 1925, *Yale Obs. Archives.*
53. G. C. Comstock, *Astrophys. Jour.,* **62**, 305, 1925.
54. C. P. Olivier, *Pop. Ast.,* **21**, 89, 1913.
55. F. Schlesinger and C. P. Olivier Correspondence, Jan. 27–Nov. 6, 1922, *Yale Obs. Archives.*
56. F. Schlesinger to C. P. Olivier, Apr. 18, 1923, *Yale Obs. Archives.*
57. C. P. Olivier, Reply to Schlesinger, Apr. 19, 1923, *Yale Obs. Archives.*
58. F. Schlesinger to F. R. Moulton, Aug. 20, 1936, *Yale Obs. Archives.*
59. F. Schlesinger and C. P. Olivier, 1934–37, *Yale. Obs. Archives.*
60. F. Schlesinger to W. Cross, June 15, 1922, *Yale Obs. Archives.*
61. C. P. Olivier, *Pub. Leander McCormick Obs.,* **2**, 199, 1919.
62. W. Cross, *Reports, Nat. Acad. Sci.,* p. 2, 1922.
63. F. W. True, Ed., *A History of the First Half-Century of the National Academy of Sciences,* 1863–1913, Washington, pp. 58, 66, 371, 1913.
64. W. Cross to F. Schlesinger, June 21, 1922, *Yale Obs. Archives.*
65. F. Schlesinger and W. Cross Correspondence, Apr. 11, 1921–June 21, 1922, *Yale Obs. Archives.*
66. F. Schlesinger to W. Cross, Apr. 12, 1921, *Yale Obs. Archives.*
67. ———, Oct. 15, 1923, *Yale Obs. Archives.*
68. H. Shapley to F. Schlesinger, June 17, 1927, *Yale Obs. Archives.*
69. F. Schlesinger and B. Wood, Correspondence, Nov. 1954, *Yale Obs. Archives.*
70. F. Schlesinger to H. Shapley, March 31, 1936, *Yale Obs. Archives.*
71. P. Millman and D. Hoffleit, *Ann. Harvard Obs.,* **105**, 601, 1936.
72. F. L. Whipple, *The Collected Contributions of Fred. L. Whipple,* Cambridge, Mass, **1**, 966 pp., 1972.
73. W. J. Fisher to F. Schlesinger, June 9 and 13, 1933, *Yale Obs. Archives.*
74. D. Hoffleit, *Proc. Nat. Acad. Sci.,* **19**, 212, 1933.
75. F. Schlesinger to W. J. Fisher, June 12, 1933, *Yale Obs. Archives.*
76. F. Schlesinger to C. P. Olivier, Nov. 6, 1922, *Yale Obs. Archives.*
77. H. Wright, *Pop. Ast.,* **47**, 256, 1939.
78. D. Brouwer, *Ann. Report* for 1954–55.
79. ———, *Ibid.,* for 1955–56.
80. P. W. Hodge and R. Wildt, *Geochemica et Cosmochemica Acta,* **14**, 126, 1958.
81. P. W. Hodge, *Interplanetary Dust,* N. Y., pp. 53–57, 1981.

14
THE QUIESCENT YEARS AND A LADY ASTRONOMER

For ten years after Elkin's retirement in 1910 as Director, no one was appointed to replace him. In 1910 F. L. Chase was appointed Acting Director and when he, too, retired in 1913 because of poor health, Mason F. Smith took over the duties of administrator until 1918, although he was not accorded the title of Acting Director. Both had collaborated with Elkin in the important parallax work, which they continued, and in the operation of the Horological (now called Time Service) and Thermometric Bureaus. In 1918 E. W. Brown, who had been Professor of Mathematics at Yale since 1907, was appointed Director but with the understanding that the Observatory was to be closed for the duration of World War I or until a new director could be found who would devote full time to the Observatory.[1]

The *Catalogue* of Yale University for 1915–16 listed the equipment and financial assets of the Observatory, adding, "Before the close of 1915 the Observatory will have ready for use a new type of instrument designed mainly for the photographic determination of stellar parallaxes," a prediction that was not to materialize. (The Loomis telescope to which this statement refers proved unsuitable for parallax work and was eventually used for variable star observations.) The financial assets comprised three major funds. The income from the Winchester Fund of $120,000 would be available for the general purposes of the Observatory. The Loomis Fund, amounting to over $300,000, was established by the will of Elias Loomis. One-third of the income was to be used for salaries of observers who would devote their whole time to the work, to the reduction of astronomical observations, and to defraying the expenses of publishing the results of such observations. The other two-thirds of the income was to be held in trust for members of the Loomis family and ultimately to become available to the Observatory. (This did not materialize until after the death of Loomis's son Henry B. Loomis in 1939.) Finally, the Observatory was to acquire two-thirds of the estate of Edward M. Reed, half of which was expected to be available soon; the final half became available in 1927,[2] the total amounting to nearly $120,000.

During this relatively quiet period between Elkin's resignation and the appointment of Frank Schlesinger in 1920, the activities at the Observatory centered around six major topics: (1) continuance of the parallax work, (2) the time and thermometer services, (3) Miss Palmer's computations of cometary orbits, (4) her work on the *Yale Index to Star Catalogues,* (5) the unique design by C. S. Hastings for the Loomis telescope and its construction by Brashear of Pittsburgh, and finally (6) Professor Brown's rigorously theoretical work in celestial mechanics and his tables of the motions of the moon, published in 1919. Of Brown's work Dirk Brouwer (Director, 1943–66) was to write that Brown's lunar theory excelled any previous studies in the field and that his lunar tables remained the basic tables against which the actual motion of the Moon was being compared.[3] The work initiated by Elkin on meteor photography ceased completely. The parallax work and that of the Horological and Thermometric Bureaus were simply continuations on a routine basis of work already in progress, as described in the previous chapters. The time service, however, became more arduous after Chase left, and Smith by 1917 was the only person except for Miss Palmer left at the Observatory. In an unpublished report to the Board of Managers dated March 20, 1918,[4] Smith commented that no clock failures had occurred within the past year. However, he wrote,

A failure here causes the stopping of all synchronized clocks and this means an immense amount of time and trouble to the two subscribers [the Telephone Company and the Fire Department] in setting them right again over their entire systems. It is not safe to leave the Observatory alone for more than a few hours at a time and I have therefore been obliged to sleep here and keep watch of the system day and night, including Sundays and holidays. In the past years when there were four or five men connected with the Observatory it was almost a certainty that there would be some person in the building at all hours.

In his first report, for 1913–14, Smith stated that the work with the heliometer had to be curtailed because he had no assistants, was occupied with plans for a new instrument, and working at the request of Asaph Hall, Jr., of the Naval Observatory on measurements of three pairs of stars to be used for determining the scale of the new Naval Observatory equatorial. Significantly Smith commented: "It was seen a few years ago that the field of usefulness of the 6 inch Heliometer was rapidly becoming exhausted and that it would soon be necessary to secure a new instrument if this Observatory was to maintain its present position in the field of parallax research." The heliometer was dismantled after the observations requested by the Naval Observatory were completed. Moreover, plans were already underway for a siderostat or coelostat telescope of 15-inch aperture, 50 ft focal length, to be mounted on a polar axis with a parallel 10-inch guide telescope. A 30-inch silvered flat mirror on an equatorial mounting would send starlight up both telescope tubes. Brashear at Pittsburgh was to furnish the optical parts and Warner and Swasey the mountings. The following year Smith expressed hope that the telescope and its housing would be completed by late 1915. The architects were instructed to make the building for this unique instrument conform aesthetically with the surrounding residential neighborhood.

The telescope had been designed by Prof. Charles Sheldon Hastings (1848–1932), Director of the Physics Laboratory of the Sheffield School. His main field of interest had been physiological and physical optics, in which he was a renowned authority. He was affiliated with Brashear who had felt the need of a mathematical specialist in this field in order to improve his optical instruments. Brashear and Hastings thus collaborated in the construction of some of the largest and most important telescopes of their time: the 24-inch Swarthmore (1911), the 30-inch Allegheny (1912), and the 26-inch Yale Southern Station (1925) refractors, all important in the determination of stellar parallaxes; and the 72-inch Victoria reflector (1918). The coelostat was erected and nearly in working order in 1916, but Smith complained, "No serious work can be undertaken until additional observers are secured." It was thus idle until after Schlesinger came to Yale.

Designed primarily for photographic parallax work, the telescope, ironically, was never used for this purpose. In 1920 it was designated *The Loomis Coelostat Telescope* as a memorial to one of Yale Observatory's greatest benefactors. It was first used for research in stellar photometry, especially by Carl Stearns for studies of the light curves of variable stars.[5] The chief advantage of the polar construction was that the observer could sit comfortably in a room at the top of the 60-foot tower, with the 15-inch photographic and 10-inch visual lenses at the foot of the tower, the 30-inch flat being capable of directing light from any star to the eyepiece, except for the relatively few polar stars being eclipsed by the tower itself (Fig. 14.1). The great disadvantage, especially for astrometry, was that varying astigmatism depending on changes in temperature could not be adequately controlled.[6] Much later, in 1956, the telescope would be moved to a new observing station in Bethany (Conn.) where it was redesigned for a study of Earth's rotation. But this astrometric effort also was unsuccessful, and the telescope was eventually permanently dismantled.

Margaretta Palmer (1862–1924), a Vassar graduate, had already been employed at the Yale Observatory as an assistant in 1889. Her first assignment seems to have been computations relative to Elkin's reductions of the asteroid Iris.[7] In 1892 she entered the Yale Graduate School, the Corporation having voted on March 3, 1892, that "beginning in the fall of 1892, the courses of the Graduate Department with the degree of Ph.D shall be open to candidates without distinction of sex."[8] For her thesis she chose the determination of a definitive orbit for Comet 1847 VI. This comet had been discovered by Miss Palmer's former teacher, Maria Mitchell, and had catapulted that lady to international fame. The previous most definitive orbit had been determined by G. Rümker in 1857,[9] according to which this comet was of great interest as it belonged to a class, comparatively rare by the standards of that time, showing a distinctly hyperpolic orbit, indicating that it came from and/or was headed out of the solar system. A redetermination of the orbit was justified by the fact that greatly improved star and sun positions had become available and that the earlier computations had not taken perturbations into account. More recent observations than those available to Rümker were, of course, out of the question, as the comet had been faint and was not expected to return. Miss Palmer's

Fig. 14.1. The Loomis Telescope Tower under construction.

thesis confirmed that the comet was indeed traveling in a hyperbolic orbit. Besides the technical reasons, Miss Palmer found philosophical reasons for having undertaken her study; she remarked

> To the woman . . . who turns her attention to astronomy this comet is conspicuous as one of the few that have been discovered by a woman, and probably the only one that has ever been discovered independently by two women. [The other was Madame Rümker, wife of the Director of the Hamburg Observatory.] But the student whose first knowledge of the heavens has been gained under the direction of Maria Mitchell has an added reason for regarding with especial interest the course through space of that body whose discovery brought before the astronomical world the young Nantucket comet-seeker, the first American woman to gain such notice.

Miss Palmer's thesis, giving the history of the discovery of the comet and details of her orbit determination were published in the *Transactions* of the Observatory.[10] For this she was awarded the Ph.D. degree in 1894. It is noteworthy that this degree, the first accorded a woman astonomer by an American university,* was awarded at Yale 31 years before Radcliffe awarded its first Ph.D. in astronomy to Cecilia Payne in 1925, and even longer, 35 years, before Harvard awarded any Ph.D. in astronomy.[11] In the preceding year 1893, however, another American woman astronomer, Dorothea Klumpke, earned the equivalent *Docteur ès Science* from the University of Paris. Other American universities before Harvard had awarded the Ph.D. in astronomy, some even to women. For example, Columbia awarded the degree to Frank Schlesinger (see Chap. 17) in 1898 and to Caroline Furness (Vassar A.B. 1891, later Director of the Vassar College Observatory) in 1900. Dr. Palmer was also to compute new orbits for three of the comets that had been discovered much earlier by Caroline Herschel (the comets of 1786 II,[12] 1788 II,[13] and 1797), but the orbit for the last she was unable to complete. Schlesinger sent her unfinished computations to C. J. Schumacher in Germany[14] but no publication seems to have ensued. Chase began a parallax investigation of 84 southern stars ($-13°$ to $0°$) according to his report of 1910–11; the following year Miss Palmer was engaged in ascertaining the proper motions of the comparison stars. Her work was not acknowledged in the final parallax results.[15]

Another massive undertaking initiated by Elkin in 1897 with the assistance of Miss J. S. Newton (one of the daughters of H. A. Newton) was the compilation of an index catalogue giving the sources of all positions that had been measured for the stars in the *Bonner Durchmusterung*. Soon Miss Palmer was to take a major part in this work, which was important in connection with Elkin's parallax and other astrometric projects. Two sets of the three-volume BD and one volume of its southern counterpart, the SD, were provided with interleaving sheets on which the references to positions measured between 1750 and 1900 for each star were inserted with meticulous care and neatness for 457,857 stars north of −23°. These priceless volumes are still in the possession of the Yale Astronomy Department, although their usefulness has been superseded by the later German publication, *Geschichte des Fixsternhimmels,* in 48 volumes (1922–52) which required more than 50 years for compilation. The early history of this undertaking was discussed by Miss Palmer in 1923.[16] In 1917 she had called attention to both the unpublished *Yale Index* and the *Geschichte* indicating that as the latter would be many years in preparation Yale would be prepared to furnish investigators the references for individual stars in which they might be interested.[17]

Professor Brown in planning the temporary closing of the Observatory during World War I was most concerned about the two people still on the payroll, Mason Smith and Margaretta Palmer. Smith was instructed to have valuable instruments moved for safekeeping to the Sloane Physics Laboratory, and to supervise the transfer of the time service then being furnished to the New Haven Railroad. Concluding these services he was to be given the opportunity either to find other employment within the University or given a leave of absence at half-salary to enable him to find a new position elsewhere. In any case, his work at the Observatory was to terminate on June 30, 1919. Brown consulted with both President Hadley and Librarian Keogh in regard to Miss Palmer, lauding the fruits of her long service, commenting, "Special arrangements will be necessary for the payment of Miss Palmer's salary it being understood that the University is unwilling to dispense with her services at the present time in view of her long connection with the Observatory and her advanced age."[18] She was 56 years old!

The upshot was that she was to work part time at the College Library until the end of the war and allowed part time to continue her comet orbit computations and work on the *Index Catalogue,* particularly in regard to filling other astronomers' requests for data. Thus, when Schlesinger arrived she was the only employee left to overlap the regimes of Elkin and Schlesinger. He immediately put her to work cataloguing stellar distances.[19] From this resulted the first *General Catalogue of Trigonometric Parallaxes,* which appeared in 1924 with Frank Schlesinger, Margaretta Palmer, and Alice Pond as coauthors. This was the last publication including Miss Palmer's name before her death that same year as the result of an automobile accident.

As the war drew to a close, Brown outlined for the Board of Managers possible plans for the future, stressing that the importance of a definite policy to be pursued through the oncoming years could hardly be overestimated, adding,

> Equally important when this policy is once settled is the choice of a director. After the appointment is made and the lines of investigation which he proposes to follow have been accepted by the Board of Managers and the Corporation, he should be given full freedom to carry them out in his own way and furnished with such financial help as may be available from the funds of the Observatory or elsewhere.[20]

Brown sketched possible choices among mainly observational work, purely theoretical investigations and a combination thereof, giving his views of the relative advantages or disadvantages for an observatory in the climate of New Haven. He also indicated how much annual income could initially be expected at the reopening of the Observatory—approximately $24,000. At that time only five American observatories (Mt. Wilson, Harvard, Dudley, Yerkes, and Lick) were known to have larger annual incomes, but only one of them (Harvard) a larger endowment. Frank Schlesinger, Director of the Allegheny Observatory and leading expert on the photographic determination of stellar parallaxes, accepted the appointment of Director of the Yale University Observatory, assuming his duties on April 1, 1920.

REFERENCES

1. G. P. Day, Acting Sec. Yale, Letter dated Apr. 11, 1918, in ms vol. *Trustees of the Winchester Observatory, Records,* p. 124.

2. Meeting of Board, Nov. 22, 1927, *Trustees of the Winchester Observatory,* p. 188.
3. D. Brouwer, *Monthly Notices Roy. Ast. Soc.,* **99**, 300, 1939.
4. M. F. Smith, in *Trustees of the Winchester Observatory,* p. 115, Mar. 20, 1918.
5. C. L . Stearns, *Trans. Yale Univ.Obs.,* **3**, 49, 1926.
6. F. Schlesinger, *Pop. Ast.,* **31**, 77, 1923; *Pub. Amer. Ast. Soc.,* **5**, 28, 1927.
7. W. L. Elkin, *Ann. Report* for 1890–91.
8. *Historical Register of Yale University, 1701–1937,* p. 24, 1939.
9. G. Rümker, *Ast. Nachrichten,* **45**, 263, 1857.
10. M. Palmer, *Trans. Yale Univ. Obs.,* **1**, 187, 1893.
11. D. Hoffleit, *Maria Mitchell's Famous Students,* AAVSO, 10 pp., 1983.
12. M. Palmer, *Ast., Jour.,* **31**, 189, 1918.
13. ———, *Ibid.,* **34**, 84, 1922.
14. F. Schlesinger to C. J. Schumacher, Dec. 23, 1924, *Yale Obs. Archives.*
15. F. L. Chase and M. F. Smith, *Trans. Yale Univ. Obs.,* **2**, 327, 1912.
16. M. Palmer, *Pop. Ast.,* **31**, 78, 1923.
17. ———, *Ast. Jour.,* **30**, 166, 1917.
18. E. W. Brown, Undated memo on M. Palmer, *Yale Obs. Archives.*
19. F. Schlesinger, *Ann. Report* for 1919–20.
20. E. W. Brown, Memo. to Board of Managers, undated. *Yale Obs. Archives.* Also in *Trustees of the Winchester Observatory,* p. 125, March 25, 1919.

*Footnote added in press:
Mercury, **21**, 8, 1992, reports that the first Ph.D. in astronomy was awarded by Columbia University to Winnifred Edgerton in 1886, eight years before Miss Palmer's degree at Yale. However, Miss Edgerton's degree was awarded by the Mathematics Department for a thesis entitled "Multiple Integrals" (*see* J. Green and J. Laduke in *Women in Science,* G. Kass-Simon and P. Farness, eds., Indiana University Press, p. 123, 1990). Edgerton did study astronomy at Wellesley and celestial mechanics at Columbia, but unlike Palmer, she did not pursue a career in astronomy.

15
THE ASCENDANCY OF CELESTIAL MECHANICS

When Ernest W. Brown (1866–1938;[1] Fig. 15.1) was a student at Cambridge University, England, he was primarily interested in mathematics. In 1888 he came directly under the influence of Sir George Darwin.[2,3] What Brown wrote of Darwin[2] constitutes the guiding principle for celestial mechanics at Yale over the next two generations:

> The influence which Darwin exerted has been felt in many directions. The exhibition of the necessity for quantitative and thorough analysis of the problems of cosmogony and celestial mechanics has been perhaps one of his chief contributions. . . . While speculations and the framing of new hypotheses must continue, but little weight is now attached to those which are defended by general reasoning alone. Conviction fails, possibly because it is recognized that the human mind cannot reason accurately in these questions without the aids furnished by mathematical symbols, and in any case language often fails to carry fully the argument of the writer as against the exact implications of mathematics. If for no other reason, Darwin's work marks an epoch in this respect. . . .
>
> To have spent an hour or two with him, whether in discussion on "shop" or in general conversation, was always a lasting inspiration.

It was Darwin who recommended to student Brown in 1889 that he study G. W. Hill's work,[4] especially on lunar theory. This was to become Brown's principal lifework. While still in Cambridge he was not yet committed to this particular application of mathematics, having also considered meteorology. However, when he was called to Haverford College in 1891 as Instructor of Mathematics and also put in charge of the observatory there, it was natural for him to select an astronomical topic for his particular application of mathematics. By the time he was invited to Yale as Professor of Mathematics in 1907 he had written *An Introductory Treatise on Lunar Theory*[3] and over twenty articles, most of them on celestial mechanics, and all but completed his voluminous treatise on *The Theory of the Motion of the Moon,* published in five parts in four volumes of the *Memoirs* of the Royal Astronomical Society between 1897 and 1908.[5] By the time Brown came to Yale he had been made a fellow of Christ College (1889–95), received his Doctor of Science degree from Cambridge in 1897, was a Fellow of the Royal Society of London, a member of the American Philosophical Society, and joint editor of the *Transactions* of the American Mathematical Society and its vice president in 1905. In 1907 he received the prestigious Adams Prize of the University of Cambridge for an essay *The Inequalities in the Motion of the Moon Due to the Direct Action of the Planets.*[6] What is said to have especially attracted him to Yale was the prospect of the production and publication of his *Lunar Tables,*[7] a monumental task completed at a cost of some $34,000 in 1919, largely with the collaboration of Henry B. Hedrick, Yale Ph.D. 1915 in mathematics, who was Chief Computer from 1910 to 1918. Hedrick, an assistant on the staff of the U.S. Nautical Almanac, in 1905 published a catalogue of 1607 zodiacal stars, intended largely for computing occultations.[8] In 1908 he began correspondence with Brown relevant to the lunar tables. He had had 22 years' experience in astronomical computations, 11 of them as assistant under Simon Newcomb, and he had enjoyed a long friendship with G. W. Hill.[9] Always interested in devising means to facilitate computations, he proffered Brown his assistance. On Nov. 9, 1908, having heard from Brown, he responded enthusiastically, "I consider your Lunar tables the most desirable work in my life and would esteem it a great honor to be connected with it from the beginning to the end. I should like to be able to devote to it the next ten years or as long as necessary—probably the best ten years of my life."[10] Hedrick's Yale appointment was approved in June 1909, to begin Jan. 1, 1910.

While still at the U.S. Naval Observatory (where he was employed 1886–1910) Hedrick had been

Fig. 15.1. E. W. Brown (1866–1938). 1929 photograph. (Courtesy AAVSO)

interested in the application of Hollerith machines to the computation of ephemerides. The Hollerith punched-card machines had been invented in 1890 to cope with census analyses (the 1880 census not having been completed by 1886, with the prospect that with the prevailing methods the 1890 census would not be completed before it was time to tackle that of 1900). Hedrick contacted the inventor, Herman Hollerith (1860–1929), head of the Tabulating Machine Company in Washington (which in 1911, after having merged with two other companies, became The International Business Machines Corporation, IBM). He communicated their discussions to Brown, who designed a machine of his own, which incorporated some of Hollerith's patents. Hollerith generously offered the loan of some of his components and permitted the use of patented information. In fact, Hollerith communicated to the Rowland Telegraph Company Hedrick's and Brown's plans, as that company was capable of building such a machine.[11] At a meeting of the American Philosophical Society, Brown presented a paper, *A Mechanical Device for the Tabulation of the Sums of Numerous Variable Functions,* published by title only,[12] but followed shortly[13] by a note, *New Plans for Tabulating the Moon's Longitude* in which he stated, "It is hoped that a machine which has just passed through the experimental stage will enable the computer to obtain the sum of the other small terms with great rapidity for half-daily intervals." Later, in 1912, F. J. W. Whipple of the Air Ministry, London, heard that Brown had "recently perfected a device for computing harmonic terms as in the lunar theory."[14] Whipple was in charge of setting up an exhibition of calculating machines at a Mathematical Congress at Cambridge, England, in August 1912, and felt that Brown's apparatus, which Brown was reluctant to exhibit, "should be all the more interesting for being home-made."[15]

Starting with a paper in 1891, *On the Determination of a certain Class of Inequalities in the Moon's Motion,*[16] Brown wrote some 150 papers, through 1938, dealing with the Moon's motion and other aspects of the mathematically complex three-body problem. Prior to Brown's work, the tables by P. A. Hansen[17,18] had been the ones that most accurately predicted the positions of the Moon and were the ones adopted for the computation of ephemerides of the Moon in most nautical almanacs. The theory of C. E. Delaunay[19] had been used for the French ephemerides. Although this theory was more complete than Hansen's, Delaunay died before convenient tables for the preparation of ephemerides could be compiled. These were subsequently prepared by J. C. R. Radau[20] but proved almost as cumbersome to use as the algebraic expressions themselves. Hansen's theory included 300 periodic terms, and fundamental constants determined from observation to as high a degree of accuracy as was available in 1857. Brown, who expressed great admiration for Hansen's early work,[21] found the least satisfactory part of Hansen's tables was in the perturbations produced by planetary action, the terms used being few in number and some erroneous.[22] Nevertheless, with a few corrections supplied by Simon Newcomb, they fulfilled the needs of navigation and astronomy for over half a century.

Brown commented that the derivation of new tables could not be justified unless they were to possess both theoretical and practical accuracy greater than what had come before. His own theory was extended to include the effects of every known force acting upon the Moon. Nearly 1,500 terms (five times as many as Hansen had adopted) were employed, yet the time required to compute an ephemeris from his tables was less than that required with Hansen's. Brown's tables were used for the first time in the computation of the lunar ephemeris in the American, British, and German Almanacs for 1923, and in the French *Connaissance des Temps* for 1926. However, the previous three volumes of the latter tabulated the differences between the lunar coordinates computed from Brown's and Delaunay's tables, the maximum differences in 1923–25 amounting to only 0.54s in right ascension, 3.5″ in declination, 7.9″ in longitude, 1.5″ in latitude, and 0.38″ in parallax.

Late in life Brown was still using the obsolescent *Millionaire* mechanical computer even after he himself had designed punch-card equipment especially for his lunar tables, and electric Monroe machines were widely in use. Nevertheless, he continued to be keenly interested in more efficient methods of computing. At the Columbia University Statistical Bureau, Professor Ben D. Wood and Mr. R. M. Mendenhall in 1930 were already applying Hollerith machines to the Brown and Brouwer *Tables for the Development of the Disturbing Function with Schedules for Harmonic Analysis.*[23] At the same time Comrie in England was using Hollerith machines for computing ephemerides from Brown's tables, carrying the computations forward to the year 2000, with the capability of also carrying them back to 1770 or beyond,[24] as Comrie commented, "at a cost of only £150 for each year of lunar positions, as compared with £500 by former methods."

In the 1930s the only known applications of punch-card machines to science were apparently in astronomy and, apart from the *British Nautical Almanac* work, only at Columbia.[25] The one person who took a special interest in these machines and eventually became the leading figure in their development was Wallace J. Eckert (1902–71),[26] of whom Ashbrook[27] was to comment, "Hardly any other astronomer of his generation influenced our science more profoundly."

After obtaining an M.A. degree from Amherst College in 1926 Eckert was employed as an Instructor at Columbia University. Soon he entered the graduate school at Yale, perhaps choosing Yale because he was already interested in celestial mechanics, particularly in the motions of the Moon, and the application of punched-card machines for their solution. Brown inevitably took an encouraging interest in Eckert's progress. After receiving his Ph.D. in 1931 Eckert returned to Columbia. In 1937 Columbia and IBM jointly established the *Thomas J. Watson Astronomical Computing Bureau*. There Eckert wrote his classical treatise, *Punched Card Methods in Scientific Computation.*[28] After describing the machine's capabilities and its ability to solve mathematical equations, the third part of the book was devoted to *Astronomical Applications,* mainly those relating to the projects in which Yale and Columbia were at the time collaborating (zone catalogues, photometry, lunar theory). From 1940 until the close of World War II, Eckert was Director of the *Nautical Almanac Office* in Washington. Then in 1945 he was appointed head of the newly established *IBM Watson Scientific Computing Laboratory* where, over the years until his retirement, first from IBM in 1967, then from Columbia in 1970, he was the one mainly responsible for the development of the complex new SSEC (Selective Sequence Electronic Calculator) which in 1948 occupied a whole skyscraper on Madison Avenue in New York and performed numerical tasks almost inconceivable up to that time. With his devotion to the Moon, celestial mechanics, and his Yale mentors, Eckert contributed immensely to numerous other Yale projects. For his work on lunar theory[29] he was awarded the James Craig Watson Medal of the National Academy of Sciences in 1966.[30] A historical survey, critical examination, and revised lunar tables are given by Gutzwiller and Schmidt in *The Motion of the Moon as Computed by the Method of Hill, Brown, and Eckert.*[31]

Brown's famous *Tables of the Motion of the Moon,*[7] published in 1919, were the basis for ephemerides of the Moon well into the 1950s. However, Eckert in 1940 found,

> the general accuracy of the theory of E. W. Brown was ample for observational purposes, but a few coefficients, particularly those giving the motions of the perigee and node, had not been obtained with sufficient accuracy. To obtain the additional accuracy by an extension of the original method would require years of highly specialized work. Then too there was the possibility of error in the original work which had required twenty years to complete. By the numerical method it has been possible in about two years to verify completely Professor Brown's results and to obtain the required additional accuracy.[32]

Accordingly, in 1954 there appeared an *Improved Lunar Ephemeris 1952–1959,* which stated in the Preface, "The increased importance of the Moon, both for precise timekeeping and for geodetic surveying, has led in recent years to so great an improvement of observational techniques that Brown's *Tables of the Motions of the Moon,* in use since 1923, are no longer capable of yielding an ephemeris of the desired precision."[33] A substantial part of the necessary computations had been carried out under Eckert's direction at the Watson Scientific Computing Laboratory.

In 1919, when he had concluded his tables, Brown commented,

> The undertaking of a complete recalculation of the moon's motions and later of tables which should make the theory available for practical and scientific use was no ambitious plan formed at the beginning but grew naturally out of the desire to continue the work as each stage in it was reached. Some part of it has always been in progress. . . . To what extent it has been worth while as a contribution to the subject must be left to the future and to others for judgment. My hope is that it will give some aid in unraveling the tangled skein of problems which our nearest celestial neighbor has never failed to present.[34]

Indeed the tables had not constituted an end in themselves. They opened the door to further investigations by Brown himself, spurred by the small unaccounted-for deviations between predicted and observed lunar positions. Of special interest are his discussions relative to the variation in the period of rotation of the Earth. As early as 1903 Newcomb had toyed with the notion that the deviations did not represent variations in the motion of the Moon but in irregularities in the rotation of the Earth; but he soon dropped the suggestion as improbable.[35] Variations in latitude had been found by various observers in the latter half of the 19th century. The ellipsoidal shape of the Earth was known and could account for some of the periodic variations in the observed position of the Moon. The effects of tidal actions on the surface layers of the Earth were already understood, but they were not sufficient to account for remaining small discrepancies: ellipticity in the figure of the sun; possible magnetic attractions between the Earth and the Moon; and shading of gravitational forces by the interposition of intervening matter (as at times of eclipses) were considered but all found wanting.[36] The Einstein effect was examined but would not resolve the discrepancies. The possibility of perturbations by the suspected inter-Mercurial planet Vulcan was even considered[37,38] and of the Leonid meteor swarm.

Already in 1914 Brown had pointed out that if the lunar discordances depended on variations in the rate of rotation of the Earth, corresponding variations should be found in the apparent motions of the sun and planets; and he found such effects for the sun, Mercury and Venus, though, as expected, to a far lesser degree than for the nearby Moon.[39] Others substantiated such observations later. Brown examined all available data as far back as 1750 (especially from lunar occultations, which he found to be more accurate than old meridian circle timings[40]). The most pronounced changes appeared to occur in about 1785 and 1898, the latter, the largest, changed the apparent length of the year by a whole second.[41] Brown concluded that the only cause for the still mysterious variations was probably in the rate of rotation of the Earth. He wrote,

> Granting the hypothesis, arguments resulting from astronomical data, mechanical laws, and geophysical evidence point to the conclusion that such variations can only arise from vertical oscillations of the whole crust of the earth, the source of these oscillations being due to oscillatory changes taking place in the isostatic layer or below it. The maximum change of the average radius needed by the astronomical data is between five inches and twelve feet, according to the hypothesis made as to the depth of the source.[42]

The deviations from a "gravitational orbit" appear to be sudden rather than gradual;[43] and the angular acceleration of the Earth, assumed to be the cause of the variations, is much greater than that produced by tidal friction alone.[44] He assumed "that the variations take place mainly in the crust of the Earth which might be assumed to slip as a whole over the nucleus to an observable extent." The cause of such slippage he was not prepared to discuss but assumed the solution must be found in the physical and chemical conditions below the outer crust of the Earth.[45]

The most accurate positions of the Moon could be derived from eclipses and from occultations, the positions of the sun and stars being known. Ancient eclipses were thus used by J. K. Fotheringham[46] for determining the secular accelerations of the sun and Moon. However, the ancient observations were not sufficiently accurate for the evaluation of the change in rotation of the Earth. With the upcoming eclipse of January 24, 1925, Brown, and L. J. Comrie of the *British Nautical Almanac,* visiting astronomer at Swarthmore at the time, inaugurated an extensive campaign for observations by both amateurs and professional astronomers, the path of the eclipse remarkably traversing ten or twelve fixed observatories.[47,48]

The spacious grounds of the Yale Observatory were strewn with equipment from the Allegheny and Sproul Observatories, and astronomers from Bermuda and California.[49,50] Professor Duncan at Wellesley College had asked to bring more than 100 students to the Yale grounds and, housing being scarce, made arrangements for the New Haven Railroad to provide a sleeping car to bring the students the night before and then wait in the RR yard until after the eclipse to take the girls back home.[51,52] Boy Scouts were engaged to patrol and protect observers and their equipment from outside molestation. Offices were reserved for the American Telephone and Telegraph Company, from which signals on weather and timing were to be transmitted to and from other stations along the eclipse path. Thus observers were alerted to the differences between predicted and observed times of contacts, important data for obtaining photographs of the flash spectrum. (The eclipse was about 4 seconds late.) Nearly all the Yale telescopes were utilized in one way or another. With the Loomis telescope, photographs five inches in diameter were obtained throughout the partial as well as the total phases. One solar prominence was photographed, showing that it arched 50,000 miles above the photospheric surface. A unique pendulum camera was devised for automatically taking pictures of totality at intervals of about one second. Film on a 25-foot-long spool was moved about two inches between exposures. Pictures from this apparatus revealed that the corona could be photographed at least 17 seconds before actual totality.[53]

In his preparations for the eclipse, in correspondence with Leon Campbell, Recorder for the AAVSO (American Association of Variable Star Observers), Brown, who was later (1933–35), to become the President of the AAVSO, recommended as further incentive to amateur observers, that those with sizable telescopes (like Butler's 10-inch which he later donated to Yale) be encouraged to look for possible novae and comets during totality. He added, however, "This may not be of very great value but I think it would add to the interest of using such instruments."[54]

Brown sent out circular forms for amateurs to fill out, giving times and places. Comrie similarly sent out instructions for occultation observations to be carried out a month before, through a month after the eclipse, in order that the relative accuracies of the two methods might be evaluated.[55] He tabulated relevant data for ten cities and gave instructions for observing. By February 26, just a month after the eclipse, Brown enthusiastically wrote Schlesinger (who at the time was in Johannesburg), "It would appear that the poor old moon, therefore, is not more than one second of arc away from its predicted position."[56] The final results he published in April 1926.[57] The weighted mean deviation from the predicted lunar orbit was only 0.28″ in longitude and 0.32″ in latitude. The results depended on 89 occultation observations, timings of contacts from twelve localities; photographs; and auxiliary observations of the transit of the sun at Washington, Greenwich, and the Cape.

The eclipse was commemorated by Chemistry Professor Philip E. Browning in a parody of Longfellow's *Paul Revere's Ride.* Among Brown's memorabilia is a copy beginning,

Listen, my nieces, and you shall hear
Of a great event in your uncle's year.

In the published version the first two lines were altered:[58]

THE 1925 ECLIPSE

Listen, my children, and you shall hear
Of a great event in the present year.
This event took place on a twenty-fourth,
In New Haven, set in the frozen north.
The sun rose clear in the east that day,
Just about in the usual way,
But shortly after the clock struck eight
We began to observe his approaching fate;
For the moon, in her regular monthly ride,
Began from our eyes his beauty to hide,
And slowly obscured his shining face,
Until, at nine, there was scarcely a trace
Of the light he so freely sheds abroad
To the system of which he is the lord.
And now on the snow strange shadows appeared
For the lines seemed curved, and the light was weird,
And the sun for an instant disappeared.
But lo, from the disc of totality,
Which had scarcely become a reality,
The corona shone forth with a glorious light.
Could there ever be a more charming sight?
The rays streamed out on every side,
For the sun refused to be denied
This chance his radiant beauty to show
To us who watched him, standing below,
Cooling our feet on the ice and snow.
With our glasses we saw from the chromosphere
The results of volcanic eruptions appear.
Great bubbles of hydrogen, bursting, took fire,
And red flames shot forth for us to admire.
Mercury, Jupiter, Venus came out,
To see what the show was all about.
Another brief instant, a flash of light,
And old Sol again shone forth in his might.

At the close of 1926 Brown issued an appeal for observations of lunar occultations, particularly by amateur astronomers, to test further the deviations of the Moon's apparent positions from its gravitational orbit.[59] Dr. Comrie offered to collaborate by collecting and reducing observations carried out in the eastern hemisphere while Brown would take care of all those for the western hemisphere. Brown especially stressed the value of contributions by amateurs as the accuracy of timing was not required to be better than the nearest second. He described his needs to the AAVSO at their annual meeting in October 1926 and received enthusiastic response from twelve members, either as observers or computers. In their first year they observed 36 occultations and reduced 30. In their tenth year they reduced 212 of the 1,483 occultations that had been observed in Europe and America in 1934, the largest number to that date.[60,61] From 1926 until his death in 1938 Brown attended nearly all the meetings of the AAVSO, giving instruction, encouragement, and progress reports. Wallace Eckert said of Brown,[62]

The appreciation of Brown's scientific work is not confined to the International Astronomical Union, but extends to amateurs the world over. He was one of the few scientists who had a popular following, not because of his popular writing but because his scientific work had in itself a popular appeal. The solar eclipse of 1925 made the results of his work known to millions, and his occultation campaign has made him familiar to many serious amateurs. He attended the meetings of the A.A.V.S.O. with as much faithfulness as those of the National Academy.

A special *Occultation Committee* was set up within the AAVSO with first the able amateur J. Ernest G. Yalden, followed by Professor Dirk Brouwer of Yale as Chairman. For giving them the opportunity to contribute significantly to his important researches the AAVSO in 1927 proudly elected Brown a "Complimentary Member," the only one besides Frank Schlesinger ever to have received this title.[63]

Members of the *British Astronomical Association* likewise responded to Comrie's request, and soon others followed. In 1928 Brown solicited the help of the armchair amateurs, the *Bond Astronomical Club* of Cambridge, Massachusetts, to reduce observations for the years back to 1923 when 40 to 50 per year had been observed. His appreciation of the amateur was clearly indicated in a 1929 article, *Practical Astronomy for the Amateur: How the Amateur Astronomer Is Helping to Observe and Calculate the Moon's Motion.*[64] This was reprinted in slightly different form in the *Journal of the Royal Astronomical Society of Canada,* whose editor hoped thereby to encourage similar cooperation in his country.[65] By 1931 Brown confessed "We do not know yet why the rotation of the earth is changing,"[66] but in 1937, shortly before his death, he prophesied that the rate of rotation of the Earth would soon alter, as its cumulative change in rotation had surpassed all previous maxima.[67] Later investigators have questioned that the changes have been sudden[68] and Brouwer,[69] in 1952, following earlier suggestions by T. E. Sterne[70] (approximately 1935) and Spencer Jones (1949)[71], investigated the hypothesis that the apparent sudden changes in the Earth's period were actually the consequence of accumulated small random changes. His analysis indicated that the observed length of the day did change in about 1937.[72]

Despite Brown's amateur following, he was not disposed to producing popular accounts of his work. Harlow Shapley, in compiling a supplement to his *Source Book in Astronomy,*[73] complained that he could find no "readable statement by E. W. Brown on his theory of the motion of the moon ... His magnificent papers in the *Memoirs* of the R.A.S. are, of course, indecently technical."[74] However, one article, *On the Completion of the Solution of the Main Problem in the New Lunar Theory,* published in 1904, served Shapley's purpose.[75] Up to that time the computations had already consumed some eight or nine thousand hours and involved "about 13,000 multiplications of series made, containing some 40,000 separate products; the whole of the work required the writing of between four and five millions of digits and *plus* and *minus* signs." And all this was before the actual construction of the famous tables. Brown concluded, "If a moderate degree of success attends these efforts it should be possible to discover whether, within the limits set by the observations, the motion of the moon shows effects which cannot be traced to the direct operation of the Newtonian law of gravitation." This early wistful hope seems to have come closest to being fulfilled, but not really so, by his and Brouwer's work on the changing period of rotation of the Earth.

The difficulty of translating Brown's theory into popular language is indicated by a short note in the *Yale Alumni Magazine*[76] shortly before the end of Brown's life. Entitled, *Professor Brown Announces New Formula,* the note reported on a paper Brown had presented at a professional astronomical meeting. The new formula, which covered two whole blackboards, would increase the accuracy of his thirty years' study of the Moon. His colleagues termed it "a remarkably rapid method of increasing precision."

REFERENCES

1. F. Schlesinger and D. Brouwer, *Biog. Mem. Nat. Acad. Sci.,* **21**, 243, 1940.
2. E. W. Brown, *The Scientific Work of Sir George Darwin,* in *Scientific Papers by G. H. Darwin,* **5**, xxxiv, 1916.
3. ———, *An Introductory Treatise on the Lunar Theory,* Cambridge Univ. Press, xvi + 292 pp., p. viii, 1896.
4. George William Hill, *Collected Mathematical Works,* Carnegie Inst. Washington., 4 vols., 1905–7.
5. E. W. Brown, *Mem. Roy. Ast. Soc.,* **53**, 39, 163, 1899; **54**, 1,1900; **57**, 51, 1905; **59**, 1, 1908.
6. ———, *The Inequalities in the Motion of the Moon Due to the Direct Action of the Planets,* Cambridge Univ. Press, xii + 92 pp., 1908.
7. E. W. Brown, with the assistance of H. B. Hedrick, *Tables of the Motion of the Moon,* Yale Univ. Press and Oxford University Press, xiv + 659 pp., 1919; Addenda and errata, *Trans. Yale Univ. Obs.,* **3**, 155, 1926.
8. H. B. Hedrick, *Astronomical Papers of the American Ephemeris,* **8**, Part III, 405, 1905; reviewed in *Pop. Ast.,* **16**, 196, 1908.
9. H. B. Hedrick to E. W. Brown, Oct. 29, 1908, *Yale Obs. Archives.*
10. ———, Nov. 9, 1908, *Yale Obs. Archives.*

11. W. G. Boyce to H. B. Hedrick, Jan. 30, 1909, *Yale Obs. Archives.*
12. E. W. Brown, *Science,* **29**, 830, 1909.
13. ———, *Ibid.,* **30**, 729, 1909.
14. ———, *Monthly Notices Roy. Ast. Soc.,* **72**, 454, 1912.
15. F. J. W. Whipple to E. W. Brown, undated, June 2 and 30, 1912. *Yale Obs. Archives.*
16. E. W. Brown, *Monthly Notices Roy. Ast. Soc.,* **52**, 71, 1891.
17. P. A. Hansen, *Tables de la Lune,* London, 511 pp, 1857.
18. ———, *Darlegung der theoretischen Berechnung der in den Mondtafeln angewandten Störungen,* Leipzig, Part 1, 497 pp, 1862; Part 2, 399 pp., 1864.
19. C. E. Delaunay, *Mem. Paris Acad. Sci.,* **28**, 1860; **29**, 1867.
20. J. C. R. Radau, *Annales du Bureau des Longitudes,* **7**, [112]+630 pp, 1911; abstract, *Comptes Rendus,* **153**, 313, 1911.
21. E. W. Brown, *Pub. Ast. Soc. Pacific.,* **32**, 90, 1920.
22. ———, Ref. 7, p. v.
23. E. W. Brown and D. Brouwer, *Trans. Yale Univ. Obs.,* **6**, Part 5, 67, 1932.
24. L. J. Comrie, *Observatory,* **52**, 199, 1929.
25. H. H. Goldstine, *The Computer from Pascal to von Neumann,* Princeton, 1972.
26. H. S. Tropp, *Dictionary of Scientific Biography,* N.Y., **15**, Supp. I, 128, 1978.
27. J. Ashbrook, "A Great American Astronomer," *Sky and Telescope,* **42**, 207, 1971.
28. W. J. Eckert, *Punched Card Methods in Scientific Computation,* N. Y., 136 pp., 1940.
29. W. J. Eckert and H. F. Smith, Jr., *Ast. Papers of the Amer. Ephemeris and Naut. Almanac,* **19**, Part II, 185, 1966.
30. Anon., *Physics Today,* **24**, 73, 1971.
31. M. C. Gutzwiller and D. S. Schmidt, *Astronomical Papers of the Amer. Ephemeris and Naut. Almanac,* **23**, Part I, 272 pp., 1986.
32. Ref. 28, p. 98.
33. H. Spencer Jones and F. A. Graf, *Improved Lunar Ephemeris* 1952–59, *Supp. to Amer. Ephemeris* and to the *British Nautical Almanac.*
34. ———, Ref. 7, p. xii.
35. *Cf.* A. Pannekoek, *A History of Astronomy,* N.Y., p. 365, 1961.
36. E. W. Brown, *Amer. Jour. Sci.,* **29**, 529, 1910.
37. W. W. Campbell to E. W. Brown, March 13, 1912, *Yale Obs. Archives.*
38. E. W. Brown, *Monthly Notices Roy. Ast. Soc.,* **70**, 342, 1910.
39. ———, "Address on Cosmical Physics," *Report Brit. Assoc. for Adv. Sci,* Section A, Australia, 11 pp., 1914; *Nature,* **94**, 184, 1914; *Science,* **40**, 389, 1914.
40. ———, *Trans. Yale Univ. Obs.,* **3,** 205, 1926.
41. ———, *Smithsonian Report* for 1937, p. 169.
42. ———, Ref 27; Ref. 29 p. 211; *Pub. Ast. Soc. Pacific,* **44**, 21, 31, 1932.
43. ———, Ref 29, p.227.
44. ———, Ref. 29, p. 231.
45. ———, *Nature,* **119**, 200, 1927.
46. J. K. Fotheringham, *Monthly Notices Roy. Ast. Soc.,* **80**, 578, 1920; **81**, 104, 1920.
47. E. W. Brown, *Science,* **61**, 10, 1925.
48. E. W. Brown, *Sci. American,* **132**, 14, 1925.
49. F. Schlesinger, *Ann. Report* for 1924–25.
50. ———, *Pop. Ast.,* **34**, 116, 1926.
51. J. C. Duncan to F. Schlesinger, Oct. 28, 1924, *Yale Obs. Archives.*
52. F. Schlesinger to J. C. Duncan, Oct. 31, 1924, *Yale Obs. Archives.*
53. E. W. Brown, *Pop. Ast.,* **33**, 180, 1925.
54. E. W. Brown to L. Campbell, Oct. 14, 1924, *AAVSO Archives.*
55. L. J. Comrie, *Pop. Ast.,* **32**, 629, 1924; **33**, 16, 1925.
56. E. W. Brown to F. Schlesinger, Feb. 26, 1925, *Yale Obs. Archives.*
57. E. W. Brown, *Ast. Jour.,* **37**, 9, 1926.
58. P. E. Browning, *Saying It with Rhymes,* p. 8, privately pub. for the New Haven Rotary Club, 1928.
59. E. W. Brown, *Ast. Jour.,* **37**, 99, 1927.
60. L. Campbell, *Variable Comments,* AAVSO, **3**, 25, 1936.
61. E. W. Brown and D. Brouwer, *Ast. Jour.,* **45**, 153, 1936.
62. W. J. Eckert, *Pop. Ast.,* **47**, 63, 1939.
63. H. M. Swartz, *Variable Comments,* AAVSO, **1**, 73, 1927.
64. E. W. Brown, *Sci. American,* **141**, 426, 1929.
65. ———, *Jour. Roy. Ast. Soc. Canada,* **24**, 177, 1930.
66. ———, *Variable Comments,* AAVSO, **2**, 41, 1941.
67. ———, *Ibid.,* **3**, 61, 1937.
68. *Cf.* B. G. Marsden and A. G. W. Cameron, *The Earth-Moon System,* N.Y., 1966.
69. D. Brouwer, *Ast. Jour.,* **57**, 125, 1952.
70. ———, *Idem.,* p. 138.
71. H. Spencer-Jones, *Smithsonian Ann. Report* for 1948–49, p. 201, 1950.
72. D. Brouwer, *The Yale Scientific Mag.,* **28**, 18, 42, 1954.
73. H. Shapley, *Source Book in Astronomy, 1900–50,* Harvard Univ. Press, xv + 423 pp., 1960.
74. Harlow Shapley to D. Brouwer, Aug. 23, 1957, *Yale Obs. Archives.*
75. E. W. Brown, *Monthly Notices Roy. Ast. Soc.,* **65**, 104, 1904.
76. Anon., *Yale Alumni Mag.,* **45**, No. 14, p. 8, 1936.

16
ASTEROIDS AND BROWN'S OTHER 3-BODY PROBLEMS

Although the Moon was E. W. Brown's chief interest, to which he devoted the greatest part of his research time, he made significant contributions to other applications of the three-body problem. The most important among these were those of the Trojan Asteroids, and the eighth satellite of Jupiter. Other papers dealt with the planetary perturbations, prediction of an unknown planet, and the stellar problem of three bodies. In addition to some 200 technical papers related to celestial mechanics, in 1933, together with C. A. Shook, he wrote a treatise, *Planetary Theory.*[1]

In an address to the American Association for the Advancement of Science on December 28, 1910, Brown presented the problems of the motions of the asteroids in relatively nonmathematical terms.[2]

> Many of their orbits are highly eccentric and inclined at large angles to the plane near to which the large planets circulate. But the most interesting problems are those which arise from the near presence of the second largest member of the solar system, Jupiter, whose mass is but little less than one-thousandth that of the sun.

> So far as we know, all the disturbances produced in our solar system by gravitation are really periodic, but some of the periods are so long, extending to thousands and ten thousands of years, that it is more convenient in the short space of time during which we desire to know the motion to treat certain of them as secular.

The long-period terms could be divided into two categories, "proper" containing slow oscillations of the eccentricities of the planets due to the attraction of one planet on another arising from the fact that the masses of the planets are small compared with the mass of the sun; and "accidental" arising from a synchronism of periods. The latter are not necessarily intrinsic but "should be regarded perhaps as a result of the defects of our mode of representing the motion in symbolic form."[3] Some periodic changes constrained to vary between finite limits are known as librations:[4]

> The familiar example furnished by the motion of the moon relative to its center of mass is of course a result of the synchronism of its period of rotation about its axis with that about the earth. The delicate balancing of the conditions necessary for a libration makes the term unusually appropriate in the cases of the motions of the asteroids.

In 1866 D. Kirkwood (1814–95) had been the first to point out that there were gaps in the asteroid belt, later named the Kirkwood gaps in his honor. He noted that the gaps appeared to occur where the periods of any objects in those positions would be roughly commensurable with the period of Jupiter. Brown in 1911 re-examined the frequencies of the known asteroid periods (about 680 at that time) and mathematically studied resonance and librational effects arising from perturbations of Jupiter and Saturn. The motion of an asteroid in a critical zone depends mainly on its apparent eccentricity (which is variable) at a given time and the difference between its instantaneous period and the period most commensurable with Jupiter's. Brown's analysis indicated that when the ratio of these periods is 2:1, stability of the asteroidal orbit becomes doubtful, accounting for the gaps. His theory for the asteroids also applies to gaps in Saturn's rings which are similarly affected by Saturn's satellites.[5,6] The problem of commensurability with Jupiter's period in the Kirkwood gaps was more fully discussed by Brouwer in 1963. In comparing the observed distributions of asteroidal motions with theory, he found gaps corresponding to the following ratios of asteroidal to Jupiter's daily motion; *2:1, 3:1,* 3:2, 5:2, 4:3, 5:3, *7:3,* 7:4, 7:5, 8:5, 9:5, and 11:7. The italicized ratios had already been noted by Brown, who also found indications for 7:2, 8:3, and 9:4. As is so often true in research, the complexity of the problem was such that Brouwer still felt that speculations he pre-

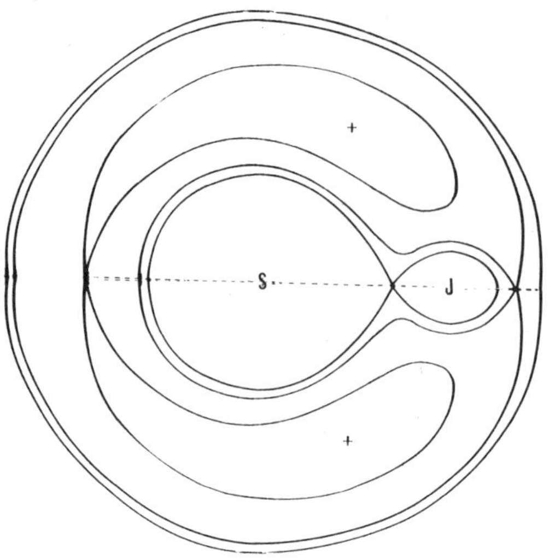

Fig. 16.1. Schematic diagram of possible "hourglass" and "horseshoe" orbits for Trojan Asteroids relative to the line joining Jupiter (J) and the Sun (S).

sented should serve merely to indicate the direction of further research.[7]

Early on, in 1772, J. L. Lagrange (1736–1813) had demonstrated theoretically that if three bodies were situated at the vertices of an equilateral triangle and that one of them had a negligible mass relative to the other two, they would form a stable configuration. It was not until 1906 that Max Wolf (1863–1932) discovered an asteroid, Achilles, that satisfied Lagrange's criterion. The asteroid was moving, not in the asteroid belt between Mars and Jupiter, but approximately in the orbit of Jupiter and almost 60° (actually 55.5°) ahead of Jupiter. The points 60° either ahead or behind the planet, are positions of stability and are named *Lagrangian Points,* or also libration points. At the Lagrangian points the gravitational pull of Jupiter and the sun just balance, so that the asteroid is not pulled toward either. Any object close to these points will show relatively small oscillations around these points.

Asteroids close to the Lagrangian points are now called *Trojan Asteroids.* By 1911 four Trojans had been found. (Now at least 70 Trojans have been discovered.) Brown set out to determine the motions of the Trojans relative to the line joining Jupiter and the sun. His amazing conclusion[8] was that they moved in horseshoe- or hourglass-shaped orbits (Fig. 16.1). (A concise description of Trojan Asteroids and Brown's orbits is given by D. Brouwer[9] in the *Encyclopaedia Britannica.*) The librational periods ranged from 140 to 200 years.[10] Thus Brown cautioned that many years must elapse before the theory could be tested with any degree of certainty. One of the most interesting of the Trojans he investigated was *624* Hektor.[11] An early discovery, Hektor (discovered by Kopff at Heidelberg in 1907) is still exceptional, librating by over 17° from its mean point relative to Jupiter. Hopefully, Brown commented,

> With the actual existence with a range so large as that of *624* it is not unreasonable to hope that other asteroids may be found which show a still greater range, and possibly one moving in a horse-shoe orbit enclosing both the triangular equilibrium points and the unstable equilibrium point in opposition to Jupiter. The limiting horse-shoe orbit is 23°.4 from Jupiter at its nearest approach to the planet.

Brown's work followed earlier incomplete investigations by G. H. Darwin,[12] which already described figure-8 orbits, but dealt only with direct orbits; and a subsequent paper by S. S. Hough[13] which also discussed retrograde orbits; and later amplifications by Darwin.[14]

Brown was warmly praised by his early mentor in an article, *On Librating Planets and a New Family of Periodic Orbits.*[15] Darwin noted that Brown "shows how to obtain the orbit of a planet which makes large oscillations about the vertex of the Lagrangian equilateral triangle." Darwin's own interest had originated in a desire to discover how a Laplacian ring could coalesce into a planet. On the current work he commented,

> These orbits throw some light on cosmogony, for we see how small planets with the same mean motion as Jove in the course of their vicissitudes tend to pass close to Jove, ultimately to be absorbed into its mass. We thus see something of the machinery whereby a large planet generates for itself a clear space in which to circulate about the sun.

Between 1923 and 1925 Brown published his extensive *Theory of the Trojan Group of Asteroids.*[16] In a summary published in the *Astronomical Journal*[17] Brown described his purpose as twofold:

> I desired to carry the developments sufficiently far so that they could be used for comparison with observation and prediction with nearly the accuracy obtained at present by observation and to indicate how they could be continued to any required degree of accuracy. Also, as we have but little knowledge of the stability of our planetary and satellite systems except that derived from the actual developments, it was hoped that some idea might be gained as to whether the stability of motion of these asteroids dif-

fered essentially from that of a major planet or from that of other well-known groups of asteroids. The gaps in the distribution of the latter may indicate some kind of instability but so far, mathematical investigation has failed to reveal it. Essential differences between the motions of the asteroids of the *Trojan* group which seem to be stable, and those of other groups whose mean motions should have exact or approximate commensurability with that of Jupiter, but which have no representatives, may possibly reveal the direction in which the instability, if it exists, should be sought.

In the Introduction to his major contribution in the *Yale Transactions* Brown stated,[16]

> The intention has been to carry the developments sufficiently far to obtain the longitude of the asteroid to 10″ at any time within a few centuries, on the assumption that the observed constants are known with the necessary accuracy. This assumption is not justified at the present time, but it is possible to so arrange tables that corrections to the constants can be made without sacrificing the tables. But my main interest has consisted in the investigation of methods for treating the problem effectively and in showing how the difficulties which may arise at any stage of the approximations can be surmounted.

The constants required were to be obtained from an osculating orbit at some given time. (An osculating orbit is the elliptical orbit in which the planet would move if there were no other forces acting upon it except the gravitational pull of the sun, starting from the planet's position, speed and direction of motion at a chosen time of observation.) With this as a first approximation the theory would provide computed positions for any future (or past) dates. To test the validity of his theory Brown used the first-discovered Trojan, Achilles. An osculating orbit had been computed by Julie Vinter Hansen, and Brown selected observations at six dates, two in 1906, two in 1907, and one each in 1913 and 1920. The differences between observed and computed ecliptic longitudes and latitudes differed by less than 10″ (except for one longitude difference of 20″), really excellent agreement, though a much longer time interval would have been desired for definitive results.

In the fall of 1927, when further observations were available from the Yerkes Observatory, Dirk Brouwer, at the suggestion of Brown, undertook a new comparison between the observations and theory.[18] He obtained "alarming results"—residuals amounting to nearly 2°. Part of the trouble was attributed to an erroneously printed constant, and Brouwer made some modifications in the theory, particularly on the perturbations by Saturn. He found that the residuals, O-C, in right ascension and declination for 18 dates between 1906 and 1930 were under 40″ except for two points, one in 1919 (−150,″ +85″) and one in 1920 (−72,″ +28″).[19] Brouwer inferred that these represented a defect of the theory. Although he had hoped the residuals would not exceed 30″ he felt that on the whole the representation of the theory was satisfactory. He thought that the large deviations indicated short-period oscillations, due to the combined effect of several short-period terms. "It is just this type of term that the theory may fail to include as exhaustively as the terms more directly connected with libration."

In the same volume of the *Yale Transactions,* W. Eckert[20] tested the theory on Hektor whose large amplitude of libration and large inclination to Jupiter's orbit made it exceptionally interesting, as well as making the analysis more complicated. The final residuals for seven positions between 1907 and 1927 were again "slightly larger than was hoped for," amounting to up to 1.7 minutes of arc in either coordinate. The residuals appeared to be systematic but again the time interval was too short for definitive conclusions.

For some years subsequently, no further work on the Trojans was carried out at Yale. Correspondence between A. O. Leuschner at the University of California at Berkeley and Schlesinger in 1936, and with Brouwer in 1941, indicated considerable outside interest in the perturbations of the Trojan group. Leuschner wrote Schlesinger,[21]

> It is a source of great disappointment to me that Brown and Brouwer feel that they cannot undertake the supervision of the development of the general perturbations of the Trojan Group in case funds for the purpose shall be made available to Yale University. I realize from my own experience that supervision of even the most competent computers trained in theory involves a great responsibility and requires much time and effort. It is, however, imperative that the remarkable contribution of Brown on the Trojan Theory be explored and applied.
>
> May I, therefore, submit to the consideration of Brown, Brouwer, and yourself the suggestion that you jointly recommend some competent young Ph.D. in Theoretical Astronomy for the award of one of Yale University's larger fellowships, with the understanding that he devote himself to the mastery of the Trojan theory, with the expectation that the following year he would be on salary from the grant which I have recommended. Such an arrangement, it would seem to me, would require only occasional conferences with Brown or Brouwer, until the man has mastered the theory.

No suitable candidate could immediately be found, and the availability at Yale of what would now be called postdoctoral fellowships was very uncertain. Moreover, both Brown and Brouwer had their hands full of other important projects. Schlesinger added, however,[22] "Surely the Trojan problem is one that will require eight, or ten, or more years of concentrated effort even with the help of an expert computer. The problem demands someone who would be attracted to this special work, and who would be willing to devote so long a term of years to it." (Computers in those days were people, not machines.)

At the time, although still active, Brown was retired, a sick and therefore despondent man whom Brouwer, quite by chance, rescued from a suicide attempt by carbon monoxide in his garage.[23]

Several years later (1941) Brouwer reactivated the correspondence with Leuschner, explaining his own previous hesitation in undertaking the general theory for the Trojan asteroids. He was nevertheless seriously interested in Leuschner's suggestions and when the German Hans Hertz came to Yale in 1937 as a Ph.D. candidate, Brouwer suggested the Trojan problem to him as a thesis project. Hertz received his Ph.D. in 1941 and continued employment at Yale to 1943. As Leuschner was Chairman of the Watson Trustees of the National Academy of Sciences, Brouwer approached him for funds to help defray Hertz's salary for the part of his work that would be devoted to the asteroids.

Hertz had started with Brown's theory as it had been revised in Brown's and Shook's *Planetary Theory* where "the Trojan Group problem was treated by a totally different method that Brown preferred decidedly over his earlier theory."[24] Hertz had then developed the theory for the asteroid Nestor in considerable detail; but a great deal still remained to be done. This work had not yet been completed when Hertz transferred to the U. S. Naval Observatory in 1943, where he did complete the project.[25] He had available 65 observations, of which four were rejected. The final opposition means of the residuals for 1908 to 1948 did not exceed 2.5′ in R.A. and 1.4′ in Dec. This accuracy was again adequate for computing ephemerides for the recovery of the asteroid but still indicated small refinements to the theory.

The overall good correspondence between observed and computed positions of asteroids in general prompted Brouwer to make use of them for improving upon the accuracy of fundamental star positions. In 1940 he wrote Paul Herget in Cincinnati,[26]

> If, during the coming few years, you wish to start on a new problem, would you be interested in taking up one of the asteroids on our program? We shall soon have accurate numerical integration orbits extending from about 1924 to the present time. The problem would be to compare such an orbit with a selection of observations, to be followed by the development of an accurate general theory. The latter is not a part of our present program, but would be desirable for a discussion of the older observations. If I am not entirely mistaken there is a great amount of latent information especially concerning systematic errors in fundamental star catalogues, in these older observations of minor planets.

Observations for checking equator points of transit instruments had been suggested by Dyson[27] in 1928, and others extended his suggestion for improving declinations determined by meridian circle of fundamental stars (stars relative to which the positions of large numbers of other stars were determined photographically). Brouwer in 1935 extended the concept theoretically to include rectification of both right ascension and declination of the fundamental stars,[28] and in 1941 initiated an observational program expected to span about ten years[29] to obtain corrections to star positions in the zone −30° to +30°. This developed into a huge project involving cooperation among the Allegheny, Leiden, and Yale Observatories. Ultimately about 7,000 photographic positions of fifteen asteroids were obtained. These were analyzed between 1966 and 1970 by one of Brouwer's former graduate students, David A. Pierce,[30] but were published in greater detail in 1978.[31]

Pierce determined systematic corrections for stars between −30° and +30° declination in both the Yale Zones and the Boss *General Catalogue* (GC). The nine Yale zones were subdivided into 54 regions, each spanning four hours in right ascension and the GC was divided into 60 regions, five zones 12° wide and two-hour intervals. The average residuals of the catalogue positions from the positions based on refined orbits of the 15 asteroids were on the order of 0.3″ in both right ascension and declination for both the Yale and GC catalogues while the average systematic corrections were well under 0.2″. In right ascension the Yale corrections ranged from 0.17″ to 0.88″, with little trend in right ascension or declination. The Boss corrections, on the other hand, ranged from, −0.12″ to +0.91″ and showed pronounced trends with right ascension, with maxima between 19 and 1 hours, and minima from 5 to 11 hours. The corrections to the declinations spanned from about $-0.3''$ to $+0.5''$ for Yale, and $-0.1''$ to $+0.5''$ for the GC. In both catalogues there were systematic trends with declination, more prominently for Yale, perhaps because the various zones had been computed at different times. Otherwise the corrections to the Yale declinations showed only minor possible trends with right ascension. For the GC, however, there was again a clear trend with right ascension, maximum corrections falling between about 3 and 10 hours, minima between 13 and 22 hours. One purpose of this type of investigation was the more precise determination of equinox and equator corrections. While estimates were given, more definitive results needed to await future observations.

In addition to his work on the Moon and asteroids, E. W. Brown spent considerable effort on the determination of the orbit of the eighth satellite of Jupiter, discovered by Melotte in 1908. Brown's first paper on this object, in 1923, was on the application of Delaunay's lunar theory.[32] The expressions for obtaining the satellite coordinates were sums of harmonic terms, whose coefficients for the Moon were uncertain to about one second of arc; for the eighth satellite of Jupiter, however, the uncertainty amounted to a whole degree. Three years later Brown commented,[33] "There seems to be no reason why the orbit should not be computed with an accuracy sufficient to give the Jovi-centric longitude within two or three minutes of arc." The satellite had been observed at all oppositions from 1908 until 1917, then lost, seen again in 1922 and 1923, and lost again until an orbit and ephemeris by Numerov and Boeva resulted in its rediscovery at Mt. Wilson in 1930. The residuals, observed minus computed positions, Brown quoted as being less than 7^m in R.A. and 45′ in Dec., discordances then considered quite satisfactory in view of the computational difficulties arising from a large eccentricity and high orbital inclination.[34]

In 1930 Brown published the first of two major contributions on the *Theory of the Eighth Satellite of Jupiter*.[35] Because of the large eccentricity, high inclination, and large ratio of its mean motion to that of Jupiter, Brown declared, "The problem of obtaining a general orbit . . . is one of the most difficult in the solar system." He gave the general theory and its application to the calculation of an intermediate orbit which included a considerable number of perturbations. From this it was expected that remaining perturbations could be found. Seven years later the second part of the theory was published jointly with Dr. Brouwer.[36] They concluded, "the theory cannot be expected to represent the motion better than 1° or 2° in Jovicentric longitude. For purposes of prediction this would, however, be amply sufficient and we should feel that our work has been fully worth while if this degree of accuracy is attained."

This paper did not include any comparison between theory and observation, which Brown left to his younger colleague, Brouwer, who had already started the necessary computations. Brown confessed, "the advancing age and other occupations of the senior author will prevent him from giving any effective assistance in the work . . . [He] can only hope that a real start has been made towards the solution of the most difficult problem in celestial mechanics which he has encountered."[37]

Brouwer's computations were evidently never completed. In July 1938 he published an ephemeris computed by Hertz,[38] which was an extension of that given by Boeva[39] in 1933. The Mt. Wilson astronomers then found that the predictions for July 27–28 differed from their observed values by 2.0^m in R.A. and 23′ in Dec.[40] No further comparisons between the observed and computed positions on the basis of Brown's elaborate theory appear to be extant. However, many years later a graduate student at the University of Paris, J. Kovalevsky, came to Yale to prepare his doctorate dissertation under the guidance of Brouwer. At a Celestial Mechanics Conference in New York in 1958, Kovalevsky summarized previous work on Jupiter's eighth satellite and discussed what problems still remained and his proposed procedures for solving them.[41] He would "apply the method of harmonic analysis in deriving a general theory from the results of a numerical integration." His Paris dissertation,[42] published in 1959, gave details of his elaborate new theory and indicated that his provisional orbit agrees with the observations to an accuracy of three minutes of arc.

Moving outward in the solar system, Brown became concerned with the problem of the predictions of the discovery of Pluto. In a very brief note in *Nature*, summarizing his current researches,[43] he expressed his doubts about the object Tombaugh had discovered: "During the past year a study of the problem of the prediction of an unknown planet by means of its effects on other planets has resulted in the discovery of a criterion which shows quite clearly why the existence of Neptune could be predicted and that of Pluto could not." Brown was convinced that the object Tombaugh had discovered in February 1930, on the basis of Percival Lowell's predictions,[44] was a pure coincidence, not the expected planet. Soon after Lowell's memoir appeared, in 1915, Brown had already considered the predictions of little value[45] because the observational data upon which they were based "seemed to be quite insufficient for the purpose in view."

In a paper presented to the National Academy of Sciences in April, 1930, Brown[46] argued, "It has always been difficult to understand why predictions of an exterior planet by Lowell and his predecessors, Galliot, Lau and W. H. Pickering, were possible from the very small residuals which the longitude of Uranus exhibits. The definiteness of these predictions appeared to be quite outside the possibilities of the material under discussion and yet it was not easy to point out any fundamental error in the arguments."

One might wonder why a majority of the arguments centered on the perturbations of the orbit of Uranus when the new planet was actually described as trans-Neptunian. There were two reasons. Uranus (discovered in 1781) had been observed over several revolutions whereas Neptune (discovered in 1846) over only half a cycle. Computations on the effects of perturbations of Neptune's orbit by Pluto revealed that a slight permissible change in the orbital elements of Neptune could equally well account for the supposed perturbations. Brown concluded,[47]

> The information concerning the newly found planet available at this moment is scanty but it appears to be sufficient to prove that it could not have been predicted from its effect on Uranus. On the most favorable assumption as to albedo (0.06) and density (5.6), the mass cannot be much greater than that of the earth or one-seventh of that predicted by Lowell. At the distance of 40 to 43 units the perturbations of such a planet on Uranus are not within the range of detection from the material available. Among the numerous solutions given by him, only those with an eccentricity of at least 0.5 appear to be possible and these will leave the unweighted sum of the squares of the residuals greater than 22. Such a solution, however, puts the planet more than 30° away from the place where it was found. This last statement refers to the perturbations as actually computed; as a matter of fact, the calculations are quite inadequate with such a high eccentricity. We must therefore regard the fact that it was found near the predicted place as purely accidental.

A note added in print on May 5, 1930, appeared to confirm his conclusions:[48] "The orbit published by Lowell Observatory for the newly discovered planet shows definitely that it cannot have any connection with that predicted." Later he provided a formula for the determination of unknown periodicities from a continuous set of observations:[49]

> The transformation is used to eliminate the unknown changes in the eccentricity and longitude of perihelion of the disturbed planet caused by the insertion of a new perturbation. The transformed observations will, in certain cases, show whether a given set of residuals can be accounted for by the existence of an unknown planet or not.

With this as criterion, the validity of the prediction of Neptune was verified, that of Pluto was not. He found that any mass less than that of the Earth would satisfy the observations of Pluto equally well, and that another century of observations of Neptune would be necessary to obtain a value of the mass of Pluto to an accuracy within a quarter of the Earth's mass.

On a lighter side of the controversy about the nature of Pluto, correspondence among H. Knox-Shaw at Oxford, H. Shapley at Harvard and F. Schlesinger at Yale indicates the diversity of opinions. While the controversy raged that Pluto was not Lowell's "Planet X," there was agreement that Pluto would turn out to be a major planet. Knox-Shaw wrote,[50]

> Jackson [J. Jackson of the Royal Observatory] expressed doubt as to its being a Trans-Neptunian planet, but I stated my opinion emphatically that in the absence of any data from which we could determine its distance we should trust the men on the spot. In fact in the reporter's presence I bet Jackson a shilling that it would turn out to be a major planet, and Jackson took the bet.

Schlesinger then wrote Knox-Shaw,[51]

> I share your views very closely. We are all agreed here that the discovery has no relation to Lowell's prediction. Just as you have bet Jackson a shilling I have bet a colleague a chocolate cake (to be consumed at our weekly colloquium) that the object is trans-Neptunian, and since this bet was laid, the evidence seems to be accumulating on my side and yours.

Although Schlesinger got his cake, the controversies over Brown's contention did not cease.

After the first orbits for Pluto were published by Bower and Whipple in April,[52] and by Crommelin in June, 1930,[53] H. N. Russell published two articles describing the orbits,[54,55] in which he expressed high praise for Lowell's work: "The orbit, now that we know it, is found to be so similar to that which Lowell predicted from his calculations fifteen years ago that it is quite incredible that the agreement can be due to accident." Table 16.1 gives the orbital elements predicted by Lowell, those quoted by Russell in 1930, and the currently accepted values.

Moreover, Russell found that "In only one particular was he [Lowell] seriously wrong—he overesti-

Table 16.1
Orbital Elements of Pluto

	Lowell Predicted	Russell 1930	Current 1990
Orbital period (years)	282	249.17	247.69
Eccentricity	0.202	0.254	0.249
Longitude of perihelion (°)	205	212.5	223.7
Perihelion date	1991.2	1989.2	1989.8
Inclination (°)	about 10	17.15	17.15

mated the apparent brightness of the planet more than tenfold." Russell inferred the mass of Pluto to be either 1.08 or 0.72 times the mass of the Earth (depending upon whether or not the earliest uncertain observation of Neptune by Lalande in 1795 was included in the solution, and concluded, "In any case it is clear that the new planet has a considerable mass and is capable of producing easily perceptible influences on the others by its attraction."

Later, however, in a letter to A. Lawrence Lowell, President of Harvard University, brother and biographer of Percival Lowell, Russell finally agreed with Brown, whom he described as the most distinguished living student of the subject,[56] "That so close a set of chance coincidences should occur is almost incredible, but the evidence assembled by Brown permits no other conclusion. Other equally remarkable coincidences have occurred in scientific experience." As an example he mentioned the announcement of the discovery of a new comet[57] for which the position had been garbled in overseas transmission. A search for the comet in the erroneous position then revealed another new comet close to that position! (The two comets were 1895 IV and 1896 I.) In regard to Percival Lowell, Russell appropriately added,

> In any event, the initial credit for the discovery of Pluto justly belongs to Percival Lowell. His analytical methods were sound; his profound enthusiasm stimulated the search, and, even after his death, was the inspiration of the campaign which resulted in its discovery at the Observatory which he had founded.

A. O. Leuschner, under whose direction his students Bower and Whipple had computed several orbits based on various assumptions, in June 1932 addressed the American Association for the Advancement of Science on "The Astronomical Romance of Pluto"[58] There he reviewed the *pros* and *cons* regarding the complex problems regarding Pluto *vs* Lowell's Planet X. His conclusions were similar to Russell's but, he cautioned, "The wisest attitude at the present time is to have an entirely open mind on the mass and consequent gravitational effects of *Pluto*."

But this did not end the controversy. In 1941 V. Kourganoff[59] refuted nearly all of Brown's objections and concluded (as translated by Gibson Reaves[60]), "The thesis of 'pure chance' in regard to the discovery of Pluto is absolutely untenable. Pluto was 'discovered' in 1915 by Lowell and re-discovered in 1919 by Pickering by the methods of celestial mechanics, before its 'physical' discovery by Tombaugh at the Lowell Observatory."

One might have expected comments by Brouwer on Kourganoff's denunciation of Brown's work. Brouwer's only citation of Kourganoff, however, appears in his *Darwin Lecture*[61] where Brouwer objects only to Kourganoff's rejection of Flamsteed's early 1715 prediscovery observation of Uranus in the determination of the mass of Pluto but does not mention the discovery *vs* prediction problems. (While Lowell's prediction of the mass of Pluto was the only parameter that differed greatly from current determinations, this had not affected his determination of the orbital parameters.) But regardless of the criticisms of Brown's conclusions as to Pluto's identification with Lowell's Planet X, Brown's methodology and arguments serve as important guides and precautions should any future orbital perturbations suggest a still more distant planet.

The question of the mass of Pluto continued to disturb astronomers until the discovery at the U.S. Naval Observatory of its companion, Charon, by James Christy in 1978.[62] Until then various attempts had been made to estimate the mass on the basis of Pluto's apparent brightness, assumptions as to its constitution, and attempts to measure its apparent diameter, as well as the unaccounted-for observed orbital perturbations of Uranus and Neptune. R. L. Duncombe (Brouwer's former student) and P. K. Seidelmann[63] at the Naval Observatory in 1980 tabulated the various mass determinations, showing an almost steady decrease in the estimates from the earliest prediscovery analyses of the perturbations of the orbits of Uranus and Neptune until the discovery of Charon. Their compilations are summarized in Table 16.2.

The discovery of Charon thus finally proved that the mass of the Pluto—Charon system was far too small to have noticeably perturbed the orbits of Ur-

Table 16.2
Determinations of the Mass of Pluto*

Date	Mass/Earth
Predicted Before Lowell:	
1848	12
1899–1909	2–9
By Lowell 1915	6.6
After Discovery of Pluto:	
1931–1955	0.5–1.0
1968	0.18
1971	0.08–0.11
1976	0.004
After Discovery of Charon:	
1978	0.002

* The 1968–71 estimates depended largely on work by Brouwer's former students, Duncombe and W. J. Klepzinsky and their collaborators at the Naval Observatory.

anus or Neptune. Hoyt, in his fine account of Planet X and Pluto, gives more details and concludes,[64]

> The discovery of Charon, and the consequent determination of Pluto's very small mass, made it certain that Pluto could not be Percival Lowell's predicted Planet X, and thus the long controversy over this question ended.
>
> It also makes certain that Pluto cannot have caused the perturbations in the motion of Uranus that Lowell, [W. H.] Pickering, Kourganov and others found over the years, and this circumstance may prove to have even more important implications for the future of planetary discovery.

Duncombe and Seidelmann similarly conclude that while Percival Lowell must be given the credit for initiating and providing the impetus in the search for his Planet X, its discovery is entirely due to the painstaking search by Tombaugh.[63] Pluto cannot account for the anomalous residuals observed in the motions of Uranus and Neptune. Consequently in recent years there have been attempts to find another "Planet X."[65] Among the investigators there have been others of Brouwer's former students: T. C. van Flandern (collaborating with R. Harrington at the Naval Observatory) and E. M. Standish at J. P. L. The latter concluded that there is no way to rule out the existence of a tenth planet, but that the orbital data for Uranus and Neptune do not need a tenth planet to account for their motions.[66] This conclusion makes Lowell's predictions and their coincidence with modern orbital data even more puzzling by effectively denying his very basis for postulating a ninth planet. Standish's computations, however, did not include the early observations of Uranus and Neptune, as he found that they do not have the requisite accuracy for precise predictions. It would seem therefore that the crucial early predictions may have been overinfluenced by the unknown errors of positional measurements. Brown's objections now appeared vindicated.

Among the ten or so papers that Brown published in the final two years of his life was a series of four in the *Monthly Notices* of the Royal Astronomical Society in 1936–37 on the stellar problem of three bodies.[67] His purpose was "to take up the stellar problem of three bodies, chiefly from the point of view of the observer, by limiting the solution to those terms which may be capable of being detected by observation at the present time and in the immediate future."

In the satellite theory of Delaunay[68] the results were useful only for the cases where eccentricities and inclinations were small. Brown extended the theory to include the higher values but still neglected some periodic terms; and in his third paper dealing with the motions of the apse and nodes he hit upon a bonus for his lunar theory: these final equations unexpectedly accounted for an outstanding discrepancy between the observed and theoretical motions of the lunar perigee, amounting to 17″ per century. In his final paper of the series Brown applied his theory of the variation of the orbital elements to the system of ζ Ursae Majoris, a quadruple system consisting of two close pairs revolving about their common center of gravity. One of the pairs is sufficiently close so that the pair could be treated as a single star, thus qualifying ζ UMa as an example of a three-body system. The period of the very close pair was 3.98 days, the other 1.832 years, and the period of revolution of the two pairs around each other 59.863 years. The eccentricities of both the latter orbits are large, 0.513 and 0.413, and the relative inclination of the orbits 37.8°. The apse—node period was about 2,000 years.

Brown's earlier interest in the stellar three-body problem is exemplified by his supervision of a 1928 Ph.D. thesis by Paul Slavenas, a graduate student from Lithuania. The thesis *The Stellar Case of the Problem of Three Bodies,*[69] gave an extensive mathematical treatment of a triple system consisting of a close binary accompanied by a distant third body. Slavenas then applied his theory to λ Tauri,[70] an eclipsing binary with a third component discovered spectroscopically by Schlesinger at Allegheny in 1914.[71] Slavenas found a primary period of 3.4529 days and the period of the third body of 34.60 days and masses of 13.8, 2.80, and 0.9 solar, and a dynamic parallax of ″.0042, in contrast to more recent determinations of 3.9540^d (SB) or 3.952955^d (EA), masses 6.8, 1.8, and 1, and a trigonometric parallax of +″.002, respectively. Slavenas pointed out that his theory could be applied to any eclipsing binary for which the spectrum of only one component is observable. In the *Masses of the Stars*[72] by Russell and Moore, λ Tauri is not included, nor is Slavenas cited.

Brown's last paper, for which he corrected proofs only a few days before his death, reverted to his favorite topic, *The Equations of the Motion of the Moon.*[73] He returned to a contemplation of the works of Hill and Poincaré, the forerunners of his own investigations of the three-body problem, and wrote,

> It is not wise, I believe, to imagine that the work of a great master in his subject cannot be improved, especially when extended calculations are involved. . . . My own experience in several cases leads me to think that it is worth while to return to the foundations even when they go so far back in time as those of Celestial Mechanics, and to examine whether valuable changes cannot be made.

A review of the procedures leading to his own final results, indicated his thinking: "For a satellite with a large eccentricity like Jupiter VIII which seems to require the ellipse as a first approximation, I hope to give elsewhere a method which appears to have considerable advantages over that which Professor Brouwer and I have used for this body and for which I was responsible." He presented revised equations and discussed the loss of accuracy due to small divisors. With these equations he found that the theory is correct to within ″.001 in longitude, "considerably better than I had hoped." In the earlier work there appeared to be no errors in longitude greater than ″.01, but he felt that further work was still necessary. Brouwer and his collaborators eventually carried on (see Chap. 21).

REFERENCES

1. E. W. Brown and C. A. Shook, *Planetary Theory,* viii + 302 pp., Cambridge Univ. Press, London, 1933.
2. E. W. Brown, *Science,* **33**, 79, 1911.
3. ———, *Idem,* p. 80.
4. ———, *Idem,* p. 82.
5. ———, *Idem,* p. 87.
6. ———, *Proc. Nat. Acad. Sci.,* **10**, 248, 1924.
7. D. Brouwer, *Ast. Jour.,* **68**, 152, 1963.
8. E. W. Brown, *Monthly Notices Roy. Ast. Soc.,* **71**, 438, 492, 1911.
9. D. Brouwer, *Encyclopaedia Britannica,* **22**, 490, 1958.
10. E. W. Brown, *Ast. Jour.,* **27**, 25, 1911.
11. ———, *Monthly Notices Roy. Ast. Soc.,* **72**, 618, 1912.
12. G. H. Darwin, *Acta Mathematica,* **21**, 99, 1897; reprinted in his *Scientific Papers,* **4**, 1, 1911.
13. S. S. Hough, *Acta Mathematica,* **24**, 257, 1901; reprinted in G. H. Darwin, *Scientific Papers,* **4**, 114, 1911.
14. G. H. Darwin, *Monthly Notices Roy. Ast. Soc.,* **70**, 108, 1909; *Scientific Papers,* **4**, 140, 1911.
15. G. H. Darwin, *Monthly Notices Roy. Ast. Soc.,* **72**, 642, 1912; *Scientific Papers,* **5**, 59, 1916.
16. E. W. Brown, *Trans. Yale Univ. Obs.,* **3**, 1, 1923; **3**, 81,1925.
17. ———, *Ast. Jour.,* **35**, 69, 1923.
18. D. Brouwer, *Trans. Yale Univ. Obs.,* **6**, 173, 1933.
19. ———, *Idem,* p. 184.
20. W. J. Eckert, *Idem,* p. 161, 1933.
21. A. O. Leuschner to F. Schlesinger, Jan. 22, 1936, *Yale Obs, Archives.*
22. F. Schlesinger to A. O. Leuschner, Jan. 29, 1936, *Yale Obs. Archives.*
23. D. Brouwer to D. Hoffleit, Private oral communication.
24. D. Brouwer to A. O. Leuschner. Oct. 10, 1941. *Yale Obs. Archives.*
25. H. Hertz, *Ast. Jour.,* **52**, 43, 1946; **54**, 128, 1949.
26. B. Brouwer to P. Herget, May 10, 1940, *Yale Obs. Archives.*
27. F. Dyson, *Trans. International Ast. Union,* **3**, 227, 1928.
28. D. Brouwer, *Ast. Jour.,* **44**, 57, 1935.
29. ———, *Ann. N.Y. Acad. Sci.,* **42**, 133, 1941.
30. D. A. Pierce, *Ast. Jour.,* **76**, 177, 1971.
31. ———, *Ast. Papers Amer. Ephemeris and Nautical Almanac,* **22**, Part 3, p. 205, 1978.
32. E. W. Brown, *Ast. Jour.,* **35**, 1, 1923.
33. ———, *Pop. Ast.,* **34**, 550, 1926.
34. ———, *Pub. Ast. Soc. Pacific,* **44**, 21, 1932.
35. ———, *Trans. Yale Univ. Obs.,* **6**, 45, 1930.
36. E. W. Brown and D. Brouwer, *Ibid.,* p. 189, 1937.
37. ———, *Idem,* p. 192.
38. H. Shapley, *Harvard College Observatory Announcement Card* No. 451, 1938.
39. N. Boeva, *Bull. Ast. Inst. Leningrad,* No. 32, 1933.
40. H. Shapley, *Harvard College Observatory Announcement Card* No. 453, Aug.1, 1938.
41. J. Kovalevsky, *Astr. Jour.,* **63**, 452, 1958.
42. ———, *Bull. Astronomique,* **23**, 1, 1959.
43. E. W. Brown, *Nature,* **128**, 216, 1931.
44. P. Lowell, *Memoirs Lowell Obs.,* **1**, 1, 1915.
45. Ref. 34, p. 24.
46. E. W. Brown, *Proc. Nat. Acad. Sci.,* **16**, 364, 1930.
47. ———, *Idem,* p. 370.
48. ———, *Idem,* p. 371.
49. ———, *Monthly Notices Roy. Ast. Soc.,* **92**, 80, 1931.

50. H. Knox-Shaw to H. Shapley, Mar. 25, 1930, *Yale Obs. Archives.*
51. F. Schlesinger to H. Knox-Shaw, Apr. 12, 1930, *Yale Obs. Archives.*
52. E. C. Bower and F. L. Whipple, *Lick Obs. Bull.* No. 421, April 1930.
53. A. C. D. Crommelin, *Jour. Brit. Ast. Assoc.*, **40**, 265, June 1930.
54. H. N. Russell, *Sci. American,* **143**, 446, 487, 1930; *Jour. Roy. Ast. Soc. Canada,* **25**, 27,1931.
55. ———, *Sci. American,* **144**, 90, 1931; *Jour. Roy. Ast. Soc. Canada,* **25**, 64, 1931.
56. H. N. Russell, *in* A. L. Lowell's, *Biography of Percival Lowell,* N. Y., Appendix I, p. 203, 1935.
57. Anon., *Ast. Jour.,* **16**, 56, 1896.
58. A. O. Leuschner, *Pub. Ast. Soc. Pacific,* **44**, 197, 1932.
59. V. Kourganoff, *Bull. Astronomique,* (2), **12**, 147, 271, 303, 1941.
60. G. Reaves, *Pub. Ast. Soc. Pacific,* **63**, 49, 1951.
61. D. Brouwer, *Monthly Notices Roy. Ast. Soc.,* **115**, 221, 1955; *Observatory,* **75**, 96, 1955.
62. J. Christy and R. S. Harrington, *Ast. Jour.,* **83**, 1005, 1978.
63. R. L. Duncombe and P. K. Seidelmann, *Icarus,* **44**, 12, 1980.
64. W. G. Hoyt, *Planet X and Pluto,* Tucson, Ariz., Chap. 11, p. 247, 1980.
65. J. D. Anderson and E. M. Standish, Jr., "Dynamical Evidence for Planet X" in *The Galaxy and the Solar System,* R. Smoluchowski, J. B. Bahcall, and M. S. Matthews, Eds., Tucson, Ariz., p. 286, 1986.
66. M. Littmann, *Sky and Telescope,* **78**, 596, 1989.
67. E. W. Brown, *Monthly Notices Roy. Ast. Soc.,* **97**, 56, 62, 116, and 388, 1937.
68. C. E. Delaunay, *Comptes Rendus, Acad. Sci.,* Paris, **74**, 17, 1872.
69. P. Slavenas, *Trans. Yale Univ. Obs.,* **6**, 35, 1927.
70. ———, *Proc. Nat. Acad. Sci.,* **13**, 833, 1927.
71. F. Schlesinger, *Pub. Allegheny Obs.,* **3**, 167, 1914.
72. H. N. Russell and C. E. Moore, *The Masses of the Stars,* ix + 236 pp., Chicago, 1940.
73. E. W. Brown, *Amer. Jour. Math.,* **60**, 785, 1938.

17
SCHLESINGER, WIZARD OF PARALLAXES

When Frank Schlesinger[1] (1871–1943) came to Yale in 1920 as the Director of the Observatory, his already distinguished career had prepared him well for his new assignment. Here he was soon to be heralded as "the father of modern astrometry."

Early in 1898, still a graduate student at Columbia, he had published a short but significant paper on astronomical refraction.[2] Its chief purpose was to show how the prevailing computations relating to the reduction of star photographs could be shortened. His Ph.D. dissertation[3] dealt with the reduction of the famous Rutherfurd plates of Praesepe taken in 1870 and 1877. For 45 stars Schlesinger determined the positions and proper motions with an accuracy ranging from $\pm''.010$ to $\pm''.051$ p.e. in either coordinate and $\pm''.003$ to $\pm''.014$ in the proper motion components—excellent results for that early epoch.

In the 1898–99 season at Columbia he was fortunate in attending (along with only two other graduate students) a series of lectures by G. W. Hill, the most renowned scholar of celestial mechanics at the time. Later, in 1936, Schlesinger was to recall this experience when he placed a bronze tablet commemorating Hill at Hill's homestead in West Nyack, New York.[4] After receiving his Ph.D. from Columbia in 1899, Schlesinger was called to Ukiah, California, to establish an *International Latitude Observatory,* one of five at approximately the same latitude (39°N) around the world, to determine changes in latitude. He may have indicated some aptitude for this type of work when, before entering upon his graduate studies, he had been a surveyor for the Department of Street Improvements in New York City. Although he was not actively engaged in this type of work after he left Ukiah in 1903, his continued interest in variation of latitude was manifested years later at Yale in his alertness to and his reviews of published papers, his collaboration with W. D. Lambert and E. W. Brown on an important survey paper,[5] his rectification of the history of the discovery of variations in latitude,[6] and his discussion[7,8] of observed changes in latitude or longitude and their interpretation. He stressed the roles of accurate stellar proper motions and seismic activities in the interpretation of the apparent changes. Moreover, at Yale he designed and had built an experimental semiautomatic photographic zenith telescope of 10-inch aperture to encourage further international work on variation of latitude. Young Jan Oort, a research assistant at Yale from 1922 to 1923, and Carl Stearns, a graduate student, took part in the testing and observations,[9] securing 144 plates on 113 nights.

During spare moments at Ukiah, Schlesinger's thoughts were occupied with improvements in the determination of stellar parallaxes.[10] He pointed out that in 1890 there were only 25 or 30 stars whose parallaxes were known to an accuracy of $\pm''.05$, a very small number in view of the fact that two fine methods were already available for better determinations: photography and the heliometer (improved since the time of Bessel's first parallax in 1838). He deemed photography to be not less accurate than the heliometer and with the additional advantage of economy of observing time.

Then in 1903 he was called to the Yerkes Observatory where Director G. E. Hale had for some years had in mind the use of the 40-inch telescope for parallax work.[11] Already Schlesinger was concerned with the magnitude-errors that would affect the results when a bright parallax star was compared with much fainter comparison stars. At first, like Russell and Hinks[12] in their early work in Cambridge, England, in 1903–07, Schlesinger tried desensitizing a small spot in the middle of the photographic plate, where the image of the parallax star would occur, so as to reduce its brightness. But this gave an uncertain diminution that could not readily be controlled for stars of different relative magnitudes. He therefore devised the rotating sector that is still being used at nearly all institutions where stellar paral-

Fig. 17.1. F. Schlesinger (1871–1943). (Photograph by Harris and Ewing, Washington, D. C.)

laxes of bright stars are measured. By adjusting the sector opening to various degrees, the brightness of the parallax-star image could be reduced by any desired amount. It is remarkable that the procedures later almost universally adopted for stellar parallax observations were formalized by Schlesinger so near the beginning of his career. Moreover, probably at about the same time, although he did not publish the details until 1911, Schlesinger[13,14] devised the computational procedure called *the method of dependences,* subsequently used, before the advent of electronic computers, at nearly all observatories where parallaxes were determined. The method was likewise applied to the determination of positions of asteroids and other transitory objects.[15] At a time when computations were being carried out with slide rules, logarithm tables, and desk calculators, it was important to reduce the amount of computing to a minimum consistent with the required accuracy of the results. Schlesinger started with the usual formulae for plate reductions involving the determination of "plate constants" before the actual coordinates of the parallax star could be determined. In the dependence method these constants, though implicit in the solutions, are not actually computed; one solved the equations only for the unknowns actually wanted.

Director George Ellery Hale was about to leave the Yerkes Observatory in 1904 for his newly founded Mt. Wilson Observatory. Hale was anxious for Schlesinger's important parallax investigations with the 40-inch at Yerkes to continue, but funds for this were somewhat in question. Hence (as a founder of both institutions) Hale suggested that Schlesinger join the staff of the Mt. Wilson Observatory but remain at Yerkes. However, in 1905 Schlesinger became Director of the Allegheny Observatory of the University of Pittsburgh where he remained until he came to Yale in 1920. At all three of these observatories it was Schlesinger who initiated the work on parallaxes, and all three have remained among those preeminent in this field, even to the present. While he was still at Yerkes, Schlesinger actually published the parallaxes for only three stars.[16] However, between 1909 and 1911 he published some 55 more, based on his Yerkes observations.[17,18]

At the Allegheny Observatory the Thaw refractor, with which Schlesinger was to do most of his astrometric work, had not yet been built. Although Brashear had already done much toward raising funds for the "new observatory" which was to acquire both a 31-inch reflector and a 30-inch long-focus refractor, the refractor, though dedicated in 1912,[19] was not operational until September 1914.[20,21] On the other hand, the new telescope, designated the "Keeler Memorial Reflector," was ready in 1906. It was put into use mainly for the spectroscopic determination of the orbits of spectroscopic and eclipsing binaries, changes in the apparent velocity in the line of sight being ascertained from the Doppler principle. In the first volume of *Publications of the Allegheny Observatory* all but five of the 23 papers published between 1906 and 1910 dealt with spectroscopic studies related to binaries by either Schlesinger or his staff (R. H. Baker, R. H. Curtiss, and F. C. Jordan), and the succeeding two volumes contained 36 further papers dealing with spectroscopic observations. Among them they discovered five new spectroscopic binaries,[22] namely the 5th magnitude stars HR 4072, 5313, 5392, 25 Ser, and ι Cr B.

On the basis of 93 spectrograms obtained between September 1906, and March 1907, Schlesinger and R. H. Curtiss[23] in 1909 derived an orbit for Algol. They stated that further observations would soon be published, but many years actually intervened. Much later (1932), in a letter to J. H. Moore at Lick Observatory, Schlesinger wrote from Yale,[24]

> I have, by the way, some three hundred spectrograms of Algol whose results remain to be published; I have a paper which I should publish very soon that shows the two year period in the photometric date [*sic*], the eclipses occurring some five minutes earlier or later than normal depending upon where the eclipsing pair is in their orbit around the center of gravity of all three stars.

The spectrograms to which he refers were presumably all taken at the Allegheny Observatory and are in addition to the 93 mentioned above. Beardsley[25] in 1969 published radial velocities based on 244 Allegheny plates taken between November 1907 and March 1912. There seems to be no record that a new orbit based on these was ever published. However, Schlesinger noticed that spectrograms taken at minimum light phase showed many lines not normally visible. After coming to Yale he turned some 250 of the spectrograms over to his assistant, Ida Barney, who identified 37 lines, mostly iron. From these she determined radial velocities which appeared in phase with the velocities of the primary eclipsing pair but with smaller amplitude. She inferred the spectrum of the third component to be later than B8, probably F or G.[26]

One important paper[27,28] with the modest title, *The Algol-Variable δ Librae,* contains the first observational discovery of stellar rotation both from the widening of lines in the spectrum outside of eclipse, in this case in synchronization with the period of revolution, but also from the larger residuals during the partial phases, when the rotational velocity from the limb of the bright partially eclipsed star is added to (or subtracted from) its orbital velocity. The same effect was observed for λ Tauri.[29] Also in this era Schlesinger considered the possibilities of determining radial velocities from objective prism plates, a subject in which Pickering at Harvard had pioneered without obtaining practical results.[30] Schlesinger discussed all the methods that had been proposed, pointing out the advantages and drawbacks of each.[31] As the length of the spectrum changes with radial velocity, he favored designing special equipment and using new emulsions that could yield spectra of nearly uniform density between the K-line at 3933 Ångstroms and the C-line at 6563. If Schlesinger could have devoted all his energies to stellar spectroscopy, that science would surely have advanced appreciably. But after the Thaw photographic refractor was completed in 1914, Schlesinger himself did comparatively little more work with the spectrograph, and after 1916 he devoted nearly all his energies to problems of astrometry.

As the new Allegheny Observatory was first being planned in 1897, it was meant to have a 30-inch visual refractor. However, there were long delays in obtaining the necessary glass disks. Fortunately, as it turned out, the telescope was not yet constructed when Schlesinger arrived; for on the basis of his experiences with the Yerkes 40-inch visual refractor, Schlesinger gave compelling arguments in favor of a photographic instead of a visual telescope. When the photographic telescope was finally ready, he gave a glowing account of its performance in comparison with visual refractors.[21] He began taking parallax plates only a few days after the objective had been installed, and achieved some 200 parallax plates per month. He published approximately 280 parallaxes before he left Allegheny in 1920; but in all nearly 500 Allegheny parallaxes were determined by Schlesinger or under his direction, many of them published after his coming to Yale.[32,33] Moreover, he had skillfully designed auxiliary equipment, including a plate-measuring engine, ventilating equipment for the telescope lens to minimize temperature variations, and improved guiding mechanisms.

With the end of World War I it had become time to select a new Director at Yale. The first choice of the Corporation seems to have been Henry Norris Russell of Princeton,[34] who was also being considered for Harvard, as was Schlesinger. After Russell declined the position at Yale and Schlesinger was selected, Russell wrote Schlesinger a hearty letter of congratulation:[35]

> You are just the man to take those large opportunities and make the most of them, and Hadley [President of Yale] told me . . . that you had plans for work in the Southern Hemisphere. I emphasized strongly to him the great importance of giving you the means of carrying them out. . . .
>
> My only regret is that you are not to go to Harvard . . . What a game of 'General Post' there will be among the observatories! Who [*sic.*] will Allegheny get to fill your place? . . .
>
> I wonder a little that you are willing to leave the splendid equipment at Allegheny . . . I hope that you will still have a chance at some spectrographic work, somehow.

The equipment at first available to Schlesinger at Yale was nowhere near equal to what he was leaving at Allegheny. However, while Schlesinger was being considered for the directorship, he had already stressed the need for a southern station. Meanwhile, Yale authorities had been contemplating the construction of a large new northern telescope, and were surprised that Schlesinger vetoed their plan.[36] In a letter to E. W. Brown,[37] Schlesinger wrote,

> It would be unwise, in my opinion, to aim to erect in or near New Haven a telescope of larger size than the new telescope just erected on campus [the 15-inch Loomis coelostat]. It should be my first endeavor to obtain funds that would enable the observatory to establish a Yale Observatory station in the southern hemisphere. The first object of such a station should be to obtain photographs for the determination of stellar parallaxes of southern objects, a field that is practically untouched.

Further, in regard to parallaxes of the northern stars, he added, "A great deal of attention is being devoted to this problem by six of the largest observatories in this country as well as by one or two abroad, while other problems of equal importance are suffering from comparative neglect."

Whole-sky coverage of parallaxes was important for investigations of galactic structure and galactic rotation. Moreover, the 15-inch coelostat, designed by Hastings and named as a memorial to Elias Loomis in 1921, did not prove satisfactory for the parallax work for which it had ostensibly been designed.[38] As early as 1912 Hastings had consulted Schlesinger at Allegheny about his plans for the coelostat and then wrote F. L. Chase[39]: "Dr. Schlesinger is quite enthusiastic about it and is quite willing to endorse my recommendations of the plan to the University. He estimates that, if the climate is as favorable to astronomical work here [in New Haven] as at Williams Bay, you may expect to determine 50 parallaxes per year." Yet after coming to Yale in 1920 Schlesinger described it as "unfortunately an instrument-maker's telescope instead of an astronomer's telescope."[36] Its disadvantage, which applies as well to all other telescopes in which oblique reflections are used, is the presence of astigmatism when the temperature is changing rapidly.[40] After some modifications the telescope was used instead for photometric work, especially on variable stars, first by Carl Stearns (between 1920 and 1925), later by John Hall and Arthur Bennett (1930 to 1940), for infrared photometry.[41]

Schlesinger's plans for the southern station were approved by the Corporation and even before his appointment was effective (April 1, 1920) he canvassed opinions on site suitability in New Zealand, starting with correspondence with C. E. Adams of the Hector Observatory at Wellington, with whom he had previously discussed the possibility of cooperation on zone-catalogue work for the determination of star positions and proper motions.[42] Then between 1920 and 1923 he considered other sites in New Zealand and elsewhere: Calamo, Chile (where the Smithsonian Astrophysical Observatory had a station), and South Africa, particularly at Bloemfontein (future site of the Harvard southern station), Kimberley, and both the Union Observatory and the University of Witwatersrand at Johannesburg. The final selection would depend on seeing conditions, percentage cloudiness, and living facilities. For site testing at Auckland, N. Z., Schlesinger, through the kindness of W. K. Green, had borrowed an 8-inch refractor from Amherst College. The tests proved very satisfactory and the authorities at Auckland were so optimistic that the Yale telescope would be erected there that other municipalities in New Zealand were distressed that Auckland had stolen the prize from them, as is evident in a newspaper cartoon![43] (Fig. 17.2) A philanthropist in New Zealand had promised a donation of £5,000 if the Yale station were to be erected on the South Island.[44] The final choice, however, was the University of Witwatersrand where the seeing conditions appeared to be exceptionally good and the University offered cooperation in the erection of the telescope, housing facilities, and office space for the Yale observers.

Meanwhile, Schlesinger had put out bids for the construction of the telescope. Brashear of Pittsburgh obtained the contract for the optical parts and was instructed to order a set of lenses 24 or 25 inches in diameter. Actually, 26-inch disks were obtained from Schott and Co. of Jena, Germany. Just two months before his sudden death in March of 1922, J. B. McDowell in Pittsburgh finished grinding the lenses, which he declared the most perfect he had ever produced.[45] Bids were also put out for the mounting, but the prices asked would have made the entire cost of the operation far exceed the total estimated expense of $50,000 for setting up the Southern Station. Hence it was decided that the mounting could be built and tested at the Mason Laboratories of the Department of Engineering at Yale. Also to save money, without jeopardizing the operation of the telescope for its major purpose, standard housing with dome was abandoned in favor of a "meridian building," a unique narrow building with sliding roof oriented north—south like a transit-instrument shed, so that stars could be observed at hour angles of only a few degrees—as they would be in any case for the primary purpose of this telescope. In the thirty years of operation of the telescope at Johannesburg there seems to have been only one occasion when there may have been disappointment about the limitation of its use so close to the meridian. Dr. van Rhijn at Groningen inquired about the use of the telescope for measuring positions to obtain proper motions relative to spiral nebulae. Harold Alden, who had been in charge of the Johannesburg station since July 1, 1925, looked into the matter. To obtain images of faint galaxies sufficiently distant to show stellar-like nuclei, he would have had to expose the plates for some three hours. Only close to the South Pole could he obtain such long exposures.[46] There was no follow-up.

Harold Alden, who had spent over ten years working on stellar parallaxes at the University of Virginia under the direction of S. A. Mitchell, came well prepared to take charge of the Yale Southern Station. Schlesinger[47] outlined the observational program for obtaining the parallaxes of nearly 1,100 bright stars and the proper motions of about 10 faint stars in the neighborhood of some 300 Boss stars[48] of well-determined proper motion. As conditions permitted, parallax plates were also to be taken of stars of large proper motion and some double stars useful for the calibration of dynamic parallaxes. The programs involved taking about 20 plates per star. Alden and one assistant were each to measure and reduce approximately 50 parallax stars a year and send the remaining plates to Yale to be reduced there. Schlesinger envisioned that the basic program would require a minimum of approximately seven years, but added, "If it should be found that the work cannot be completed in seven years it is understood that you will have fulfilled your contract and may if you so desire return to this country after ten years." Alden eventually remained in South Africa for twenty years, returning to the University of Virginia as S. A. Mitchell's successor as Director of the Leander McCormick Observatory in 1945.

By December 1924 Schlesinger was ready to go to South Africa to make the definitive selection of the observatory site and then supervise the erection of the telescope. As a safety precaution he took the lens along as his personal luggage, arriving at Johannesburg on February 8, 1925. Ground was broken for the building ten days later, the construction completed, and the telescope mounted in June (Fig. 17.3 and 4). Alden had already arrived at Johannesburg by the end of May. When the operation was completed Schlesinger reported that the total cost to Yale

Fig. 17.2. New Zealanders fought for Yale Southern Station. Cartoon from a local paper, probably *Auckland Weekly,* Oct. 11, 1923. (From a xerox copy, courtesy Yale Archives)

had been $57,765, less than $8,000 more than the original estimate made six years before.[49]

As the Prince of Wales happened to be in Johannesburg at the time, he was asked to open the Observatory officially on June 22. He unlocked the door and pulled the cord to start the driving mechanism of the telescope, "as a mark of respect for the United States and one of her great institutions."[50]

The first plates that Alden sent to New Haven, Schlesinger found disappointing. Alden complained of apparent astigmatism of the lens and of mechanical vibrations. Fortunately, the problems were either not serious or were easily rectified. However, not until October 1926 did Schlesinger express complete satisfaction with the plates he had just received, saying "I have examined these with much care and pleasure. They contain the first excellent images I have seen."[51] In the first year of its operation, 2,840 plates suitable for parallax determination were secured with the new telescope. In 1945, at the time of Alden's departure, 57,253 plates had been secured. In 1936 parallaxes were published for 851 stars[52] and in 1943 for 472 more.[53]

When Alden was first appointed at Johannesburg, he had been told that after seven years he and his family would be entitled to return to the States on vacation at University expense. Meanwhile the Great Depression of 1929 intervened, and by March 5, 1932, Schlesinger had the unhappy duty of informing Alden that if they did go to the States they would run the risk that their return trip to Johannesburg could not be financed.[54] Moreover, there was talk that the Corporation might reduce all salaries by 10 percent until the Depression was over. Jobs at Yale, as elsewhere, were being curtailed: if anyone left, there would be no replacement. Ultimately Alden did return alone for the summer of 1932 to attend the meetings of the International Astronomical Union in Cambridge, Massachusetts.[55]

Schlesinger was not only concerned with the actual determination of new trigonometric parallaxes; already in 1924 he had compiled, with the assistance

Fig. 17.3. The 26-inch building at Johannesburg, South Africa.

Fig. 17.4. The 26-inch refractor mounted at Johannesburg.

of Margaretta Palmer and Alice Pond, his first *General Catalogue of Stellar Parallaxes.* This contained 1,870 entries, the majority (1,684) being trigonometric parallaxes from all available sources. Also included were 61 labeled "Clus" based on moving cluster data; six designated "OrbV" were dynamic parallaxes computed from well-determined orbits based on radial velocities; and finally, designated "Ceph," were 139 parallaxes for Cepheid variables determined by Shapley from the period-luminosity relation, for 15 of which trigonometric values were also available. Moreover, in a separate column the authors indicated by an *s* or *d* for which of the stars spectroscopic parallaxes were available, or for which dynamic parallaxes were or might become available. The second edition, published in 1935 with the collaboration of Louise Jenkins, contained 7,534 stars with either or both trigonometric and spectroscopic parallaxes, all reduced to absolute values on a homogeneous system. Trigonometric parallaxes were available for about half of these stars and spectroscopic for three-fourths. In this edition, dynamic parallaxes are given in a separate section for nearly 2,450 pairs, a column being added to indicate by *t* or *s* when trigonometric or spectroscopic parallaxes are also included in the main part of the catalogue. The third edition was compiled by Louise Jenkins in 1952, after Schlesinger's death. It contained 5,822 stars, all with trigonometric parallaxes only, spectroscopic and dynamic now being omitted. In 1963 she published a *Supplement* containing updated information on about 600 of these as well as some 600 additional stars. A fourth edition now (1991) in progress by William van Altena contains about 7,700 stars. Fig. 17.5 shows the increase in numbers of parallaxes since the first three reliable values were obtained in 1838. This and Table 17.1 bring out the spectacular increase in productivity in this field when Schlesinger established Yale's Southern Station, and the decline in rate of increase after his retirement.

At the Southern Station, Alden's primary responsibilities concerned the parallaxes and zone catalogues (to be described in the following chapter). In addition he also obtained photographs of satellites of Jupiter, Saturn, and Neptune and of the Moon. In 1928 Dirk Brouwer had written as his Ph.D. dissertation at Leiden a discussion of occultations of the Galilean moons of Jupiter[56] carried out by Innes at the Union Observatory in Johannesburg from 1908 to 1926. De Sitter at Leiden had both preceded and extended this investigation.[57] Alden then carried out further observations in 1927 and 1928 with the Yale 26-inch, yielding more precise longitudes.[58] He also obtained 66 plates of the satellites of Saturn in 1926 and 37 more in 1927, publishing precise coordinates of the five satellites, Titan, Rhea, Dione, Enceladus, and Japetus relative to the satellite Tethys.[59,60] Of this work Schlesinger wrote,[61] "I have read with a good deal of pleasure your paper on the satellites of Saturn. I suppose this must be the most important single contribution to our knowledge of the relative positions of these bodies."

In 1939 Schlesinger[62] instructed Alden to make a series of photographs taken at or near full moon, and timed as accurately as possible, "for the purpose of determining mechanical libration." He indicated the project should extend over a period of eighteen years, but that "good use can be made of a series that lasts three or four years." Schlesinger, of course, did not live long enough to see this project to completion. However, Alden continued to take (or have taken) plates until he left Johannesburg in 1945, when Cyril Jackson continued the program through

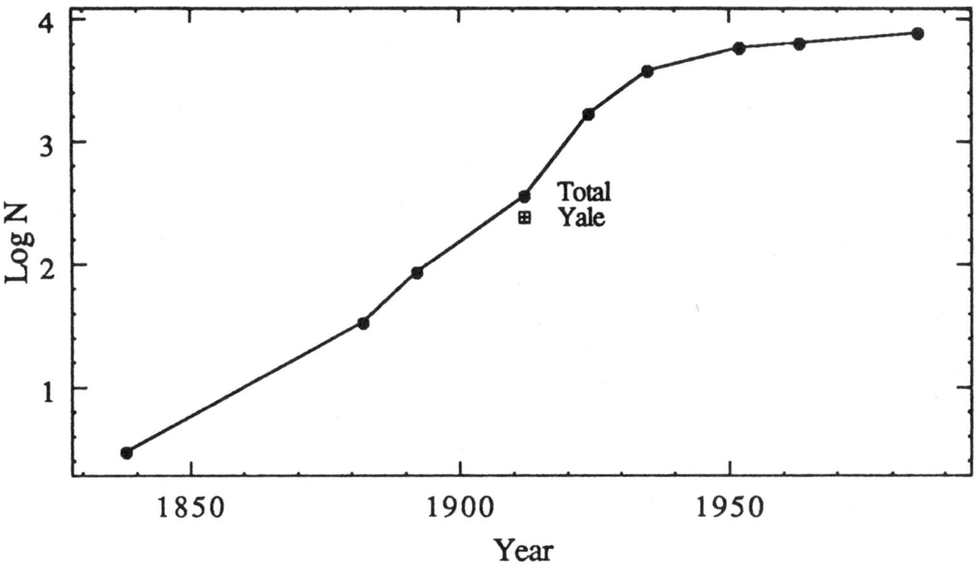

Fig. 17.5. Increase in numbers of parallaxes 1838–1989.

Table 17.1
Number and Rate of Increase of Trigonometric Parallax Determinations

t	N	ΔN/Δt
1890	30	
1910	365	17
1924	1680	94
1935	4100	220
1952	5800	100
1963	6400	73
1988	7700	54

1952. Thus 126 plates were obtained and these, along with 14 earlier plates taken between 1927 and 1934, were sent to C. B. Watts at the U.S. Naval Observatory. Together with over 650 plates taken at the Naval Observatory between 1947 and 1957, and 27 at Flagstaff in 1927–28, they constituted the basis for Watts' massive analysis, *The Marginal Zone of the Moon.*[63] The measurements of the plates were carried out on a specially designed automatic polar photoelectric micrometer[64,65] at the Naval Observatory.

The projects on stellar parallaxes that Schlesinger initiated at Yerkes, Allegheny, and Yale continue to be carried out at many institutions worldwide. His work on zone catalogues for the determination of proper motions has been greatly expanded to investigations on the determination of proper motions relative to external galaxies, initiated at Lick Observatory for stars available to northern observatories and now being extended at Yale to the South Pole. While astrometric projects constituted the highest priority at Yale under Schlesinger's directorship, in a greatly expanded Department of Astronomy they now (1991) have equal or sometimes lower priority than numerous modern theoretical and astrophysical projects, for example, those involving the internal structures and ages of stars and galaxies.

REFERENCES

1. D. Brouwer, *Biog. Mem. Nat. Acad. Sci.*, **24**, 105, 1945.
2. F. Schlesinger, *Ast. Jour.*, **18**, 180, 1898.
3. ———, *Ann. N.Y. Acad. Sci.*, **10**, 189, 1898; *Contr. Rutherfurd Obs., Columbia Univ.*, No. 15, 1898.
4. ———, *Pub. Ast. Soc. Pacific*, **49**, 5, 1937.
5. W. D. Lambert, F. Schlesinger, and E. W. Brown, *Bull. Nat. Research Council*, **78**, 245, 1931.
6. F. Schlesinger, *Observatory*, **59**, 191, 1936.
7. ———, *Ast. Jour.*, **34**, 42, 1922.
8. ———, *Monthly Notices Roy. Ast. Soc.*, **97**, 696, 1937; **98**, 203, 1938.
9. ———, *Ibid.*, **87**, 506, 519, 1927; *Pop. Ast.*, **33**, 115, 1925.
10. ———, *Astrophys. Jour.*, **10**, 242, 1890.
11. E. B. Frost, *Pub. Yerkes Observatory*, **4**, 1, 1917.
12. H. N. Russell, *Pub. No. 147, Carnegie Inst. Washington*, 1911.
13. F. Schlesinger, *Astrophys. Jour.*, **33**, 161, 1911.

14. ———, "Seeliger Festschrift": *Probleme der Astronomie,* Berlin, p. 422, 1924.
15. ———, *Ast. Jour.,* **37**, 77, 1926.
16. ———, *Astrophys. Jour.,* **20**, 123, 1904.
17. ———, *Ast. Nachrichten,* **182**, 359, 1909.
18. ———, *Astrophys. Jour.,* **33**, 172, 234, 353, 418, 1911; **34**, 26, 1911.
19. F. Schlesinger and others, *Misc. Sci. Papers of the Allegheny Obs.,* n.s., **2**, No. 2, 39 pp., 1913.
20. F. Schlesinger, *Pop. Ast.,* **23**, 638, 1915.
21. ———, *Pub. Allegheny Obs.,* **4**, 1, 1918.
22. ———, *Ibid.,* **1**, 121, 1909.
23. F. Schlesinger and R. H. Curtiss, *Idem.,* p. 25.
24. F. Schlesinger to J. H. Moore, Oct. 10, 1932, *Yale Obs. Archives.*
25. W. R. Beardsley, *Pub. Allegheny Obs.,* **8**, No. 7, 1969.
26. I. Barney, *Ast. Jour.,* **35**, 95, 1923.
27. F. Schlesinger, *Pub. Allegheny Obs.,* **1**, 123, 1909.
28. ———, *Monthly Notices Roy. Ast. Soc.,* **71**, 719, 1911.
29. ———, *Pub. Allegheny Obs.,* **3**, 23, 1913.
30. E. C. Pickering, *Ann. Harvard College Observatory,* **26**, xxi, 1891.
31. F. Schlesinger, *Proc. Amer. Philosophical. Soc.,* **52**, 175, 1913.
32. ———, see esp. *Pub. Allegheny Obs.,* **4**, and **5**, 1918–20.
33. ———, *Trans. Yale Univ. Obs.,* **6**, 19, 1927.
34. Board of Managers of the Observatory to President Hadley, June 1919, *Yale Obs. Archives.*
35. Henry Norris Russell to F. Schlesinger, Nov. 18, 1919, *Yale Obs. Archives.*
36. F. Schlesinger to J. S. Kapteyn, Dec. 24, 1920, *Yale Obs. Archives.*
37. F. Schlesinger to E. W. Brown, Oct. 25, 1919, *Yale Obs. Archives.*
38. F. Schlesinger, *Pop. Ast.,* **31**, 251, 1923.
39. C. S. Hastings to F. L. Chase, July 15, 1912, *Yale Obs. Archives.*
40. Ref. 38, pp. 77, 251.
41. *See* Chap. 19.
42. F. Schlesinger to C. E. Adams, Jan. 2, 1920, *Yale Obs. Archives.*
43. New Zealand newspaper clipping, "Yale Gift Telescope," undated. Probably Auckland Weekly, Oct. 11, 1923. *Yale Obs. Archives.*
44. F. Schlesinger to T. F. Cheeseman, Curator of the Auckland Museum, April 25, 1923, *Yale Obs. Archives.*
45. *Winchester Observatory Records,* p. 165, 1924.
46. H. Alden to F. Schlesinger, Oct. 20, 1937, *Yale Obs. Archives.*
47. F. Schlesinger to H. L. Alden, Sep. 3, 1924, *Yale Obs. Archives.*
48. L. Boss, *Preliminary General Catalogue,* Washington, 1910.
49. *Winchester Record Book,* p. 182.
50. F. Schlesinger, *Annual Report of the Yale Observatory,* for 1924–25.
51. F. Schlesinger to H. L. Alden, Oct. 21, 1926, *Yale Obs. Archives.*
52. F. Schlesinger, *Trans. Yale Univ. Obs.,* **8**, 1936.
53. ———, *Ibid.,* **15**, Part 1, 1943.
54. F. Schlesinger to H. L. Alden, March 5, 1932, *Yale Obs. Archives.*
55. H. L. Alden to F. Schlesinger, Nov. 2, 1932, *Yale Obs. Archives.*
56. D. Brouwer, *Leiden Ann.,* **16**, Part 1, 99 pp., 1928.
57. W. de Sitter, *Idem.,* Part 2, 94 pp, 1928; Part 4, 80 pp., 1929.
58. H. L. Alden, W. de Sitter, and G. Pels, *Bull. Ast. Inst. Netherlands,* **5**, 55, 121, 1929.
59. H. L. Alden and W. C. O'Connell, *Ast. Jour.,* **38**, 53, 1928.
60. H. L. Alden, *Ibid.,* **40**, 88, 1930.
61. F. Schlesinger to H. L. Alden, July 7, 1927, *Yale Obs. Archives.*
62. ———, Feb. 8, 1939, *Yale Obs. Archives.*
63. C. B. Watts, *Astronomical Papers of the Amer. Ephemeris and Nautical Almanac,* **17**, 951 pp., 1963.
64. ———, "The Moon," *I.A.U. Symposium* No. 14, p. 59, 1962.
65. C. B. Watts and A. N. Adams, *Ast. Jour.,* **55**, 81, 1950.

18
POSITIONS, PROPER MOTIONS, CATALOGUES

Equally as important as his parallax work was Schlesinger's other major astrometric project, namely zone catalogues for the determination of large numbers of proper motions, a project indispensable for the eventual analyses of galactic rotation and solar motion. (They were called "zone catalogues" because successive volumes were prepared for discrete zones of declination.) In an address as retiring President of the American Astronomical Society in 1922, Schlesinger[1] pointed out the necessity for obtaining large numbers of proper motions. At the time there were two theories on the motions and shape of the Milky Way galaxy. Kapteyn in 1904 had proposed the theory of two star streams; while Karl Schwarzschild in 1907 presented his ellipsoidal hypothesis. Although numerous investigations followed, the differences were not yet resolved. Schlesinger felt that only with large numbers of proper motions could a definitive decision between them be reached. For this, large numbers of determinations were more important than high accuracy.

The importance of proper motions (as well as of radial velocities) was accentuated by investigations carried out by the young Jan Oort who spent the years 1922–24 at Yale, and Jan Schilt in 1926–31. While at Yale, Oort wrote two important papers long predating his 1927 and Lindblad's 1926 discovery of galactic rotation. One of Oort's Yale papers[2] indicated the similarities in motion and distribution of globular clusters and high-velocity stars; the other[3] dealt with velocity differences between absolutely bright and absolutely faint stars, showing that the differences disappeared when the relatively few stars of high velocity were rejected in the comparison.

Jan Schilt was invited to Yale primarily to work on photometry for the zone catalogues. However, he also wrote several papers based on already known stellar motions and dealing with galactic rotation and star streaming.[4–10] Here he noted that giants have space motions different from those of main sequence stars. Previously at Leiden, Schilt and Oort had jointly received the Bachiene Foundation prize offered in 1921 for an essay discussing solar motion and velocities of the stars in the Boss *Preliminary General Catalogue.*[11,12] In 1928 Schilt[7] extended this investigation and pointed out the need for further investigations of radial velocities, to compare with the proper-motion results.

Schlesinger was also interested in the Taurus moving cluster. He speculated that there should be numerous other such clusters still undiscovered in the Milky Way, which could be identified only by the common proper motions of the components. He reported one such find,[13] showing in this case that high accuracy was the critical factor for determining the convergent point, so necessary for the determination of cluster parallaxes. Similarly, high accuracy was important for the discovery of common proper-motion stars which might form physical systems; Proxima Centauri was cited as an example. For stars of well-determined parallax he recommended that a few plates of the parallax field be taken ten years after the parallax series in order to obtain accurate proper motions of the faint reference stars, which were much fainter than the limits of the *Astronomische Gesellschaft Catalogues.* The probable errors of the proper motions of the faint stars he estimated would be as small as about $\pm''.002$. With typical altruism he concluded:[1]

> I venture to guess that fully 100,000 highly accurate proper motions are required to settle the problems that are at present urgent, to say nothing of the additional developments that the immediate future is likely to reveal . . . We owe it to our successors to take immediate advantage of newer and more accurate methods, and in each age to set down as accurate positions of the stars as our means permit. In doing this we shall merely be passing on the legacy bequeathed to us by the observers of the nineteenth century.

Table 18.1
Decrease in Probable Errors from Hipparchus until 1943

Hipparchus (190?–120? B.C.)	± 240″
Tycho (1546–1601), mural quadrant	30
Bradley (1693–1762), meridian circle	2
Modern meridian circle (to 1929)	0.35
Bessel (1784–1846), heliometer	0.2
More recent heliometers	0.1
Rutherfurd (1816–92), photographs	0.08
Modern (to 1929) long-focus photographs	0.025
Meridian circle *ca* 1940	0.15
Wide-angle cameras, scale about 100″/mm	0.10

In a paper, *Some Pioneer Observers,* Schlesinger[14] indicated the progressive decrease in errors of the determinations of stellar positions from the time of Hipparchus until 1929. His table is reproduced here (Table 18.1) with the additions of the dates of the astronomers cited, and the last two lines representing the approximate errors involved in Schlesinger's own zone catalogues published by 1943.

In every aspect of Schlesinger's work he was intensely interested in determining both the accidental and systematic errors of measurement and their sources. In regard to the latter he quoted Kapteyn[1] as having said, "I know of no more depressing thing in the whole domain of astronomy, than to pass from the consideration of accidental errors of our starplaces to that of their systematic errors." (Quoted probable errors have generally referred to discordances between successive measurements of the same quantity, "internal errors of measurement," whereas the actual errors including still unrecognized systematic differences between independent sources are often appreciably larger.) For example, meridian circles could yield coordinates with probable errors of ±″.2 or ±″.3 while the real errors based on a thousand observations from the best observatories indicated errors of half a second or more.

Schlesinger's meticulous attention to all possible sources of error was exemplified in his investigation of the relative positions of stars in globular clusters. At Yerkes Observatory Professor E. E. Barnard made visual filar-micrometer measurements at the 40-inch of the relative positions of stars in the three clusters, M3, M13, and M92, while Schlesinger measured plates taken on the same clusters by G. W. Ritchey, also at the 40-inch refractor.[15] This work was carried out in 1903–05 but not published until 1934 because Schlesinger was awaiting the publication of Barnard's discussion of his measurements. The surprising conclusion of the intercomparison of the results was that the probable errors of the photographic measurements were only a third as large as those of the micrometer measurements of the same stars, while the amount of time spent on the micrometer measures was a hundredfold that required for the photographic.[13]

Schlesinger then pondered the reasons for the differences. Most of the discrepancies he found could be attributed to rapid variations in atmospheric refraction. While he pointed out that Von Nusl and Fric at Prague had previously noted that the position of the pole star fluctuated about its mean position in a period too long to be consistent with unsteady seeing, Schlesinger obtained trails of Alcyone which showed deviations from the mean declination of up to ″.4 in a period of only about 80 seconds. Then he obtained simultaneous trails of Alcyone and Merope, separated by about 18 minutes of arc. The two trails showed identical fluctuations in cycles on the order of a minute of time. On photographs of a cluster, the relative positions of the various stars would not be affected by these slow anomalous refraction effects. In the case of the filar micrometer measurements, on the other hand, the individual stars could not be measured simultaneously and their relative positions would therefore be subject to errors sometimes amounting to a second of arc in zenith distance.

By 1884 strong support for photography as a means for obtaining better positions of large numbers of stars had been provided by two French astronomers, the brothers Paul and Prosper Henry,[16,17] using telescopes of their own design and construction. The Director of the Paris Observatory, Admiral Mouchez, in 1887 therefore proposed an international arrangement whereby different observatories using similar telescopes constructed by the brothers Henry (13-inch lens, 11-ft focal length), or comparable equipment, would photograph different zones on plates covering 2×2°, ultimately embracing the whole sky.[18–20] Charts were to be published for stars to 14th magnitude and catalogues to 12th. This resulted in the *Carte du Ciel* (more commonly called the *Astrographic Catalogue,* not to be confused with the AG, or *Astronomische Gesellschaft Catalogues*), a project that bogged down because of its magnitude. The voluminous catalogues (some 150 volumes) were completed in 1963, giving, with but few exceptions, only rectangular coordinates on the plates and the constants necessary for their reduction, but not the final equatorial coordinates so essential to future applications.[21,22] Only about 75 per cent of the charts were published.

The German organization, the *Astronomische Gesellschaft,* published three series of catalogues, the AGK1, AGK2, and AGK3. The first of these included stars between +80° and −2,5° to a visual magnitude limit of about 10.0, and southern observatories extended the work to the South Pole.[23] The AGK1 catalogues, some 17 volumes published between 1890 and 1910, gave positions based on meridian-circle observations carried out at numerous observatories according to precepts outlined by Argelander as early as 1867.[23] For the determination of proper motions the *Astronomische Gesellschaft* contemplated repeating the meridian observations in about half a century; but later it was decided that the positions for the new epoch could be obtained more rapidly by photography than by meridian circle. However, this suggestion was not made until 1921, and it was based upon conclusions already reached by Schlesinger in 1914.[24] The plans for the AGK2 were approved in 1927, and all the plates, each covering 5 × 5 square degrees, were to be obtained between 1928 and 1932.[25] The catalogues, published between 1951 and 1958, again gave only positions, for 183,520 stars, as it was found that the errors of the earlier catalogues were not well enough established to make proper motion determinations feasible. The AGK3, giving both new positions and proper motions was not completed until 1973. By then, however, Schlesinger and his associates and successors, had already made significant progress along similar lines.

While still at Allegheny in 1913, Schlesinger had made experimental measurements on some Harvard plates taken with the 16-inch Metcalf telescope (plate scale 90″/mm) and a 4-inch Cooke (scale 193″/mm) to test the feasibility of using wide-angle cameras for positional work.[24,26–28] The preliminary results indicated measuring errors of ±″.29 and ±″.36, respectively, as compared with about ±″.50 for AG positions. This was sufficiently encouraging for him to plan new equipment especially for stellar positional determinations. With plates covering 25 square degrees, he estimated that only a sixth as many comparison stars would be needed as for the 4-square-degree areas being photographed for the *Carte du Ciel.* His first "catalogue camera" was a 3-inch doublet, designed by Hastings and built by Brashear, with a focal length of 163 cm (scale 126″/mm). Schlesinger was probably the first to use plate glass carefully selected for flatness to be sent to Eastman Kodak Company for coating with photographic emulsion. He also designed a long-screw measuring engine for measuring coordinates on 8 × 10-inch plates.

The first zone photographed was the equatorial zone, the plates covering 4° by 6°, followed by two zones 5° wide between +50° and +60°. The measurements of the stellar positions were determined relative to meridian-circle positions of selected stars provided by observers at the Lick and Leiden observatories.

At a meeting of the American Astronomical Society in 1917, Schlesinger reported on the results for 5,954 stars between −2°10′ and +1°10′ in the Nicolaieff Zone of the AG for which 120 plates each covering 25 square degrees had been taken, and 602 comparison stars with meridian-circle positions were provided.[29] The probable errors of the final positions were ±″.21 in R.A. and ±″.19 in Dec., an accuracy indicating that the weight of a photographic determination is about four times that of a visual. However, the actual catalogues for the zones measured at Allegheny were not completed until he came to Yale. They comprise the first four[30] of the twenty zone catalogues published at Yale mainly by Schlesinger and Dr. Ida M. Barney (Fig. 18.1), who was associated with him from 1921 until his retirement and then continued the work through 1959.

While the work on the Allegheny zones was progressing, Schlesinger prepared for further zones to be photographed at Yale. At first he contemplated repeating all the *Gesellschaft* zones, but eventually limited his plans to −30° to +30°. Instead of the Allegheny 3-inch he obtained a Ross-McDowell 5.6-inch camera, with a field 14° in R.A. by 10° in Dec., scale 101.86″/mm, photographed on 0.25-inch thick plate-glass plates 17 by 23 inches. A new measuring engine was built to accommodate measurements on such large plates.[31] To reduce both eye-strain and accidental errors due to the observer's bumping his nose on the microscope, Schlesinger and A. L. Bennett in 1933 devised a projection method for measuring photographs which they first tested on the smaller parallax plates; viewing and bisecting images on a small screen proved easier than peering through a microscope, and this increased both the speed and accuracy of the measurements.[32]

For the −10° to −30° zones, Schlesinger acquired a better 5-inch Ross lens to be installed piggyback on the 26-inch at Johannesburg. Now the use of 17-inch square plates having a scale of about 100″/mm became standard. Four catalogues for these

Fig. 18.1. Ida M. Barney (1886–1982).

zones were published between 1939 and 1943 by Schlesinger and I. M. Barney, in volumes 11 through 14 of the *Transactions of the Yale University Observatory.* These were the first catalogues for which IBM punch-card machines were used for a large part of the necessary computations, services provided by the Watson Bureau of IBM, mainly through the good offices of Dr. Wallace Eckert. The measurements and reduction of the plates for the subsequent volumes, 16 (1945) through 27 (1959), became the responsibility of Dr. Barney. These included the north polar cap for stars north of +85° and new measurements for the +50 to +60° zones, as well as the completion of the remaining zones of the originally planned −30° to +30° belt. For the northern zones an 8-inch Ross lens replaced the 5-inch in 1936 (vols. 18, 1947, and 19, 1948); and another 8-inch was installed in 1950 for the north polar region, both again having the standard plate scale of about 100″/mm. Meridian-circle positions for the reference stars in the various zones were contributed, in addition to those from Lick and Leiden, by the U.S. Naval Observatory in Washington and the Greenwich and Cape observatories. Apparent magnitudes and spectral classes were also important to the reductions; these, however, were not independently determined at Yale. Most of the spectral classes were furnished by Harvard Observatory; many for the red stars were classified by Vyssotsky at the Leander McCormick Observatory of the University of Virginia. Visual magnitudes were generally taken from the AG catalogues; photographic magnitudes were given for only a few zones (+20° to +30°, and the 1959 edition of the +50° to +60° zones) and were determined by or under the direction of Dr. Jan Schilt at Columbia from plates taken especially for the purpose at Yale.[31,33]

By the time of Schlesinger's death, ten volumes of Yale zone catalogues had been published, including 92,329 stars. Under the directorship of his successor, Dirk Brouwer, the work was continued by and under the supervision of Ida Barney who, by 1959, added thirteen more volumes and brought the total number of stars to over 151,000. Miss Barney's final work on zone catalogues dealt with the reduction of the new measurements of the +50° to +60° zones. The measurements themselves were carried out at the IBM Watson Scientific Laboratory by Rebecca B. Jones with a new electronic engine designed by Wallace Eckert.[34] The automatic image-centering device which replaced the conventional eye-estimate bisections of the images resulted in higher positional accuracy (from a probable error in either coordinate of ±″.115 by the older method, to ±″.090), and less eye fatigue.

When Dirk Brouwer became Director in 1941 he planned to redetermine the positions of stars in the Cape Zone Catalogues from −30° to the South Pole for obtaining improved proper motions.[35] The plates for −40° to −50° and −60° to −70° were obtained at Johannesburg by Alden between 1941 and 1942. Wartime conditions then disrupted operations; it became difficult to ship photographic materials, and no photographic plates were manufactured in South Africa. The observatory at Johannesburg was dismantled in 1952 and the equipment shipped to Australia. There the 8-inch catalogue camera was temporarily mounted at the Sydney Observatory where in 1955–56 Brouwer himself took some of the plates on the −30° to −40° and −70° to −90° zones, and Harley Wood and assistants at the Sydney Observatory completed the program. (The −50° to −60° zone was to be reserved by the Sydney Observatory.) The meridian-circle positions of the reference stars for these zones were furnished by Dr. R. H. Stoy, the author of most of the Cape Zone catalogues to be remeasured. After Miss Barney's retirement, the general supervision of the reductions of the southern zones was assigned to Dorrit Hoffleit. Ultimately five more volumes of the *Transactions* were published,[36–39] bringing the total number of stars with Yale proper motions to over 227,000.

The electronic machine at the Watson IBM Laboratory was used for measuring the plates of the −40° to −50° zone by Mrs. Dorothy A. Eckert and an assistant, Mrs. Meredith B. Weddle.[37] Unfortunately, the machine suffered premature aging, relays having to be replaced frequently during the measuring. The most accurate machine ever available up to that time had to be abandoned after the completion of that zone. For the −60° to −70° zone the now antique long-screw engine was again used, in an unfavorable environment where the temperature could not be controlled. (The length, or scale, of the measuring screw changes with temperature, vitiating the stellar coordinates measured on the photographic plates.) Finally, in 1964, while the measurements for the −70° to the South Pole regions were in progress, a new Mann engine was acquired from the Mann Instrument Co. of Boston. For these zones more profuse plate-overlap was obtained than for any of the previous zones. The necessarily more elaborate computations were carried out by Phillip Lu according to procedures advocated by consultant Heinrich Eichhorn.[38]

Schlesinger always took an active interest in the needs of astrometrists elsewhere. Thus in 1926 he designed and had constructed a blink comparator for Harvard Observatory, to accommodate 14 by 17-inch plates for W. J. Luyten's *Bruce Proper Motion Survey.*[40–42] Similarly he had built blink microscopes of his own design for the Leiden Observatory in 1933, and previously for other institutions.[43]

At the Southern Station, besides the plates for the parallax and zone proper-motion programs, plates for proper motions were also secured on southern *Selected Areas,* for proper motions of long-period variables and, at the request of the Leiden Observatory, of Cepheid variable-star fields.[44]

Always aware of the special importance of both nearby stars and of apparently bright stars, Schlesinger compiled general catalogues of all available data on stars with known parallaxes (see previous chapter) and of the naked-eye stars. E. C. Pickering at Harvard had compiled a catalogue of stars to effectively 6.5 visual magnitude, for which he had determined accurate visual magnitudes. This was called the *Harvard Revised Photometry,* or HR for short.[45] The catalogue also gave positions for 1900, spectral classes, and star names or designations. Schlesinger found a need for an updated catalogue of this sort, and compiled the *Yale Catalogue of Bright Stars,* BS for short. The first edition of this came out in 1930, containing only the HR stars and their companions, with exactly the same numbering as in the HR. To the types of data Pickering had given were now added precession, proper motions, parallaxes, and radial velocities. The second edition, by Schlesinger and Louise Jenkins in 1940, contained many more parallaxes and radial velocities, but no new categories. In 1964 the third edition, compiled by D. Hoffleit, contained many more cross-references to designations in other catalogues, gave positions for both 1900 and 2000, and added columns for variable stars, galactic coordinates, B-V color indices, and double-star data, the last having occurred only among the *Remarks* in previous editions. Finally a fourth edition in 1982, by Hoffleit and C. Jaschek (the title now abbreviated to *The Bright Star Catalogue*), added U-B and R-I colors, and rotational velocities ($v\sin i$). The successive catalogues of Bright Stars proved probably the most quoted of any star catalogues ever published, being used by both professional astronomers and astrophysicists, and by amateurs.

In his address at the dedication of the new Allegheny Observatory in 1912 Schlesinger had half-facetiously stated that he was tempted to mount a plaque over the door, "ABANDON HOPE OF MAKING DISCOVERIES, ALL YE WHO ENTER HERE"[46] What he meant was that it would take so long to accumulate the necessary data for solving the problems of galactic properties that only future generations could benefit from the fruits of current efforts. His own lifelong efforts to provide data for future research were eminently successful. Finally, in his *Annual Report* for the year 1940–41, Schlesinger wrote, "In retiring from the directorship in July, 1941, it is with a feeling of satisfaction that can be rarely matched, in the confidence I have that the work will be carried forward well and efficiently, and that it will be broadened as circumstances permit."

Although priorities and major interests of successive directors have changed progressively within the department, Schlesinger's confidence was well founded. In all four of his major specialties while he was Director at Yale, work has continued: the determination of new parallaxes; zone catalogue work (now in progress for extending the Lick proper-motion survey for proper motions relative to distant galaxies to the South Pole); and the continual updating of both the *General Catalogue of Trigonometric Parallaxes*; and of the *Bright Star Catalogue.*

REFERENCES

1. F. Schlesinger, *Pop. Ast.,* **30**, 535, 1922.
2. J. Oort, *Proc. Nat. Acad. Sci.,* **10**, 256, 1925.
3. ———, *Idem,* p. 253.
4. J. Schilt and I. Barney, *Ast. Jour.,* **37**, 181, 1927.
5. J. Schilt, *Pub. Ast. Soc. Pacific,* **39**, 203, 1927.
6. ———, *Ast. Jour.,* **38**, 149, 1928.
7. ———, *Ibid.,* **39**, 17, 1928.
8. ———, *Idem,* p. 143.
9. ———, *Ibid.,* **41**, 97, 1931.
10. ———, *Idem,* p. 105.
11. L. Boss, *Preliminary General Catalogue of 6188 Stars for the Epoch 1900,* Washington, 1910.
12. W. J. Luyten, *My First 70 Years of Astronomical Research,* privately published, Minneapolis, p. 18, 1986.
13. F. Schlesinger, *Monthly Notices Roy. Ast. Soc.,* **87**, 506, 1927.
14. ———, *Pub. Ast. Soc. Pacific,* **41**, 212, 1929.
15. ———, *Ast. Jour.,* **43**, 97, 161, 1934; **44**, 21, 1934.
16. F. W. D., *Mon. Notices Roy. Ast. Soc.,* **64**, 296, 1904.
17. ———, *Ibid.,* **65**, 349, 1905.
18. A. E. B. Mouchez, *Ann. Report, Obs. Univ. Paris for 1887.*
19. B. Baillaud, *Histoire de l'Astronomie de Position,* **I**, Chap. I, pp. 5ff, 1933? Typewritten copy at Yale Obs.
20. *cf.,* A. Pannekoek, *A History of Astronomy,* p. 337, N.Y. 1961.
21. J. Lankford, *The General History of Astronomy,* **4**, 29, 1984.
22. H. Eichhorn, *Astronomy of Star Positions,* p. 279, N.Y., 1974.
23. ———, *Idem* p. 227.
24. F. Schlesinger and C. J. Hudson, *Pub. Allegheny Obs.,* **3**, No. 9, 59, 1914.
25. R. Schorr and A. Kohlschütter, *Introd. to AGK2,* **1**, E1, 1951.
26. F. Schlesinger, *Proc. Am. Philosophical Soc.,* **53**, 303, 1914.
27. ———, *Observatory,* **38**, 129, 1915.
28. D. Brouwer, *Biog. Mem. Nat. Acad. Sci.,* **24**, 105, 1945.
29. F. Schlesinger, *Pub. Amer. Ast. Soc.,* **3**, 339, 1918.
30. F. Schlesinger *et al., Trans. Yale Univ. Obs.,* **3**, Part IV, 1925; **4**, 1925; **5**, 1926; **7**, 1930.
31. F. Schlesinger and I. Barney, *Trans. Yale Univ. Obs.,* **9**, 1933.
32. F. Schlesinger and A. L. Bennett, *Monthly Notices Roy. Ast. Soc.,* **93**, 382, 1933.
33. J. Schilt and S. J. Hill, *Contr. Rutherfurd Obs., Columbia Univ.,* No. 30, 1937; No. 31, 1938.
34. W. J. Eckert and R. B. Jones, *Ast. Jour.,* **59**, 83, 1954.
35. D. Hoffleit, *Ast. Jour.,* **67**, 696, 1962.
36. ———, *Trans. Yale Univ. Obs.* **28**, 1967; **29**, 1968.
37. D. Hoffleit with D. Eckert, P. Lu and K. Paranya, *Ibid.,* **30**. 1970.
38. P. K. Lu. *Ibid.,* **31**, 1971.
39. F. W. Fallon, *Ibid.,* **32**, Part II, 1983.
40. F. Schlesinger, *Ast. Jour.,* **37**, 45, 1926.
41. Ref. 12, p. 24.
42. W. J. Luyten, *Bruce Proper Motion Survey,* Minneapolis, 2 vols., 1963.
43. F. Schlesinger, *Ann. Report* for 1933–34.
44. ———, *Ibid.,* 1926–1938.
45. E. C. Pickering, *Ann. Harvard Coll. Obs.,* **50**, 1908.
46. F. Schlesinger, *Miscellaneous Scientific Papers of the Allegheny Observatory,* n.s., **2**, No. 2, p. 16, 1913.

19
PHOTOMETRY 1920–1966

Stellar photometry had not been one of Schlesinger's primary interests when he first came to Yale. But when he found that the Loomis telescope was not suited to the astrometry for which it had ostensibly been designed, it occurred to him that its very deficiency, namely not giving adequately sharp images, could be an asset in photographic photometry. Carl L. Stearns, who came to Yale as a graduate student in 1920, was assigned the task of testing the equipment for this purpose.

While extrafocal photographic images had previously been used successfully elsewhere for magnitude determinations, such images at the Loomis would not reach sufficiently faint stars. Stearns found, however, that only slightly out-of-focus images would yield the desired results for stars to nearly 10.5 photographic magnitude without unduly long exposures. He made measurements of both the diameter and density of the images of stars in the Pleiades, calibrating his measurements against magnitudes previously determined by Hertzsprung.[1] For measurements of the density, Stearns used a Hartmann photometer[2] with modifications designed by himself and Schlesinger, and built at Yale's Sloane physics laboratory.[3] The photometric wedge used in this instrument was supplied by the Harvard College Observatory.

With preliminary tests completed, Stearns turned his attention to two variable stars,[3] the RR Lyrae-type star, RR Ceti, and the β Lyrae eclipsing-type star, VV Orionis. For RR Ceti he obtained 131 exposures on 47 plates taken on 17 nights between November 1921 and December 1922. The measures of image density of the variable, calibrated against five comparison stars, yielded a more precise light curve than any previously published. RR Ceti varied from 9.17 to 10.39. For VV Orionis he made 502 exposures on 84 plates taken on 24 nights between November 1922 and February 1923. This star varied from 5.08 to 5.36. In both cases the probable error of a single observation uncorrected for systematic night and plate errors was ±0.04pg, and after correction for such effects, ±0.030pg. For VV Orionis, Stearns also computed an orbit based on these observations. His dissertation based on these data, for which he was awarded his Ph.D. in 1923, proved that the Loomis telescope was eminently valuable for variable-star observations over a wide magnitude range. Stearns continued working on variable stars at Yale until 1925 when plans for the solar eclipse interrupted the use of the telescope for variable stars. Soon thereafter he left Yale for an appointment at his alma mater, Connecticut Wesleyan University.

In the fall of 1926, Schlesinger invited Dr. Jan Schilt, a former student and disciple of J. C. Kapteyn, to come to Yale. Schilt's primary assignment was the determination of photographic stellar magnitudes for the zone catalogues. Previously, the magnitudes given in the zone catalogues were visual only and came either from Harvard or were taken from older positional catalogues. As an assistant at the Leiden Observatory, Schilt had designed and built a thermopile-microphotometer, later named the *Schilt photometer,* for the determination of magnitudes from photographic plates.[4] At Yale he designed such a photometer to accommodate Yale's largest zone catalogue plates, 19 by 23 inches and weighing 10 lbs each.

At first it was decided to use the same plates for magnitude determinations as for the astrometric measurements. However, this did not prove feasible and Schilt experimented with various lenses, finally deciding upon a 3.5inch Ross.[5] In 1928 Schlesinger was to report that all future zone catalogues would contain photographic magnitudes. At that time Schilt had already determined some 8,500 magnitudes for the +20° to +30° zones, with a probable error of one determination of ±0.07mag, or ±0.05 for the mean from two plates. Schilt resigned on June 30, 1931, to become Director of the Rutherfurd Observatory of Columbia University. However, he took the photometer with him and continued deriving magnitudes for the Yale zone catalogues. Dr. Arthur Bennett replaced Schilt at Yale and took the plates for

Schilt to measure, as did also V. Goedicke in 1941–46. Goedicke also determined photo-red magnitudes (measured on a photometer built by Bennett) in the +10° to +20° zone for the purpose of deriving color indices.[6–8] For some zones, however, no photographic magnitudes became available. In all, Schilt provided photographic magnitudes for about 89,000 stars (Table 19.1). Years later (1948–52) at Johannesburg, Schilt and Cyril Jackson were to determine photoelectric V-magnitudes and colors of 2,031 bright stars.[9,10]

Before coming to Yale, Schilt had spent some time at the Mt. Wilson Observatory where he obtained photographs of variable star fields. He brought 97 plates on the globular cluster, M3, to Yale and turned them over to graduate student P. Slavenas, who determined the light curves of 17 of the cluster variables and, according to Schlesinger,[11] expected to determine the light curves for 30 more, and elements for an additional 30. None of this seems ever to have been accomplished.

Slavenas was also interested in verifying the variability of the light from the planet Uranus. From 77 plates obtained with a small camera attached to the catalogue camera, he ascertained a period of 0.4497 days, in good agreement with earlier spectroscopic determinations of the planet's period of rotation, 10.75 hrs by P. Lowell[12] and 10^h50^m by Slipher.[13] However, the amplitude of variation was smaller than what others had found.[14]

Schilt used his photometer for determining the distribution of light in the globular cluster ω Centauri.[15] Following the example of Hertzsprung, who had similarly measured the distribution of light in M3,[16] Schilt used an out-of-focus plate supplied by Spencer-Jones at the Cape Observatory. The diameters of star images were 2mm, and the central region of the cluster was uniform gray, not black. Previously (1893) Solon Bailey at Harvard had determined the distribution of light in this cluster from star counts. Schilt measured the intensity at 3,240 equally spaced intervals, the total area scanned being 40 by 40mm, or 20′ by 20′. Schilt's results indicated greater increase in density toward the center of the cluster than did Bailey's star counts.[17] The discrepancy Schilt attributed to the overcrowding of images near the center of the cluster, making the star counts there unreliable. The isophotes Schilt found indicated ellipticity of the cluster in agreement with that found by Shapley in 1919 from star counts on Harvard plates.[18]

Another investigation by Schilt while he was at Yale did not involve new photometric measurements but depended on his questioning some interpretations of variable-star results by others. W UMa type stars showed periodic variations in the widths of their spectral lines attributed at that time to rapid rotation of these stars. Schilt, however, found that the lines showed greater dispersion in light than could be accounted for by rotation alone. Moreover, if interpreted as binaries, none of these stars indicated a total eclipse. He interpreted these variables as being in a stage intermediate between single and double stars, attributing their spectral and light variations to single dumbbell-shaped objects in the process of fission.[19]

Another group of variables that were formerly considered to be spectroscopic binary stars were the Cepheids. This decision stemmed erroneously from Belopolsky's discovery[20] in 1894 that their radial velocities varied in synchronism with their apparent magnitudes. However, the variations, as Belopolsky realized but others were reluctant to admit, were not such as would have been expected from simple binary motion, neither in the shapes of the light and velocity curves nor in the relative timings

Table 19.1
Photographic Magnitudes

Zone		Number	Reference
+25°	to +30°	10358*	Trans. Yale U. Obs., **9**, 1933
20	25	8703*	Trans. Yale U. Obs., **10**, 1934
55	60	6902	Contr. Rutherfurd Obs., No. 30, 1937
50	55	7280	Contr. Rutherfurd Obs., No. 31, 1938
10	20	16379	Contr. Rutherfurd Obs., No. 32, 1952
30	50	39294	Contr. Rutherfurd Obs., No. 32, 1952

*Number of stars in the Catalogue, for a small percentage of which photographic magnitudes could not be determined.

of light and velocity phases. In 1914 Shapley,[21] then at Mt. Wilson, proved that the Cepheids were in fact single stars and he proposed pulsation as the most likely reason for the variations in both light and radial velocity.

Although it was 1952 before Baade[22,23] announced that the Cepheids belonged to two separate "Populations," the classical Cepheids to Populations I, the W Virginis stars to Population II, Schilt[24] already in 1928 had foreseen the distinction that there must be more than one group. The Cepheids with periods under 10 days all had c-type spectral classes, indicative of high luminosity. But among the apparently fainter Cepheids there was a surprising percentage of stars with longer periods, which seemed contradictory to the accepted period-luminosity relation. A large dispersion in spectral class of stars of the same period indicated to him "the possibility that we have to do with more than one group of stars, which having the same order of periods, are of different surface temperatures, and which may have rather different masses and absolute magnitudes." He concluded, "This kind of variability does not seem to be restricted to the stars of high mass or high luminosity known as c-stars, and a close examination not only of the period, but also of the spectral type and the shape of the light curve, is necessary to justify conclusions of equality of absolute magnitude as used in the estimates of the distances of star groups." This paper did not receive the early recognition it deserved, having anticipated Baade's discovery of Populations I and II by nearly two and a half decades.

Possibly the most significant of Yale's contributions to stellar photometry was the original work of then graduate student John S. Hall (Fig. 19.1), now Director Emeritus of the Lowell Observatory in Flagstaff, Arizona. He began working with photoelectric photometers at the Loomis telescope in 1931. With the polar telescope eyepiece in a fixed position in an enclosed room, this telescope seemed ideal for his purpose, as the relatively heavy photoelectric equipment could also stay in a fixed position. Hall experimented with the use of a caesium oxide photoelectric cell, designing his own photometer. He found it would work well only when cooled to −40°C, and to keep it that cool he cleverly mounted it inside a thermos bottle to protect it from heat and keep it dry chemically. The color curve of the cell ranged from under 3000 to over 12,000 Ångstroms, with maximum sensitivity at about 7500Å. With the use of different filters he was able to derive color indices.

Fig. 19.1. John S. Hall (center) with British astronomer F. J. M. Stratton at IAU in Stockholm, 1938.

For calibrating his instrument, Hall made measurements of the Pleiades. Comparing the magnitudes determined without filter, with those using filters giving effective wave lengths of 8100 and 8800Å, he determined red indices (not on the present standard system) for 347 stars brighter than 5.0 visual magnitude.[25] These compared well with Bottlinger's[26] indices obtained with a blue sensitive cell and blue and yellow filters. This was at a time when Joel Stebbins and associates in America, and other astronomers in Europe, had already been using photocells at visual and photographic wave lengths, but it was a decade before Stebbins and Whitford[27] initiated their six-color photoelectric system.

Neglected and now almost forgotten in variable-star literature, the first photoelectric determination of an infrared light curve of a variable star was obtained by Hall in 1932 for the Cepheid ζ Geminorum.[28] He acquired measures on 33 nights between January 28 and November 27, 1932. He compared his light curve at effective wave length 7400Å with Guthnick's[29] at 4500Å for 1919–20 and Luizet's[30] at 5500Å for 1898–1918. The amplitudes of the three curves are 0.40 in the red, 0.55 visual, and 0.81 in the blue, thus extending the trends with wave length previously noted between B and V for numerous variables. At successive wave lengths the observed

Table 19.2
Variable Stars with Early IR Light Curves

Name	ζ Gem	Algol	AO Cas	S Sge	RT Aur
BS	2650	936	65	7609	2332
Max	3.62V	2.12V	5.96B 6.07V	5.79B 5.24V	5.00V
min	4.18	3.37	6.11 6.24	7.05 6.04	5.82
Type	δ Cep	EA	Ell	δ Cep	δ Cep
Per.days	10.15	2.87	3.52	8.38	3.73
Spect.	F7 – G3	B8 + B8	O9 + O9	F6 – G5	F4 – G1
PEP amp:					
red	0.40	0.9	0.143	0.459	0.521
vis	0.56	1.28	0.17	0.76	0.87
blue	0.81	1.25	0.15	1.26	1.18
λ Eff	7400	8660	8000	8000	7500

maxima fall later, the IR falling 0.024 period later than the visual.

Hall was granted the Ph.D in 1933 for his thesis on *Photo-Electric Photometry in the Infra-Red with the Loomis Telescope*. After leaving Yale he went first to Swarthmore, then to his *alma mater*, Amherst. From similar observations in the infrared (8660Å) carried out at both of those observatories he published a light curve of Algol which he compared with a blue photoelectric curve obtained by Stebbins.[31] Again the amplitude in blue was appreciably greater than that in the red.

Dr. A. L. Bennett (Fig. 19.2) came to Yale late in 1931. He was familiar with photometric problems, having written his Ph.D. thesis at Princeton in 1928 on *A Photometric Investigation of the Brightness of 59 Areas of the Moon*, for which he had developed his own equipment. The results of this project were not published until he was encouraged to do so by Schlesinger after Bennett came to Yale.[32] Here he made improvements over Hall's equipment and built his own.[33] He introduced a rotating-sector timing device for automatically making star readings through a red and an infrared filter and corresponding sky-background observations at a rate of one reading a minute, or five minutes for each completed set of measures per star, with a probable error of only $\pm 0.010^m$ for a single observation on a good night.[33,34]

Bennett determined color indices for 254 stars of all spectral classes with visual magnitudes between 5 and 9.[35] Such data were important for their relationships to the spectra and temperatures of the stars. He also carried out infrared observations for variable stars. Although he observed some thirty long period variable stars which F. Hollander was almost simultaneously observing photographically in blue light with the 5-inch Ross camera,[36] they published no results on these long period variables.

Infrared observations for the following shorter-period stars were likewise carried out: S Sge and RT Aur (Cepheids); Pleione (type unknown at the time, now γ Cas type); 44i Boo, VV Cep, U Oph, and VV Ori (eclipsing) and AO Cas (then called eclipsing, now Ell).[37] However, Bennett's light curves for only three of these were published: AO Cas,[38] S Sge,[39] and RT Aur.[40] AO Cas, a spectroscopic binary, Bennett found to have a symmetric light curve from which he was the one to deduce that the observations could be represented by ellipticity alone, without eclipses.

Both S Sge and RT Aur were observed at the request of Martin Schwarzschild, then at Harvard, who was investigating the astrophysical properties of Cepheids. S Sge was of special interest because it showed a pronounced hump on the photographic light curve shortly after maximum, as Hertzsprung had earlier pointed out[41] when he observed the star in 1910–11. Bennett found that the hump had nearly the same magnitude as the primary maximum and therefore described the curve as having a double maximum. The red magnitudes were considered important for helping to determine the star's bolometric magnitude.

Table 19.2 summarizes both Hall's and Bennett's infrared results on the variable stars and compares them with the data given in the most recently available *GCVS*,[42] the *Bright Star Catalogue*,[43] or other sources. The effective wave lengths were not uniform for all the variables and are noted in the bottom

Fig. 19.2. Arthur L. Bennett, December 1936, at Loomis Telescope.

line. It may be pointed out that none of these coincides exactly with the modern standards of Stebbins and Whitford,[44] whose R is for effective wave length 7190 and I for 10,300Å. The three Cepheids again all show decreasing amplitude from blue through red.

In November 1940 Bennett was called away for war work for the Navy, regrettably not to return to astronomy. The bulk of his observations on variable stars has remained unpublished, although he had converted them to magnitudes. Comprising twenty notebooks for the years 1933 through 1940, they are stored in the Archives of the Yale University Library. A sampling of these for the two variables, W And and R Lyn, with S-type spectra and periods of 395.93 and 378.75 days, respectively, is illustrated in Figs. 19.3 and 19.4.[45] Bennett's red observations are plotted in the upper parts of the figures while the middle curves are the visual magnitudes by AAVSO observers made in the same time interval.[46] The lowest curves in the figures represent the color indices, infrared *minus* visual, which range from slightly less than four to over eight magnitudes. These estimates are comparable with results of a similar investigation by Hetzler[47] who also compared his observations at about 8500Å with visual AAVSO curves. He found infrared *minus* visual indices ranging from five at maximum to eight near minimum for W And and three to six for R Lyn.

In his annual report for 1935–36 Schlesinger had already observed, "The useful way in which photoelectric cells have been employed on the Loomis Telescope since 1930, first by Mr. Hall and now by Mr. Bennett, is a source of satisfaction to all the friends of the Observatory." Later, in 1940, Ejnar Hertzsprung was to write Schlesinger,[48]

> Though I do not know if Dr. Arthur L. Bennett is still active at your observatory I want to put down a few remarks concerning his interesting work on color indices observed photoelectrically with the Loomis telescope. The fact that Dr. Bennett has obtained good results down to nearly 9th visual magnitude opens possibilities for further applications of this promising method.

Alas! World War II put an end to this fruitful research at Yale. After the war Bennett remained at the Navy's Michelson Laboratory at Inyokern, California, and no one was hired to replace him at Yale.

Bennett's observations at the Loomis telescope were the last infrared observations carried out at Yale. After World War II the telescope was refurbished for an entirely different project (D. Brouwer's investigations into the variation of the pole of rotation of the earth) for photographing exclusively the polar region. From July 1946 until June 1953 Dr. Joseph Ashbrook made both visual and photographic estimates of variable stars at Yale, but with other instruments and none photoelectrically or at red wave lengths. Still later, Dr. Harlan J. Smith used the 10-inch Butler refractor at Yale for photoelectric observations of short-period variables, but only in UBV.

Ashbrook, who replaced Goedicke, came to Yale in 1946 bringing with him the enthusiasm for variable stars he had experienced as a graduate student at Harvard where he received his Ph.D.in 1947. At Yale he began taking plates with the 5-inch Ross lens of the eclipsing variable, Y Sex, for which he had already made some 850 observations on Harvard plates. With the same instrument he observed the Cepheid, SV Vul, for ascertaining its changing period. Later he made use also of the 3-inch and 8-inch Ross, deriving periods and changes in period of δ Cepheid and RR Lyrae stars, observations of novae, and times of minima of eclipsing variables. He found changing periods for the pulsating stars AP Her, HU Cas, and SW Her.[49] For much of the work he published after coming to Yale he combined his Yale observations with earlier or contemporary measurements he had made at Harvard, either visually or with the extensive Harvard plate collection. In addition to his photographic studies he also made visual use of binoculars and of the 8-inch Reed telescope.[50,51]

An eclipsing star with large apsidal motion, AG Persei, intrigued Ashbrook. The period of the apsidal motion was not well determined, 53 or 20 years having previously been suggested. In the fall term of 1948 Ashbrook took a leave of absence and, under an ONR Grant, carried out 96 photoelectric observations at the Amherst College 18-inch refractor.[52] He obtained a period of the line of apsides of 72 years, in good agreement with a more recent value of 76.55 years quoted in the *Eighth Catalogue of the Orbital Elements of Spectroscopic Binary Systems.*[53] This investigation may be what stimulated Brouwer's efforts to obtain modern photoelectric equipment for Yale. In his *Report* for 1949–50 he wrote,

> While photographic photometry can still be usefully employed on certain types of investigations, the introduction of powerful multiplier tubes for photoelectric photometry direct at a telescope has largely replaced the older methods for many applications. The limitations placed upon the Observatory by the lack of this newer equipment will be overcome during the next few months with the installation of photoelectric equipment employing a 1P21 photomul-

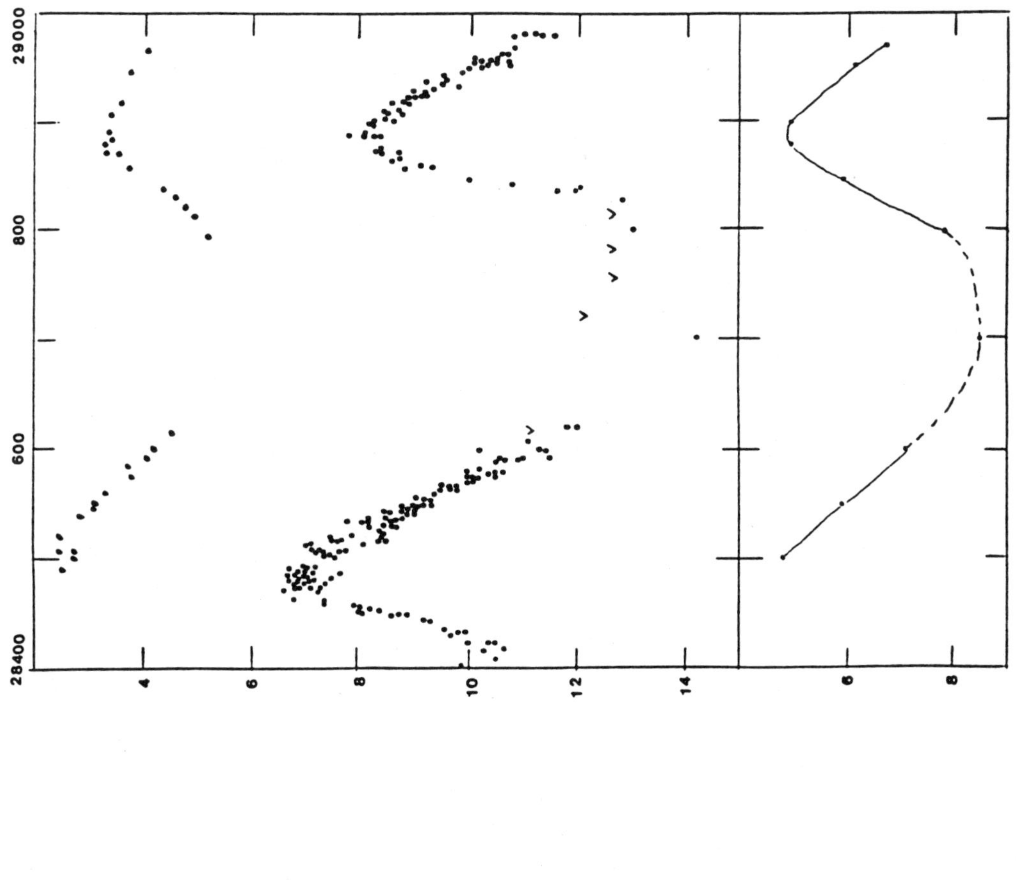

Fig. 19.3. Light and color curves for W And. Upper plot, Bennett's infrared (IR) observations; middle, corresponding AAVSO visual; bottom, color curve, visual *minus* IR. Ordinates are magnitudes, abscissae, J.D.-2400000.

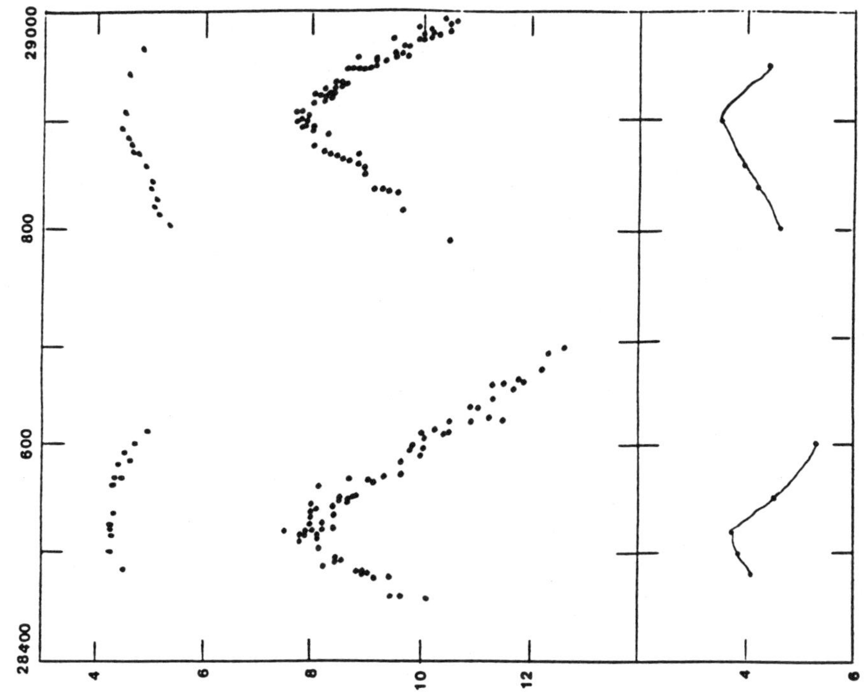

Fig. 19.4. Light and color curves for R Lyn. Upper, Bennett's IR observations; middle, corresponding AAVSO visual; bottom, visual *minus* IR.

tiplier tube for use with the 8-inch Reed refractor. While the telescope is too small to enable the Observatory to compete with institutions that have much more powerful telescopic equipment available, it will enable Mr. Ashbrook to turn to this equipment for investigations for which the photographic method is clearly outdated.

Meanwhile a new two-year project was being started for obtaining 45-minute exposures on a field in Cepheus every clear night with the 8-inch Ross lens for the detection and study of the variable stars.

When the 1948 edition of the *General Catalogue of Variable Stars* came out, Ashbrook and graduate student Raynor L. Duncombe studied the distributions of the classical Cepheids in order to determine the best position of the north pole of the galaxy.[54] Their result (R.A. 191.9°, Dec. 27.4°) confirmed, rather than improved upon, earlier determinations carried out elsewhere over the preceding decade.

Besides variable stars, Ashbrook observed the planet Uranus[55] with an opera glass from 1936 to 1947. The purpose was to check an eight-year period found by W. Becker for observations compiled from 1864 to 1932.[56] Ashbrook's light curve showed an amplitude of $0.4^{m.}$ and approximately confirmed the period which, however, appeared from the earlier data not to be constant. The variations he attributed to effects in the atmosphere of Uranus and he called attention to the need for spectra at both maximum and minimum light phase.

Aside from his strictly photometric investigations, in 1953 Ashbrook made a new determination of the period of rotation of the planet Mars.[57] He supplemented old observations of central meridian transits of Martian features with his own observations at the 10-inch Butler telescope. Of this Brouwer in his *Annual Report* for 1951–52 stated, "the period of rotation obtained has an accuracy far surpassing that of previous determinations."

Ashbrook was not reappointed and left Yale the end of June 1953 to assume the Assistant Editorship of *Sky and Telescope* magazine. In regard to his work on variable stars Brouwer commented,[58] "This branch of astronomy has been very well represented by Mr. Ashbrook's constant and diligent activity. It permitted him to make excellent use of our modest telescopic equipment in New Haven. It is not likely that after Mr. Ashbrook's departure this type of activity will be continued at anywhere near the level of the past seven years." So valuable a researcher was ostensibly dropped because a minor speech handicap prevented his being a clear speaker for the undergraduate classes he was required to teach. His share in the teaching of graduate students, on the other hand, had been a major one and they found him stimulating. How unlike Ashbrook's experience was that of H. A. Newton of whom Timothy Dwight had written,[59] "He had a certain hesitation of speech and slowness in utterance. Whether this interfered in any measure with his success in teaching I do not know. . . . We who knew him well were accustomed to it, and his pupils thought little of the matter when they came to have familiar acquaintance with him as an instructor."

Harlan J. Smith, like Ashbrook whom he replaced on July 1, 1953, also came from Harvard well equipped to carry on variable star research. He was particularly interested in the RRc-type pulsating variables with periods less than 0.2day. In his Ph.D. thesis he found that most of these were a new class of variables which he called "dwarf Cepheids,"[60] with CY Aqr and AI Vel as prototypes, now renamed SX Phe and δ Sct types, respectively. This work Smith completed after coming to Yale. Of the stars with such short periods that he studied, two, FG Hya and UZ Leo, turned out to be W UMa-type eclipsing binaries.[61,62] In 1954, A. A. Disco, an instrument technician knowledgeable in electronics, came to the Yale Astronomy and Physics Departments from the Netherlands. One of his first tasks was to equip the 10-inch Butler refractor for photoelectric work. Smith and his undergraduate students made considerable use of the new equipment, which was intended, in part, for just such undergraduate training. They observed especially stars with short periods. Smith observed DY Her and YZ Boo, and with undergraduate Paul Hodge (now head of the Department of Astronomy at the University of Washington) investigated the eclipsing binaries CW Cep and V822 Aql. With J. McCullough (now at the Woods Hole Oceanographic Institution) and R. Wing (an expert on red variable stars at Ohio State University), Smith demonstrated that the dwarf Cepheid V703 Sco shows beat-period phenomena. W. Tifft (now at Steward Observatory of the University of Arizona) and Smith analyzed photoelectric and spectroscopic data on another dwarf Cepheid, T Sex, and attempted a theoretical interpretation of the complexities of the observations.[63]

When Mt. Wilson astronomers in 1961 first reported that the radio source 3C48 was coincident with a faint optical star, Smith, Hoffleit, and their graduate students made a pilgrimage to Harvard College Observatory to search the extensive photographic plate collection there for possible variations of the 16th magnitude star. They surmised that the peculiar 3C48 might be a supernova remnant, somewhat like the Crab Nebula, and that an outburst might be revealed on the Harvard plates.[64] No variation was found for 3C48, but two years later a similar case arose in regard to 3C273, and again a trip to Harvard was undertaken, this time with gratifying results: this object was indeed variable.[65] Russian astronomers independently and simultaneously made the same discovery, but based on far fewer observations than were available from the Harvard plate collection.[66,67]

In his report for the two years ending June 30, 1961, Brouwer reported that a 20-inch reflector had been built and was being equipped with a photoelectric spectrophotometer designed by Smith and a graduate student, J. P. Rodman. (The optical parts of the telescope had been constructed by Tinsley Laboratories of Berkeley, California; the mounting was adapted from the original mounting of the Loomis Telescope.) This equipment was intended for studying possible auroral activity on Jupiter at H-α wave length, solar flares, and flare activity on nearby stars.[68] Rodman (currently, 1991, head of the Physics and Astronomy Department at Mt. Union College, Alliance, Ohio) wrote his Ph.D. thesis on this instrumentation, its function, and such observations as had already been obtained.[69] Smith contemplated further observations, but resigned as of August 1, 1963, to accept the post of Director of the McDonald Observatory and a professorship at the University of Texas.

Another graduate student, E. Milone, under the direction of Rodman who had been appointed Research Associate, continued a search for flare stars, but no new ones were discovered. Ultimately Milone completed his Ph.D. dissertation[70] under the guidance of Dr. A. Wesselink, successor to Harlan Smith. Wesselink with the assistance of Milone made improvements at the 20-inch, with which Milone made observations of eclipsing and peculiar Cepheid variables for his thesis. The completion of his thesis and most of the variable-star investigations carried out by or under the supervision of Dr. Wesselink, relate to a later time, after Dr. Brouwer's directorship, the ostensible conclusion of this particular history.

Photometry and variable stars did not constitute one of the major fields for research in the administrations of Schlesinger and his successor, Brouwer. Nevertheless, as noted, much pioneering and productive work along these lines was actually accomplished. However, the various successive investigators did not take over from their predecessors but pursued their own preferences. Hence there was little continuity or trend from one investigator to the next, especially in regard to the particular variable stars or types of variability to be observed.

REFERENCES

1. E. Hertzsprung, *Ast. Nachrichten,* **186**, 177, 1910.
2. J. Hartmann, *Astrophys. Jour.,* **10**, 321, 1899.
3. C. L. Stearns, *Trans. Yale Univ. Obs.,* **3**, 49, 1926.
4. P. van de Kamp, *Physics Today,* **35**, 67, June 1982.
5. J. Schilt, *Trans. Yale Univ. Obs.,* **9**, (27), 1933.
6. D. Brouwer, *Ann. Reports* for 1940–46.
7. V. Goedicke, *Ast. Jour.,* **50**, 145, 1943.
8. ———, *Ibid.,* **51**, 168, 1945.
9. J. Schilt and C. Jackson, *Ast. Jour.,* **55**, 9, 1948.
10. ———, *Ibid.,* **56**, 209, 1952.
11. F. Schlesinger, *Ann. Report* for 1927–28.
12. P. Lowell, *Bull. Lowell Obs.,* **2**, 17, 1912.
13. V. M. Slipher, *Idem.,* p. 19, 1912.
14. P. Slavenas, *Ast. Nachrichten,* **233**, 125, 1928.
15. J. Schilt, *Ast. Jour.,* **38**, 109, 1928; *Pop. Ast.,* **36**, 296, 1928.
16. E. Hertzsprung, *Ast. Nachrichten,* **207**, 89, 1918.
17. S. I. Bailey, *Astron. and Astro-Phys.,* **12**, 689, 1893.
18. H. Shapley and M. B. Shapley, *Astrophys. Jour.,* **50**, 42, 1919.
19. J. Schilt, *Pub. Ast. Soc. Pacific,* **39**, 160, 1927.
20. A. Belopolsky, *Ast. Nachrichten,* **136**, 282, 1894.
21. H. Shapley, *Astrophys. Jour.,* **40**, 448, 1914.
22. W. Baade, *Trans. International Ast. Union,* **8**, 397, 1952.
23. ———, *Pub. Ast. Soc. Pacific,* **68**, 5, 1956.
24. J. Schilt, *Ast. Jour.,* **38**, 197, 1928.
25. J. S. Hall, *Astrophys. Jour.,* **79**, 145, 1934.
26. K. F. Bottlinger, *Veröff. Berlin-Babelsberg Obs.,* **3**, Heft 4, 1923.
27. J. Stebbins and A. E. Whitford, *Astrophys. Jour.,* **98**, 20, 1943.
28. J. S. Hall, *Proc. Nat. Acad. Sci.,* **18**, 365, 1932.
29. P. Guthnick, *Ast. Nachrichten,* **214**, Jubiläumsnummer, p. 10, 1921.
30. E. Hertzsprung, *Bull. Lyon Obs.,* **11**, No. 33A, 1929.
31. J. S. Hall, *Astrophys. Jour.,* **90**, 449, 1939.
32. A. L. Bennett, *Astrophys. Jour.,* **88**, 1, 1938.
33. ———, *Pub. Amer. Ast. Soc.,* **8**, 209, 1935.
34. F. Schlesinger, *Ann. Report* for 1935–36.
35. A. L. Bennett, *Astrophys. Jour.,* **85**, 257, 1937.
36. F. Schlesinger, *Ann. Reports* for 1937–40.
37. F. Schlesinger, *Ann. Report for* 1939–40.
38. A. L. Bennett, *Ast. Jour.,* **47**, 104, 1938.
39. ———, *Astrophys. Jour.,* **90**, 289, 1939.
40. ———, *Ibid.,* **93**, 52, 1941.
41. E. Hertzsprung, *Ast. Nachrichten,* **205**, 281, 1917.
42. P. N Kholopov, *Gen. Cat. Var. Stars,* Moscow, 1985–87.

43. D. Hoffleit and C. Jaschek, *The Bright Star Catalogue,* Yale Univ. Obs., 1982.
44. J. Stebbins and A. E. Whitford, *Astrophys. Jour.,* **98**, 26, 1943.
45. D. Hoffleit, *Jour. A.A.V.S.O.,* **16**, 29, 1987.
46. L. Campbell, *Ann. Harvard Coll. Obs.,* **104**, Nos. 4–6; **107**, Nos. 1–3, 1936–39.
47. C. Hetzler, *Astrophys. Jour.,* **83**, 372, 1936.
48. E. Hertzsprung to F. Schlesinger, March 11, 1940, *Yale Obs. Archives.*
49. D. Brouwer, *Ann. Report* for 1951–52.
50. J. Ashbrook, *Ast. Jour.,* **53**, 104, 1948.
51. ———, *Ibid.,* **56**, 87, 1951.
52. ———, *Ast. Jour.,* **55**, 2, 1949.
53. A. H. Batten, J. M. Fletcher, and D. G. MacCarthy, *Pub. Dominion Astrophys. Obs.,* **17**, 153, item 219, 1989.
54. J. Ashbrook and R. L. Duncombe, *Astr. Jour.,* **56**, 204, 1951.
55. J. Ashbrook, *Pub. Ast. Soc. Pacific,* **60**, 116, 1948.
56. W. Becker, *Die Himmelswelt,* **44**, 47, 1934.
57. J. Ashbrook, *Astron. Jour.,* **58**, 145, 1953.
58. D. Brouwer, *Ann. Report* for 1952–53.
59. T. Dwight, *Memories of Yale Life and Men,* N.Y., p. 397, 1903.
60. H. J. Smith, *Ast. Jour.,* **60**, 179, 1955.
61. D. Brouwer, *Ann. Report* for 1953–54.
62. H. J. Smith, *Ast. Jour.,* **68**, 39, 1963.
63. W. G. Tifft and H. J. Smith, *Astrophys. Jour.,* **127**, 591, 1958; *Ast. Jour.,* **63**, 54, 1958 (Abstract).
64. H. J. Smith and D. Hoffleit, *Pub. Ast. Soc. Pacific,* **73**, 292, 1961; *Ast. Jour.,* **66**, 295, 1961 (Abstract).
65. H. J. Smith, *Nature,* **198**, 650, 1963; *Ast. Jour.,* **68**, 292, 1963 (Abstract).
66. A. S. Sharov and Y. N. Efremov, *Inform. Bull. Var. Stars,* No. 23, 1963.
67. L. Detre, *Ibid.,* No. 27, 1963.
68. J. P. Rodman and H. J. Smith, *Ast. Jour.,* **67**, 585, 1962.
69. J. P. Rodman, *An H-Alpha Spectrophotometer and Some Astronomical Applications,* Dissertation for the Degree of Doctor of Philosophy, Yale University, August, 1962.
70. E. F. Milone, *Photoelectric Photometry of Some Selected Variables,* Dissertation for the Degree of Doctor of Philosophy, Yale University, 1967.

20
SCHLESINGER, PUBLIC-SPIRITED HUMANITARIAN, TEACHER

Apart from his concentration on astrometry, Schlesinger invariably concerned himself with the more humanitarian problems of astronomers, and with public relations. As early as 1922, at the end of the Russian Revolution, Schlesinger, as then President of the American Astronomical Society, was involved in sending food to starving Russian astronomers. He resolved visa problems for Jan Schilt of the Netherlands and in 1928 assisted Brouwer in his efforts to bring his fiancée to America. During Hitler's regime and World War II, Schlesinger was active in getting refugee astronomers out of Germany, particularly Gustav Land, whose original name was Deutschland, to which the Nazis had objected! Land came to America in 1939, first to the Sproul Observatory and in 1942 to Yale where his most important work dealt with analyses of the accuracy of astrometric measurements. Another refugee aided by Schlesinger was Herbert Luft, ace sunspot observer, whose observations ultimately spanned more than sixty years. Luft had spent several years in Brazil before coming to New York in 1946.

On January 2, 1937, thirteen gentlemen met at the home of Donald S. Kimball in Hamden, a suburb of New Haven, to plan the organization of an Amateur Astronomical Society of New Haven.[1] A week later the group met again and elected Kimball as its first president, and voted to drop "Amateur" from the name of the society. Also, "The matter of admitting ladies to our society was briefly discussed but no action taken." This problem was resolved without comment when a lady was elected to the membership committee a few weeks later. Meanwhile, Kimball had discussed the newly organized society with Schlesinger, who made constructive comments, including that meetings should be limited to once a month. Starting on February 20, the meetings were held at the Yale Observatory and continued to be held in Yale buildings until 1975.[2] For his interest and helpfulness, Schlesinger was elected an Honorary Member. Eventually, members of this Society were to carry out observations of lunar occultations needed by Dr. Brouwer in his continuation of the work begun by E. W. Brown. Much later, in recent years, the Society became actively involved in refurbishing worn-out instruments at the Bethany station and making observations with them. Several of the members are (1991) also voluntarily identifying variable stars and measuring plates for more accurate positions than the available published coordinates. This work is important relevant to space telescope projects under the guidance of Dr. van Altena. These amateurs have over the years kept alert for opportunities to help the professional.

Schlesinger (like astronomers at several other institutions) concerned himself with the importance of public awareness of astronomy. Already in his first annual report for 1920–21, he described plans for using the 8-inch Reed telescope for public open nights, and he wrote Anson Phelps Stokes, Secretary of the University, requesting that the dome be fitted for visitors two nights a week, possibly as many as three or four.[3] In the first full year of operation, 1921–22, open nights were conducted on 51 nights attended by a total of 1,192 visitors. In other years, until the closing of the Observatory on Canner Street in 1955, anywhere from 7 to 36 nights were conducted each year, with an average attendance of 17 to 25 visitors per night—the numbers being limited by the capacity of the dome.

The importance of informal communication among astronomers was early recognized by Schlesinger when, beginning in June 1920, he started the group called *The Neighbors*. Astronomers primarily from the northeastern states planned to meet about four times a year, generally in New Haven, "to exchange ideas." There was to be no formal organization, no rules, no constitution, no dues. The meetings began on Friday afternoons and most participants left on Saturday, some on Sunday. Schlesinger reported,[4]

"We spend all this time together except the few hours that must be wasted in sleep." Reading of papers was strictly forbidden. He summarized, "In a small group like this, meeting so frequently, our knowledge of each other has gotten to be intimate and has given rise to a network of friendships which in themselves, aside from any questions of scientific results, justify the existence of the Neighbors."

In regard to the participation of women, Schlesinger was sometimes harshly antagonistic, at other times ambivalent. Dr. Ida Barney once said she spent the happiest years of her life working with Schlesinger at the Observatory. Nevertheless, she was not at all happy about the prevalent discrimination against women in regard to rank and pay. Early in his career (1901) Schlesinger had written Professor Hale at Yerkes Observatory,[5]

> I am thoroughly in favor of employing women as measurers and computers and I think their services might well be extended to other departments. Not only are women available at smaller salaries than are men, but for routine work they have important advantages. Men are more likely to grow impatient after the novelty of the work has worn off and would be harder to retain for that reason.

The success of his *Neighbors* was bound to attract professional women astronomers. Cecilia Payne of Harvard Observatory was particularly anxious to participate and correspondingly irate at being rejected. Schlesinger commented,[6] "Oh, the entering wedge," expressing the opinion that admitting the women would constitute "imposing a burden on hosts."

The value of reading the history of astronomy was always apparent to Schlesinger. In history one would find encouragement for continuing work that could not possibly be concluded in a single lifetime; there one could see that old observations, even when of low accuracy by modern standards, could be extremely valuable just because of their age. In a long article published near the end of his career[7] he pointed out the importance of observing changes in the sky, changes that might take centuries to be revealed. He concluded,

> Some of the greatest names in the history of science are those of men who foresaw the possibility of such changes and who set about making as accurate observations as they could, knowing well that they could not live to see these observations fulfill their purpose, but conscious of the value that their work would take on after the lapse of years. It is this characteristic of his work that is at once the despair and the inspiration of the astronomer. He often has occasion to envy the physicist and the chemist who are able to complete their own experiments. On the other hand, his work is not so easily superseded as theirs; for although it will happen to him as often as to anyone else that later improvements will make his own observations relatively inaccurate, these observations may still continue to be useful because they were made early, the element that preserves them from decay being their epoch. And as a corollary to this, the history of astronomy contains (and will continue to contain) many examples of men who have been able to make advances because they were acquainted with this history and have known where to look and what to look for in the records.

On March 11, 1927, Schlesinger was chosen by the Royal Astronomical Society to deliver its first George Darwin Lecture.[8] Here also he introduced his topic, *Some Aspects of Astronomical Photography of Precision,* with a brief historical survey. This paper, written over sixty years ago, could still be considered profitable compulsory reading for anyone considering a career in astrometry, for it described in adequate detail the successive problems that arose in determinations of stellar parallaxes and proper motions and the steps taken for their solution. He concluded this dissertation with a description of the new equipment erected in Johannesburg, indicating how an electric fan had been introduced to clear the air inside the telescope tube. Exposures for parallax plates were short, a new plate being exposed at intervals averaging eight minutes. The fan was turned on inside for about two minutes between exposures. With humor he ended,

> Experience with this device shows that it actually does improve the character of the images, and that, moreover, it obviates changes in focus that may otherwise be very annoying. If these precautions are not as effective as we hope, it is at least a satisfaction to know that we have done our best not to spoil the quality of starlight in the last few feet of so vast a journey.

As Director of the Observatory and member or officer of numerous scientific organizations, Schlesinger was sometimes called upon to recommend other speakers. Thus, when asked by the Dean of the University to recommend a speaker for one of the Silliman Lectures he, perhaps somewhat reluctantly, nominated Edwin Hubble in place of Ejnar Hertzsprung about whom he commented,[9] "Hertzsprung, a Dane long resident in Holland, has an important message to convey, but he is a very poor speaker in at least four languages, of which English is one. This is a great pity because he comes as close to genius as anyone now active in astronomy." Hubble was a fortunate choice. His lecture resulted in his classic, *The Realm of the Nebulae.*[10]

In a biographical article Dirk Brouwer said of Schlesinger,[11]

> He did not lecture a great deal, but when he did, he preferred to speak about a subject close to his own special fields of interest, and succeeded in lifting his audience to an appreciation of the objectives of astronomical research. . . .
>
> It may be regretted that a scientist who can expound his subject so clearly has so few students. Schlesinger's activity as a teacher of undergraduate students was always very limited. For graduate students he offered a number of subjects, with astronomical photography and the theory of errors as his favorite topics. . . . he succeeded in stimulating the initiative of his students. Especially when they were engaged in independent research . . . he would always be ready to drop his own work and concentrate with them on finding the best approach to the solution of a problem that had arisen.

Although he may not have had time to give many undergraduate courses at Yale, Schlesinger had a lifelong concern in the importance of teaching. After he earned his B.S. degree in astronomy from the College of the City of New York in 1890, he taught three years in the Public Night School System of New York City. In 1901 he was being considered for a professorship at the University of the State of Missouri. In his application for that post he dreamed as follows of founding a graduate school in astronomy:[12]

> There is an excellent opportunity for the establishment somewhere in the West, of a graduate school in the *new* Astronomy, that is in Spectroscopy and Astronomical Photography. These subjects would not require any expensive additions to the present equipment of the Laws Observatory; this is especially true of Astronomical Photography, which by the way, is taught in only one institution in this country. Should my application be favorably acted upon by the Curators, it would be my endeavor to establish such a graduate school as soon as the condition of the department would warrant, and it would be my expectation that the University authorities would not withhold the proper amount of encouragement towards this end.

Nothing seems to have come of this application, but while he was at the Allegheny Observatory Schlesinger also became the first head of the graduate school at the University of Pittsburgh.[13] At Yale, conditions were ripe for increasing emphasis on undergraduate as well as graduate education. In 1915 E. W. Brown, as chairman of a committee of the American Association for the Advancement of Science, had reported the committee's consensus on the training of students, especially the gifted, for research careers.[14] Subsequently he wrote several articles making suggestions for improving the curriculum, especially in mathematics, and comparing American education with that at Cambridge, England.[15–17] In a particularly perceptive address as retiring president of the American Mathematical Society, *The Relations of Mathematics to the Natural Sciences,*[18] Brown stressed that for applied mathematics a man "cannot usually be solely a physicist or solely a mathematician but must have received a sufficient education in both directions."

In his first annual report, for 1920–21, Schlesinger noted that, even though desirable, no undergraduate courses had been given that year. He commented that there was a tendency of institutions to devote little or no time to the training of astronomers, adding, "the situation presents a duty to Yale and an opportunity for science." He felt that the chief subjects needed were mathematics, physics, and modern languages, mathematics and physics being "as indispensable for the pursuit of astronomy as astronomy itself." But it was not until the 1923–24 term that, for the first time in many years, a course in elementary astronomy was offered. Five students attended, and twenty-five enrolled the following year.

By 1927 Schlesinger made arrangements for qualified students to use the Reed telescope, and in 1928 when the 9-inch Sheffield telescope was mounted under a new dome on Bingham Hall, it was put to use in connection with the undergraduate astronomy class. In 1930 Schlesinger relinquished the teaching of this class to Brouwer who, for the first time, also introduced an advanced undergraduate course on instrumentation, using a transit instrument as well as the Sheffield telescope. Schlesinger had introduced graduate courses in astronomy earlier; in 1920–21 he taught one course on astronomical photography, another in astronomical spectroscopy, and the following year one on the history of astronomy. Later other courses were offered, on least squares, celestial mechanics, and astrophysics, according to the interests of the faculty and needs of the students.

To keep staff and students informed of work in progress at the Observatory and discuss important papers in current journals, Schlesinger started a weekly meeting he called the *Journal Club.* This survived intermittently and was finally replaced by regular colloquia by outside speakers.

During Schlesinger's regime, an east wing was added to the Observatory's administration building. It was completed in 1928, providing six new offices and a small lecture room (Fig. 20.1) which was to

Fig. 20.1. Yale Observatory *ca* 1930 looking north; from left, the heliometer dome, transit room, original office building with 1928 addition to east, top of Loomis Tower in background, scaffolding for sliding roof of Catalogue Camera building (hidden behind foliage), and the Reed telescope dome. (Photo by Samuel Kravitt, New Haven.)

be used not only for classes but also for public open-night lectures when the weather was unsuitable for observing.[19]

The Yale Divinity School was built, partly on Observatory property, in 1932. This seemed like a culmination of a conflict of interest between science and religion. When the Divinity School plans were first proposed, Schlesinger had prepared a memorandum,[20] citing the disadvantages to observational astronomy, particularly to photographic astronomy, by the erection of a row of buildings so close to the telescopes, particularly to the south where most observations were conducted. His objections were to no avail, and prospects for setting up major new astronomical equipment on Observatory grounds had to be abandoned.

Until 1932 the Observatory, unlike other departments of the University, was under the jurisdiction of a Board of Managers as set up in 1879. The Board consisted of the Director, *ex officio,* as Chairman, the President of the University, also *ex officio,* and other members appointed by the Corporation. When this system was established, the Observatory was primarily a research institution financed by its own endowment. Now, with increased teaching loads, it was time for astronomy to be considered a separate department of the University, just as the Sloane Laboratory was a part of the Physics Department. Upon the recommendation of Schlesinger this plan was approved by the Corporation and, with the unanimous approval of the Board of Managers, that body was disbanded, its functions being taken over by the faculty of the Department of Astronomy. This was a triumph for Schlesinger who summarized the situation as follows:[21]

> It is hoped and confidently expected that this change will be especially beneficial with regard to undergraduate instruction, which has sometimes been casual or even haphazard in the past. There is no reason why a strong undergraduate department should not be established at Yale whenever, in the opinion of the administrative officers of the University, the time has come for doing so. It is significant that in spite of the fact that Yale's educational policy is precisely of the kind that fosters progress in pure science, the last graduate of Yale who entered astronomy as a career was of the Class of 1850, eighty-two years ago.

That graduate was H. A. Newton. During Schlesinger's directorship at most three undergraduate courses in astronomy were offered. Two students who graduated from Yale in this era became highly esteemed professional astronomers: Thornton Page, Yale 1934, Director of the Van Vleck Observatory, 1959–81, and Lyman Spitzer, Yale 1935, Director of the Princeton University Observatory, 1947–79.

REFERENCES

1. J. Marks, *Fifty Years Ago This Month,* a brief history of the Astronomical Society of New Haven, prepared for the members, 1987 (unpublished).
2. J. D. English, *A History of the Astronomical Society of New Haven, Conn., from 1950 to the Present,* privately duplicated for the members, 1979.
3. F. Schlesinger to A. P. Stokes, Apr. 18, 1921, *Yale Obs. Archives.*
4. F. Schlesinger, *Science,* **64**, 596, 1926.
5. F. Schlesinger to G. E. Hale, July 13, 1901, *Yale Obs. Archives..*
6. F. Schlesinger, May 1, 1936, *Yale Obs. Archives.*
7. F. Schlesinger, *Pop. Ast.,* **48**, 120, 1940.
8. ———, *Monthly Notices. Roy. Ast. Soc.,* **87**, 506, 1927.
9. F. Schlesinger, Apr. 24, 1934, *Yale Obs. Archives.*
10. E. Hubble, *The Realm of the Nebulae,* New Haven, xiii + 210 pp., 1936.
11. D. Brouwer, *Biographical Mem., Nat. Acad. Sci.,* **24**, 105, 1945.
12. F. Schlesinger to President R. H. Jesse, May 10, 1901, *Yale Obs. Archives.*
13. J. T. King, Sec. to the Dean, Univ. Pittsburgh, to K. Burns, May 15, 1945, *Yale Obs. Archives.*
14. E. W. Brown, R. G. Harrison, G. A. Hulett, and W. Lindgren, *Science,* n.s., **42**, 368, 1915.
15. E. W. Brown, "Reading Term at Cambridge," *Yale Daily News,* June 1, pp. 1, 6, 1915.
16. ———, *Ibid.,* April 7, pp. 1, 4, 1916.
17. ———, *Yale Alumni Weekly,* **28**, 310, Dec. 13, 1918.
18. ———, *Bull. Amer. Math. Soc.,* **23**, 213, 1917.
19. F. Schlesinger, *Ann. Reports* for 1827–28, 1928–29.
20. ———, undated, *Yale Obs. Archives.*
21. ———, *Ann. Report* for 1931–32.

21
PREEMINENCE OF CELESTIAL MECHANICS IN THE BROUWER ERA

Schlesinger had reason to be happy with the appointment of his successor, Dirk Brouwer (1902–66; Fig. 21.1), whom he had recommended as Director and who conscientiously carried forward the astrometric projects that Schlesinger had brought to such perfection. Scientifically, however, Brouwer was more significantly the successor of E. W. Brown, the greatest contributions of both having been in celestial mechanics. Brouwer, a Ph.D. from the University of Leiden, had first come to Yale in the fall of 1927 as a Fellow of the International Education Board. E. W. Brown was much impressed with Brouwer's work, and Schlesinger, in a letter to President Angell of Yale, stressed the importance of Brown's wish to have Brouwer as his assistant:[1]

The subject of Celestial Mechanics has, for reasons that I need not take your time to go into here, been much neglected of late, and has always been a subject that has been cultivated by solitary workers. Very seldom indeed is the work of any master in this subject continued by a pupil. Thus Simon Newcomb cannot be said to have had a pupil in any sense, and the same is true of G. W. Hill, who, as you doubtless know, never occupied an academic position. Brown might be said to be a pupil of Hill's, but only in the sense that he has continued his work after becoming acquainted with it through his publications. Brown himself now occupies a unique position in this branch of Science. There are few others who are concerning themselves with Celestial Mechanics, but they are doing so more from the point of view of Pure Mathematics than of devising methods by which the positions of the heavenly bodies can be predicted with accuracy and facility. Mr. Brouwer, whom Professor Brown recommends for the appointment, and who has been with us for half a year, has made a most favorable impression. He is not only competent and enthusiastic about Celestial Mechanics, but I have been delighted to see that he maintains a broad view of Astronomy as a whole. This is a particularly valuable asset for one who proposes to attack the problems in Celestial Mechanics that now need attention. If he should be appointed as Professor Brown recommends and should continue to work with him for a further period of two or three years, there is a good chance that he will absorb a knowledge of Brown's methods in a way that would hardly be practicable otherwise, and so the University would lay the foundation for the continuation of Brown's work by a very promising pupil.

In 1939 Brouwer had been offered the directorship of the U.S. Nautical Almanac as successor to James Robertson who was retiring. Brouwer was being offered an annual salary of $6500 at a time when Schlesinger was trying to persuade Yale University to increase Brouwer's salary to $4500! Schlesinger was about to retire in two years' time and he wrote President Seymour that he had Brouwer in mind as his own successor as Director of the Yale Observatory. Tempting as the Nautical Almanac offer was, Brouwer decided to remain at Yale because of its "better research possibilities."[2] There were other promising candidates for the Nautical Almanac position: for example, F. L. Whipple at Harvard was interested in the opening and wrote Schlesinger, asking to be recommended for the post.[3] But Wallace Eckert (Yale Ph.D. 1931) was in the forefront. Harlow Shapley, however, was not in favor of Eckert at that time, commenting that with Eckert's too great devotion to Hollerith machines "the Nautical Almanac is likely to sink into quiet respectability."[4] When Eckert was nevertheless appointed, Shapley remarked "I have a feeling his directorship will be more competent than inspired, but certainly we should feel confident that the job of the Nautical Almanac is not in blundering hands."[5]

Five years later Eckert resigned in order to become Director of the newly expanded IBM-Watson Computer Center affiliated with Columbia University, which was already collaborating with Yale on computational projects. Brouwer was again approached to head the Nautical Almanac Office, again with a more lucrative salary than he was earning at Yale, and with other attractions as well. But he wrote Provost Furniss,[6]

Fig. 21.1. Dirk Brouwer (1902–1966).

> The temptation of the appointment is that in the organization of the Naval Observatory I would be able to cooperate with several capable men in my own field, whereas at Yale Observatory I stand rather alone. In this respect the conditions at Washington are considerably more favorable than in 1939. In addition, the equipment and resources of the Naval Observatory are far beyond those available here. On the other hand, the director of the Nautical Almanac is burdened with administrative duties of considerable extent, and finds his freedom of action in the organization of his office restricted by Civil Service regulations.
>
> Without going into the arguments pro and con, I can say that I am inclined not to accept the appointment to the Directorship of the Nautical Almanac. The principal reason is that the new Thomas J. Watson Bureau will open up such excellent opportunities for our research here. However, as I have expressed at various occasions, the loss of our Southern Station would limit our research possibilities very greatly, and it seems to me that I should not turn down the offer without having more definite assurance with regard to the future arrangements concerning the Southern Station.

The Southern Station was suffering both for financial reasons and from city lights and smoke pollution from the nearby railroad. Schlesinger would have been distressed that his cherished Station was in difficulties hardly a year after his death. (But these problems will be more fully treated in Chap. 24.)

Like Schlesinger, Brouwer did not fail to encourage research in other branches of astronomy, particularly, in addition to astrometry, in astrophysics, photometry, and radio astronomy. However, partly for financial reasons and partly because of University regulations beyond his control, Brouwer was less successful in increasing the overall scope of research activities at Yale, as exemplified especially in the case of radio astronomy (see Chap. 22). But in the field of celestial mechanics he achieved worldwide preeminence.

Also like Schlesinger before him, Brouwer was ever ready to help those handicapped or in need, hiring them for whatever tasks he considered them capable; but he did not supervise their work himself. Usually they turned out very helpful. If some of them did not succeed in their assignments he still gave them the benefit of the doubt and held the immediate supervisor to whom he had assigned them responsible for whatever went wrong!

Another instance of Brouwer's helpfulness arose during World War II. Dr. W. Dieckvoss of the Hamburg Observatory, an officer in the German army in the invasion of North Africa, had been captured and spent May 1944 to November 1945 at the military prison camp at Como, Mississippi. From there he contacted various American astronomers, asking for current astronomical literature and also volunteering for work that he might be able to do in prison. Many responded by sending journals or texts, and after the war he wrote an appreciative review, "Die Astronomie in Amerika 1944–1945," for the popular German journal *Die Himmelswelt.*[7] But evidently, during the war only Brouwer considered the psychologically more important problem of supplying professional work. Others, most on leave from their observatories for military duties, could not, or were wary of, proffering this type of mutual benefit. Brouwer, however, in his annual report on his Asteroid Program for 1944–45 wrote:[8]

> Mr. Land also computed eight ephemerides of asteroids, and prepared data for the computation of twenty-two ephemerides by Dr. W. Dieckvoss, a German astronomer, now a prisoner of war in this country. In the summer of 1944 this prisoner of war wrote to Yale Observatory, asking to be enabled to perform voluntary astronomical calcula-

tions for the Observatory. Notwithstanding difficulty of communication we succeeded in having Dr. Dieckvoss do a considerable amount of work useful in the asteroid program. Unfortunately, few types of calculation lend themselves to being carried out at a great distance with a minimum amount of computing facilities.

In the decade before World War II, in many universities celestial mechanics played but a minor role in the astronomy curriculum, theoretical astrophysics having attained supremacy. Hence, after the war, potential graduate students interested in obtaining a Ph.D. for a thesis in celestial mechanics faced a dilemma. One dedicated advocate was Morris S. Davis, in 1947 an M.A. candidate at the University of Missouri. He had written Brouwer of his interest, but with a growing family he would require financial support, which at the time was not available at Yale. He canvassed numerous other institutions to no avail, and finally wrote Brouwer again:[9] "I received the impression that, except for the work you are doing (and have been doing), Celestial Mechanics is almost a defunct field as far as fruitfulness for research is concerned." At Yerkes Observatory Davis "received the overall impression that Celestial Mechanics was not the thing to be studied these days but rather some one of the more up-and-coming fields in vogue today." He was accordingly interested to know Brouwer's views, asking, "Has Celestial Mechanics been milked dry and is there no future until some mathematical tools are invented to handle the unmanageable problems left unsolved? Is it desirable for people to specialize in Celestial Mechanics these days? Your ideas will certainly help me to form an opinion."

Evidently Brouwer was impressed by the enthusiasm and perseverance of this young man, and did succeed in obtaining an assistantship at Yale for Davis who then earned his Ph.D. in 1950 for a thesis entitled *A Study of the Method of Rectangular Coordinates in General Planetary Theory.* Helping Davis achieve his goal in the long run proved advantageous for the Yale Astronomy Department as Davis eventually (1957–66) became the head of the Yale Computer Center and devoted considerable talent to the solution by high-speed computers of the problems previously too cumbersome for old-fashioned electric desk calculators.

In May of 1947 Yale had been awarded a contract with the Office of Naval Research for research in celestial mechanics whereby Yale, the U.S. Naval Observatory, and the newly founded IBM Watson Scientific Computing Laboratory in New York would collaborate on projects of mutual interest. This contract enabled Yale to acquire new IBM machines. Initially seven projects were jointly undertaken:[10]

1. *Revision of the motions of the principal planets.* Many precise observations had accumulated since the most recent determinations by Simon Newcomb some fifty years before;[11] the theories of motion had been and were undergoing refinement; and it was now possible to carry out the computations to higher degrees of accuracy than ever before. Results of this investigation, giving the positions for the outer planets from 1653 to 2060, were published in 1951 by Eckert, Brouwer and Clemence.[12] In 1953 Clemence[13] (Director of the U. S. Nautical Almanac) published a supplement giving the heliocentric coordinates of the combined center of mass of the sun and the five outer planets. These results were again supplemented by determinations of the perturbations acting upon the outer planets by the four inner[14] and on the secular variations of the orbital elements of the principal planets.[15] The heliocentric positions for Venus for the years 1800 to 2000 were published by Herget[16] in 1953; and R. L. Duncombe[17] in 1958 analyzed the motion of Venus from meridian-circle observations for 1750 to 1949. (This, with minor revisions, is Duncombe's Yale Ph.D. thesis of 1956.) The motions of Mercury, 1765–1937, had already been computed by Clemence before the advent of the joint ONR contract.[18] The motions of Mars were discussed by Clemence[19] and R. Laubscher.[20]

2. *Measurement of photographic plates of Saturn's satellites for the purpose of evaluating the mass of the system.* The mass of Saturn had earlier been obtained both from observations of the periods of revolution and separation from the planet of its satellites, and from the perturbations of Saturn on the orbits of the other planets, especially Jupiter. Redeterminations were in order as the results from the two types of determinations had yielded somewhat discordant results.

In an early Yale Ph.D. dissertation, in 1889, Asaph Hall, Jr.,[21] from an analysis of the orbit of Saturn's satellite Titan, had determined the reciprocal mass of Saturn (i.e. the ratio of the mass of the sun to the mass of Saturn) as 3,500.5. This value was in good agreement with Bessel's[22] widely accepted value of 3,501.6 ± 0.8, also based on observations of the satellite Titan. However, this result was found to be subject to systematic errors and a discussion by H. Jeffreys[23] involving six of Saturn's satellites yielded 3,494.8 ± 1.1. Meanwhile in 1953 H. Hertz[24] utilized available observations of Jupiter, 1884–1948, subsequent to those G. W. Hill[25] had used in 1895 for his tables of Jupiter for the determination of the mass of Saturn. Hertz's result was the most precise to date and effectively removed the discrepancy of the results from the satellites. Hill had found the reciprocal mass of Saturn to be 3,502.20 ± 0.53. Hertz obtained 3,497.64 ± 0.27. Later Clemence[26] found that discrepancies between observed and predicted coordinates could be removed by correcting the reciprocal mass to 3,499.7. The most recent accepted value in the current *Nautical Almanac,* 3,498.5, was determined by B. G. Marsden in the course of an analysis of the motion of the asteroid Hidalgo.[27] At Allegheny and Johannesburg, 465 plates had been taken, with over 7,000 satellite images for a new determination of Saturn's mass.[28] Although Brouwer and Clemence[29] indicated that the plates taken by Alden at the Yale Southern Station had been measured by F. H. Hollander and were being reduced by M. S. Davis for the re-determination of Saturn's mass, no results of this investigation appear to have been published.

3. *Search of the literature for observations of eclipses of the Galilean satellites of Jupiter.* An analysis of the observations was expected to help improve the theory of the system and establish a method for obtaining accurate positions of planets from photographs of their satellites. Brouwer remarked[30] "It is hoped that a discussion of the eclipse material will throw new light upon the discordances found by the late Prof. W. de Sitter[31] between the mean motions obtained from observations since 1890 and the average mean motions during the past three centuries."

Dr. Joseph Ashbrook undertook the tremendous task of compiling the old observations, amounting to some 20,000, the first 1,500 already found when the contract had been operative only one month.[10,32] Dating back to 1653, the observations were the reason for extending the calculated positions for the major planets back that far. Ashbrook also supplemented this material with visual observations at the Reed 8-inch refractor of current eclipses of Jupiter's satellites. However, although Brouwer's periodic reports to ONR indicate work in progress, there is no specific record of any published analysis of this massive accumulation of data. In 1966 in his Ph.D. thesis on the Galilean satellites, Brian Marsden[33] commented, "Eclipse observations certainly leave a lot to be desired," estimates of the times varying with the size of the telescope used, transparency of our atmosphere, and changes in the refractive properties of Jupiter's atmosphere. He concluded that "observations of eclipses cannot determine the time with an accuracy better than $\pm 10^s$, and that, since they contain errors of a systematic nature that cannot be allowed for, there is little improvement no matter how many observations are combined. Bradley had come to this same conclusion in 1728."

4. *Work on the secular perturbations of Pluto.* The effect of an unknown planet upon the motions of Uranus and Neptune was the major incentive for the search that led to the discovery of Pluto by Clyde Tombaugh at the Lowell Observatory in 1930. In an article, *Current Problems of Pluto,* Brouwer[34] reviewed the then current determinations of the mass of Pluto, calculated from its perturbation of Neptune's orbit. The most likely value then appeared to be about 0.9 times the mass of the Earth. More modern determinations have reduced this to only 0.002 Earth masses. The discrepancy is attributed to the fact that the orbit of Neptune was not sufficiently well established and that Charon, the companion of Pluto, crucial to the modern determination, was not discovered until 1978. Brouwer in 1950 had deplored the fact that Pluto had no known companion which, by its orbital motion, would have clinched the determination of the mass of the planet.

5. *Work on a new mathematical theory of the motions of Jupiter and Saturn.* In 1951 Clemence and Brouwer[35] summarized the results of the far more accurate calculations of the major planets stemming from both refined theory and the use of the new SSEC (Selective Sequence Electronic Calculator) recently completed by IBM. The residuals between observations and computed positions were greatly reduced. The maximum deviations for Jupiter, Saturn, and Uranus changed from $\pm 2''$ to $\pm 1''$, and for Neptune drastically from nearly $6''$ to less than $1''$ for observations after 1840.

6. *Development of accurate orbits for the first four asteroids (Ceres, Pallas, Juno, and Vesta).* The definitive orbits were to be carried out at the Cincinnati Observatory by P. Herget.[36] In 1949 and 1962 the *Astronomical Papers of the American Ephemeris and Nautical Almanac* contained the rectangular coordinates of these four minor planets at intervals of twenty days from 1920 to 1960 and 1960 to 1980, respectively.[37,38] In 1956 Herget reported[39] that perturbations by planets Venus to Saturn at ten-day intervals from 1920 to 2000 enabled the accurate computations of the ephemerides beyond 1960. The purpose of these investigations was the better determination of the position of the equinox among the

stars. This problem had been tackled by E. S. Jackson[40] at the Naval Observatory. However, R. Duncombe[41] found discordances between theory and observation which could be reduced by adopting a more recent determination of the mass of Mercury than that adopted by Jackson. Duncombe then computed more precise heliocentric coordinates for Ceres, Pallas, Juno, and Vesta for the years 1928 to 2000.

7. *Calculations for developing the theory of the motions of Mars by the Hansen method.* Clemence in two extensive contributions[19] stated that his goal was to attain a precision of ″.01 in the predicted geocentric position of Mars for approximately 500 years before and after 1850. Although there were indications that this goal was not quite achieved, he nevertheless concluded, "The theory is, however, considerably more precise than any other general planetary theory yet constructed." In a 1970 doctoral thesis, R. E. Laubscher[20] applied Clemence's theory to all the available meridian-circle and radio observations of Mars. He concluded, "Clemence's new theory of Mars is completely adequate for representing the available meridian observations of the planet."

In addition to the aforementioned seven major projects, Brouwer and the other celestial mechanicians in his department were interested in a variety of other topics. He himself became intensely involved in artificial satellite theory. Under the sponsorship of the Air Force his first paper, *Solution of the Problem of Artificial Satellite Theory without Drag,* was published in 1959.[42–43] Soon Japanese visiting astronomer Gen-ichiro Hori collaborated in this project,[44] extending the theoretical analysis to include the effects of atmospheric drag and the oblateness of the Earth.[45–48] These factors resulted in producing secular and long-period terms stemming from the short periods found in the drag-free solutions. Brouwer's theory did not hold for orbits with inclinations near a critical value of 63.4°. For such orbits Hori[49] and Garfinkel[50] developed the theory farther. Garfinkel believed that his solution clearly revealed all the essential features of the motion. In a later note[51] it was concluded that, if the mean inclination were close to the critical value, the perturbations in the coordinates would be of the order of 25 times greater than otherwise.

In the mid-1920s E. W. Brown had discussed apparent irregularities in the motion of the Moon, attributing the changes rather to irregularities in the rotation of the Earth (Chap. 15). Brouwer reactivated this problem, both theoretically[52–55] and observationally. According to his frequent collaborator, Gerald Clemence,[56] "He established that the fluctuations in the rate of rotation of the Earth are of a statistical character that would be produced by random disturbances in the interior of the Earth, and in 1950 he proposed *ephemeris time,* which has been adopted throughout the world for the astronomical measure of time that is free from such fluctuations."

The history of time-keeping, culminating with ephemeris time, Brouwer described in largely nontechnical terms in *Physics Today.*[57] In attempting to obtain observational measurements of the variations in the rotation of the Earth he was less successful. When the 15-inch Loomis telescope was moved to Bethany it was mounted as a polar telescope to photograph only the north polar sky. Throughout the night on a single plate, successive 10-minute guided exposures were made at 15-minute intervals (Fig. 21.2). A superposition of plates taken on different nights should then show the relative motion of the pole. Unfortunately, the results could rather be correlated with local seismic activity than terrestrial rotation. In her dissertation on the performance of the telescope, Carol Williams[58] found residuals between the observed and the theoretical star positions which could have arisen from a systematic change in the orientation of the axis of the instrument, which in turn might have been due to tides in the solid Earth.

Victor Szebehely, who was at Yale from 1963 to 1968, was primarily concerned with the three-body problem in celestial mechanics. In this short interval he published no fewer than 33 articles[59,60] and a monograph *Theory of Orbits, the Restricted Problem of Three Bodies.*[61] The problems that he addresses deal with two bodies revolving around their center of gravity and the relative motion of a third body of comparatively negligible mass. The shapes of orbits of the third companion are illustrated in Fig. 21.3 for the case of equal masses for the two primaries.[62] To the mathematically uninitiated such diagrams seem nothing short of fantastic; that the orbits are so complicated is less surprising than the fact that they can be convincingly represented mathematically.

In his work in celestial mechanics Brouwer seldom deviated from applications within the solar system. In two highly mathematical papers, however, he dealt theoretically with the problems of close binary systems. He first dealt with a particle of negligible mass revolving about a rigid spheroid .[63] In the second paper[64] he considered two rigid spheroids of about the same mass in which the axes of rotation

Fig. 21.2. A Loomis plate planned for analyzing the motions of the pole of the Earth's rotation.

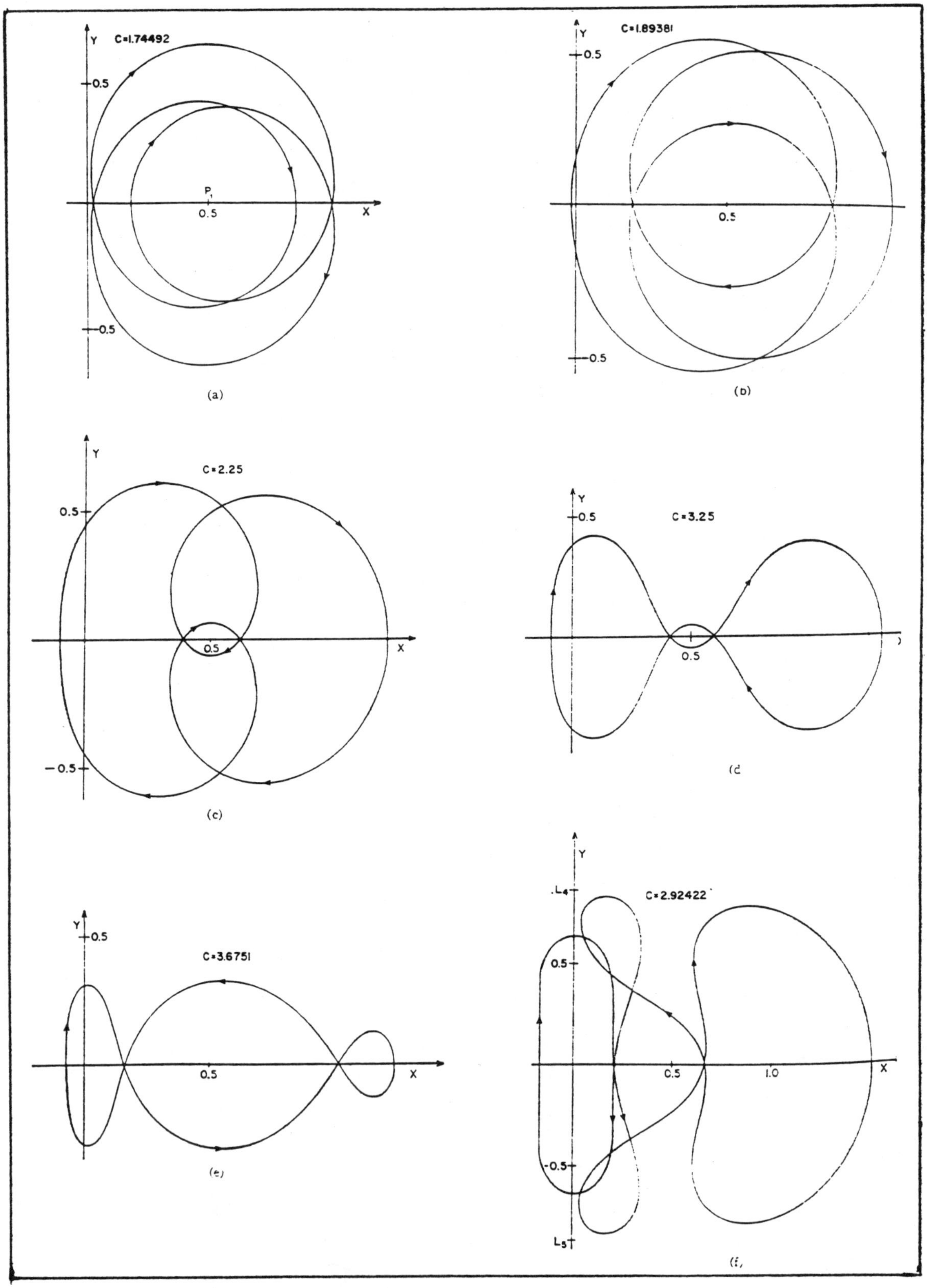

Fig. 21.3. Diagrams of possible orbit-shapes of a small satellite under the gravitational attraction of a binary star having equal mass components (Ref. 62).

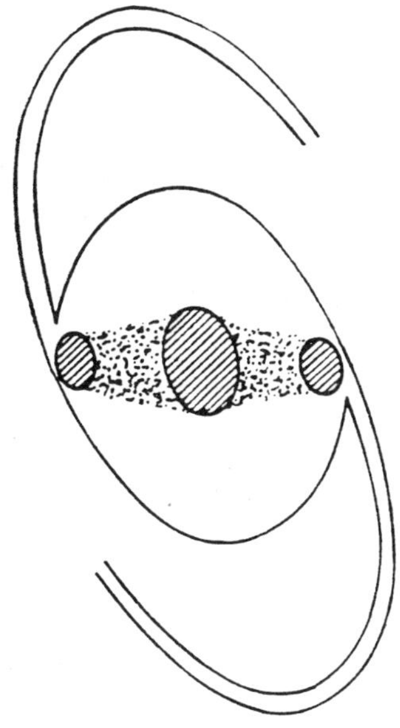

Fig. 21.4. Danby's sketch of NGC 2950. Reproduced by permission of author and Editor of *The Astronomical Journal*.

are not perpendicular to the orbital plane. For this case he found that "the motions of the orbital plane and the equatorial planes are expressed by two fundamental periods, both of which are short compared with the period of revolution of the line of apsides." He indicated that the light curves of many eclipsing binaries might not reveal the inferred motions because of the low inclinations of the orbital planes. However, "A study of the periodic terms is expected to lead to a dynamical explanation of the observed fluctuations in phase in numerous eclipsing binary systems." He cited discussions by Dugan and Wright of eclipsing binaries to which his theory might be applicable,[65] but Brouwer himself made no specific comparisons between this theory and such observations as were then available.

A member of the celestial-mechanics faculty, J. M. A. Danby, meanwhile was carrying out meticulous theoretical solutions in the difficult problem of the complex orbital motions in a three-body system.[66–68] This led him to a still more complex problem relating to barred-spiral galaxies,[69–70] suggested to him by Lynden-Bell.[67] A barred spiral such as NGC 2950 (Fig. 21.4), having pronounced condensations at the two ends of the bar, he found to be an unstable configuration. Particles escaping from the condensations would then form thin spiral arms.

ONR support for celestial mechanics continued throughout Brouwer's regime. With the perfection of ever faster and more versatile IBM equipment under Eckert's direction, the joint projects of the three institutions also increased in number and scope. Dr. Paul Herget of the Cincinnati Observatory voluntarily joined in the investigations and astronomers from other observatories, notably Lowell and Lick, contributed valuable observational data. The bulk of the results stemming from the ONR contract, both mathematical theory and extensive positional catalogues for both the major and many minor planets, were published in the *Astronomical Papers of the American Ephemeris and Nautical Almanac,* mainly between 1949 and 1969 (*see* Appendix E). Many other relevant papers were published in the *Astronomical Journal* and miscellaneous other journals. In much of the ONR-supported work numerous Yale graduate students took part, resulting in about twenty-five Ph.D. dissertations in celestial mechanics alone (Appendix B).

In his biographical memoir on Dr. Brouwer, Gerald Clemence[56] wrote that the launching of *Sputnik* in 1957 "led to an upsurge of interest in dynamical astronomy unprecedented in the history of the subject." Meanwhile Yale, under Brouwer's leadership, had become recognized as a world center for celestial mechanics. He attracted to Yale not only graduate students but also experts in the field as either distinguished visiting professors or to the Yale faculty. (These included, for periods ranging from one semester to several years, Drs. G. A. Wilkins, P. J. Message, A. Fletcher, T. B. Slebarski, and A. Hewish from Great Britain; J. Meffroy, J. Kovalevsky, and A. Deprit from France; G. Contopoulos from Greece; V. G. Szebehely from Hungary; G. Hori from Japan; C. A. Altavista from Argentina; and P. C. Choudhuri from India.) Brouwer himself was called upon as a consultant by military and industrial organizations concerned with preparations for launching vehicles into space and predicting their orbits.

As a consequence of his many commitments to celestial mechanics, Brouwer envisioned the establishment of a summer institute in dynamical astronomy at Yale. In response to his initiative, the National Science Foundation and the National Aeronautics and Space Administration in 1959 jointly sponsored the first experimental four-week *Summer Institute in Dynamical Astronomy at Yale.* The object was twofold: first to give instruction to college teachers who wished to introduce a course in dynamical astronomy in their own institutions; and second, to give instruction to government and industrial research workers who lacked formal training in the dynamical astronomy they needed for orbit and trajectory research. The courses were given by Yale astronomers and invited specialists from various institutions in the United States and abroad, including in its first year Brouwer (Director); Morris Davis (Associate Director); G. M. Clemence and R. L. Duncombe (U.S. Naval Observatory); B. Garfinkel (Ballistic Research Laboratories, Aberdeen Proving Ground); P. J. Message (Cambridge University, England); and J. P. Vinti (National Bureau of Standards). Of the approximately 100 participants, 20 per cent were college teachers and 80 per cent were nearly equally divided among government and industrial scientists.[71] The proceedings of the first summer,[72] covering fifteen sessions, were published in 1959 and, because of demand, were reprinted in 1960. The experiment was sufficiently successful for the program for 1960 to be extended to a six-week period, and it continued to be supported by the NSF for the next six years, bringing in many foreign experts as summer-staff members. It is estimated[73] that some 600 college teachers and government or industry employees were trained under this program which was so vital to the national defense and space-exploration projects. During this era, it should also be noted that Brouwer and Clemence[74] wrote a much-needed text, *Methods of Celestial Mechanics,* which is still a standard. Another text, *Fundamentals of Celestial Mechanics,* by J. M. A. Danby,[75] was intended primarily for advanced undergraduates or first-year graduate students.

Meanwhile Brouwer planned the creation in the Astronomy Department of the *Research Center for Celestial Mechanics,* to increase activity in both applied and the more mathematical aspects of the subject. This was sponsored jointly by the Office of Naval Research and the Air Force Office of Scientific Research, beginning July 1, 1962, with an expected annual budget of $350,000. The grant was largely in response to Pentagon concern that the increasing interest in the modern field of radio astronomy tended to deplete the supply of researchers in celestial mechanics.[76] The new Center at Yale was expected to play a vital role in solving the problems involving earth-circling satellites, lunar and interplanetary probes, and the determination of trajectories for planned interplanetary space vehicles. The grant enabled Brouwer to increase his staff of celestial-mechanics professors by three (G. M. Clemence, G. Hori, and V. G. Szebehely, in addition to Brouwer, Davis, and Danby), doubling its size. Quarters in the departmental building at 135 Prospect Street now became intolerably crowded, and a former residential building on nearby Edwards Street was acquired for the new Research Center, with an additional room in another building. Dr. Brouwer commented unhappily,[77] "Scattering a department's activity over three buildings, even if only a few blocks apart, is not an ideal solution. An important difficulty is that everyone in the Department wishes easy access to the Departmental library. Efforts toward finding a solution for this particular problem continue . . ."

No happy solution was found; nevertheless scientific productivity did not slacken. The publication of technical papers proliferated. From the inception of the Center until 1968, when, after Brouwer's decease, government support of the Center declined, members of the Center had published well over a hundred scientific papers related to celestial mechanics, producing theories, tables, and texts of lasting value.

REFERENCES

1. F. Schlesinger to J. R. Angell, Jan. 12, 1928, *Yale Obs. Archives.*
2. Schlesinger Files, June 1939, *Yale Obs. Archives.*
3. F. L. Whipple to F. Schlesinger, Oct. 9, 1939, Schlesinger files, *Yale Obs. Archives.*
4. H. Shapley to F. Schlesinger, 1939, *Yale Obs. Archives.*
5. ———, Jan. 9, 1940, *Yale Obs. Archives.*
6. D. Brouwer to E. S. Furniss, Nov. 8, 1944, *Yale Obs. Archives.*
7. W. Dieckvoss, *Die Himmelswelt,* **55**, 32, 1947.
8. D. Brouwer, *Ann. Report* for 1944–45.
9. M. S. Davis to D. Brouwer, Feb. 10, 1947, *Yale Obs. Archives.*
10. D. Brouwer, *Ann. Report* for 1946–47.
11. S. Newcomb, *Ast. Papers Amer. Ephem. and Naut. Almanac,* vols. **5–7**, 1895–98.
12. W. J. Eckert, D. Brouwer, and G. M. Clemence, *Ibid.,* **12**, XXXI + 327 pp., 1951.
13. G. M. Clemence, *Ibid.,* **13**, Part 4, viii + 41 pp., 1953.
14. ———, *Ibid.,* **13**, Part 5, 42 pp., 1953.
15. D. Brouwer and A. J. J. van Woerkom, *Ibid.,* **13**, Part 2, 29 pp., 1950.
16. P. Herget, *Ibid.,* **15**, Part 3, x + 297 pp., 1955.
17. R. L. Duncombe, *Ibid.,* **16**, 261, 1958.
18. G. M. Clemence, *Ibid.,* **11**, Part 1, 1, 1943.
19. ———, *Ibid.,* **11**, Part 2, 223, 1949; **16**, Part 2, 261, 1961.

20. R. E. Laubscher, *Ibid.,* **22**, Part 4, 363, 1981; *The Motion of Mars, 1751–1969,* Yale Ph.D. Dissertation, 1970.
21. W. L. Elkin, p. 18, *Ann. Report* for 1888–1889.
22. F. W. Bessel, *Ast. Nachrichten,* **11**, 17, 1833.
23. H. Jeffreys, *Monthly Notices Roy. Ast. Soc.,* **114**, 433, 1954.
24. H. G. Hertz, *Ast. Papers Amer. Ephemeris and Naut. Almanac,* **15**, Part 2, 167, 1953.
25. G. W. Hill, *Ibid.,* **7**, 1, 1898.
26. G. M. Clemence, *Ast. Jour.,* **65**, 21, 1960.
27. B. G. Marsden, *Bull. Amer. Ast. Soc.,* **2**, 249, 1970.
28. D. Brouwer and G. M. Clemence, *Report to ONR,* No. 17, June 15, 1951.
29. ———, "Orbits and Masses of Planets and Satellites," *in* G. P. Kuiper and B. M. Middlehurst, Eds., *Planets and Satellites,* **3**, Chap. 3, p. 31, 1960.
30. D. Brouwer, *Trans. International Ast. Union,* **7**, 226, 1950.
31. W. de Sitter, edited by D. Brouwer, *Bull. Ast. Inst. Netherlands,* **8**, 213, 1938.
32. D. Brouwer, *Ann. Report* for 1949–50.
33. B. G. Marsden, *The Motions of the Galilean Satellites of Jupiter,* Ph.D. Dissertation, Yale Univ., 1966.
34. D. Brouwer, *Sky and Telescope,* **9**, 103, 1950.
35. G. M. Clemence and D. Brouwer, *Ibid.,* **10**, 83, 1951.
36. G. M. Clemence, Ref. 30, p. 77.
37. P. Herget, G. M. Clemence, and H. G. Hertz, *Ast. Papers Amer. Ephemeris and Naut. Almanac,* **11**, Part 4, 521, 1950.
38. P. Herget, *Ibid.,* **16**, 335, 1962.
39. ———, *Monthly Notices Roy. Ast. Soc.,* **116**, 218, 1956.
40. E. S. Jackson, *Ast. Papers Amer. Ephemeris and Naut. Almanac,* **20**, Part 1, 1, 1968.
41. R. Duncombe, *Idem,* Part 2, 133, 1969.
42. D. Brouwer, "Solution of the Problem of Artificial Satellite Theory without Drag," *Air Force Cambridge Research Center,* Contract TN-59–638, 1959.
43. ———, *Ast. Jour.,* **64**, 378, 1959.
44. D. Brouwer and G. Hori, *Ibid.,* **65**, 342, 1960.
45. ———, *Ibid.,* **66**, 39, 1961.
46. ———, *Idem,* p. 193, 1961.
47. ———, *Idem,* p. 264, 1961.
48. D. Brouwer, "Research on the Theories of Motion of Artificial Satellites," *Air Force Command Control* Contract AF-19(604)-4137, 1961.
49. G. Hori, *Ast. Jour.* **65**, 291, 1960.
50. B. Garfinkel, *Idem.,* p. 624, 1960.
51. P. J. Message, G. Hori, and B. Garfinkel, *Observatory,* **82**, 168, 1962.
52. D. Brouwer, *Proc. Nat. Acad. Sci.,* **38**, 1, 1952.
53. ———, *Ast. Jour.,* **57**, 125, 1952.
54. ———, "The Polar Motion and Changes in the Earth's Orbit," *in* H. Shapley, Ed., *Climatic Change,* Cambridge, p. 159, 1953.
55. ———, *Ast. Jour.,* **64**, 97, 1959.
56. G. M. Clemence, *Biog. Memoirs Nat. Acad. Sci.,* **41**, 69, 1970. Reprinted with permission of the National Academy Press, Washington, DC.
57. D. Brouwer, *Physics Today,* **4**, No. 8, p. 6, 1951.
58. C. A. Williams, *An Analysis of the Performance of the Loomis Polar Telescope,* Yale Ph.D. Thesis, 1967.
59. D. Brouwer, *Ann. Reports* for 1963–65.
60. R. Wildt, *Ibid.,* for 1965–68.
61. V. G. Szebehely, *Theory of Orbits, The Restricted Problem of Three Bodies,* 668 pp., N.Y., 1967.
62. V. Szebehely and P. Nacozy, *Ast. Jour.,* **72**, 184, 1967.
63. D. Brouwer, *Ast. Jour.,* **51**, 223, 1946.
64. ———, *Ibid.,* **52**, 57, 1946.
65. R. S. Dugan and F. W. Wright, *Contr. Princeton Univ. Obs.,* No. 19, 1939.
66. J. M. A. Danby, *Ast. Jour.,* **69**, 165, 294, 1964.
67. ———, *Ibid.,* **70**, 136, 1965.
68. ———, *Ibid.,* **72**, 198, 1967.
69. ———, *Ibid.,* **70**, 320, 501, 1965.
70. J. M. A. Danby and R. V. M. Zahar, *Ibid.,* **72**, 919, 1967.
71. D. Brouwer, *Ann. Report* for July 1959 to 30 June 1961.
72. G. E. Bowden and J. Flis, Eds., *Notes of the Summer Institute in Dynamical Astronomy at Yale University,* 1959 and 1960.
73. J. Ashbrook, *Sky and Telescope,* **31**, 146, 1966.
74. D. Brouwer and G. M. Clemence, *Methods of Celestial Mechanics,* Academic Press, N.Y., xii + 598 pp., 1961.
75. J. M. A. Danby, *Fundamentals of Celestial Mechanics,* The Macmillan Co., N.Y., xiii + 348 pp., 1962.
76. Anon., New Haven *Journal Courier,* May 24, 1962,
77. D. Brouwer, *Ann. Report* for July 1, 1961, to June 30, 1963.

22
RADIO ASTRONOMY: GREAT EXPECTATIONS AND HARD TIMES

Research in radio astronomy at Yale had but a brief albeit productive life, from about 1956 to 1965. Soon after coming to Yale in 1953, Harlan Smith wrote *A Prospectus for Radio Astronomy at Yale*[1] in which he reviewed the then current status of radio astronomy. He stressed the importance to Yale of taking on modern projects to supplement the ongoing older lines of research; the desirability of training students in the newer fields; proposing what he considered the most feasible equipment and projects to be undertaken at Yale; and finally, he provided cost estimates for parabolic dishes with or without steerable mountings. At Harvard, as a graduate student under B. J. Bok, Smith had come to appreciate the importance of this new field of endeavor. In 1958 Brouwer was to comment.[2] "It was his [Smith's] initiative that led to the inclusion of radio astronomy in our graduate teaching program and to a research project on planetary long-wave radio emission." Further, Brouwer added, "It is probable that with improved facilities, and with a strong activity in the new field of radio astronomy the department will attract more graduate students."

In his report for the year 1955–56, Dr. Brouwer noted that as soon as land for the Observatory at Bethany was assured, H. J. Smith and Rupert Wildt pointed out the desirability of providing space there for the possibility of undertaking research in radio astronomy. Brouwer remarked that the site requirements for radio astronomy are nearly the opposite of those for optical astronomy. For radio astronomy he felt the best location would be in a hollow, shielded on all sides by hills or mountains, whereas optical astronomy requires an elevated location without obstructions. After studying contour maps of the proposed area in Bethany, Brouwer communicated with owners and prospective donors to secure additional land to satisfy both requirements. But funding for equipment was also needed. In a fund drive in 1956 Brouwer[3] remarked that only four U.S. universities and about the same number abroad had already combined traditional and radio astronomy in the same department, adding,

> Yale has the opportunity to begin this now, before the lead of other institutions becomes too great. Yale has strong Physics and Electrical Engineering departments which are much interested in the new developments and with which fruitful cooperation will be possible. This past autumn, in 1956, two outstanding students have entered the graduate school with the specific intention of obtaining their degree in Physics and Astronomy combined, that is, in radio astronomy, and other students have expressed great interest.

Until substantial funds should be forthcoming, relatively modest "home made" equipment would suffice for studying radio signals from the planets Venus and Jupiter. Moreover, such studies would be very appropriate in view of the department's interest not only in planetary motions (celestial mechanics) but in their astrophysical properties, Yale's Dr. Wildt having been recognized as one of the world's leading scientists in the physics of planets.

The two students to whom Brouwer referred, though not by name, were James N. Douglas and Kenelm W. Philip, who had come to Smith asking that courses be offered in radio astronomy. Philip is the one mainly credited with suggesting that Yale observe Jupiter for radio variability, this having already been indicated by observations by Burke and Franklin[4,5] at the Mills–Cross antenna of the Carnegie Institution of Washington. Smith and his two students tackled the problems of observing Jupiter not so much in a typical student—professor relationship, but rather as a cooperative research team in which each of the three had knowledge or expertise in different aspects of the problem. As Douglas was to recall years later,[6] "We were three very different people—Harlan the astronomer, Ken the iconoclast, skeptic, and incredibly creative jack-of-all-trades, and me the hardware man. We all grew to partake of each other's characteristics, and the sum of the three was greater than three."

All three were "simply trying to learn the subject from the ground up." They had no money to buy equipment, but a friend, presumed to be Brouwer, contributed a few hundred dollars for buying component parts such as resistors, capacitors, and Yagi antennas. Douglas then solicited the help of the Yale Radio Club (W1YU) of which he was president at that time, and in 1956–57 built a simple two-element interferometer operating at 21 megahertz (i.e. 21 million cycles per second), to be erected on the roof of Hendrie Hall in downtown New Haven. Observations began in March 1957. Meanwhile Smith and Wildt received a grant of $10,000 from Research Corporation to begin a research project in planetary radio astronomy.[7]

The initial results of their efforts were presented at a meeting of the American Astronomical Society in August 1957.[8] In the first month of their operation they had concentrated on Jupiter, recording nine probable "Jupiter events" (storms). Saturn, more favorably located, was followed for the next two months, yielding thirteen events which appeared to indicate a period of 10^h22^m. (Its currently accepted rotational period is $10^h39.9^m$.) The results, which the authors granted were inconclusive pending further observations and better discrimination against terrestrial atmospheric and galactic background effects, were disputed by Hewish and Pausey.[6] Later, Douglas believed that the supposed early Saturn events were more likely terrestrial static.[6]

Soon Smith was awarded a $21,000 NSF grant for two years to continue his radio astronomy research.[9] Now three interferometer pairs of antennas were operating at 23.0 mc/s: one on Hendrie Hall, another at Pound Ridge, N.Y., about 35 miles west of New Haven, and the third at Bethany, about 11 miles north of Hendrie Hall. The purpose of the configuration of stations was to confirm decametric storms on Jupiter by disentangling the observations from terrestrial ionospheric effects. At Bethany also, another pair of Yagis was operating at 22 and 23mc/s.[10] With these, Smith and his students were in fact to confirm storms on Jupiter, with pulse elements on the order of one-second duration. Nevertheless, they found that "Saturn remains a tantalizing, though still doubtful, possible source."[11] Indeed, an unsigned memorandum on radio astronomy for 1963 in the Yale Observatory archives, reporting on the principal results of the past five years, states specifically that no other planet besides Jupiter had been proved to be a decametric source, although Saturn had been suspected by several groups. The same memorandum also notes that Jupiter was found not to be a conspicuous source of H-α emission associated with radio emission. And perhaps the most daring speculation was that the radio observations suggested that the core of Jupiter rotates 13 seconds faster than the mantle, and that changes in rotation of the mantle may arise from its sliding over the more massive core.

For their planetary studies, by 1959 Smith and Douglas were operating at six different frequencies (13.25, 16.5, 19.2, 20.2, 22.2, and 23.0Mc).[12] Already in 1961, in his report to the President and Fellows of the University,[13] Brouwer had written enthusiastically,

> The research so far has concentrated on radio storms from Jupiter and the sun. Such studies can be made only at the lowest frequencies accessible to radio astronomers. The efforts of the last several years have made the Yale Observatory one of the four important observatories in the world active in low frequency radio astronomy research. Recently . . . experimentation has begun on medium-frequency equipment designed for the cosmological problem concerned with positions and counts of galaxies that are observable as radio sources.

Graduate students J. P. Rodman and F. A. Bozyan shared in the design and construction of a radio polarimeter and corner reflectors, as did undergraduates Barry Lasker and Michael M. Davis in electronic developments for the various projects. By 1961 large Yagi antennas had also been set up at Middletown and at the Pomfret School in Connecticut. In 1962 Bozyan was awarded a Fulbright Fellowship to Kodaikanal, India, where he installed a Yagi interferometer to be used in cooperation with the Yale programs intended for studying fine structure of the ionosphere, the magnetosphere, and possible other sources.[6] Equipment was also being built at Yale in 1963 for establishing four stations equally spaced in longitude for a project sponsored by the National Aeronautics and Space Administration for observing radiation from Jupiter and the sun. Bozyan's work in Kodaikanal was a part of this project. Douglas was also constructing a broad-band interferometer for obtaining high positional accuracy.[14]

Douglas earned his Ph.D.[15] in 1961 and was appointed Assistant Professor. Over a span of about ten years (or until both had left Yale for the University of Texas), Smith and Douglas published some fifteen papers or abstracts dealing with radio observations of Jupiter. Appreciating the importance of duplicate observations as well as large base-lines between stations, they succeeded in getting the cooperation of other institutions to strengthen their results, and progressively improved the facilities at Yale itself. In an international colloquium held in Liège in 1962, *La Physique des Planètes,* Douglas and Smith[16] summarized their conclusions in regard to four aspects of their problem. *First,* the Jovian storms indicated a period of rotation of the planet of $9^h\ 55^m\ 29.37^s$ with a probable error estimated at only 0.1^s, more than three orders of magnitude greater accuracy than from optical determinations.[17] This depended upon observations of a total of nearly 700 separate observed storms, compiled from published sources and from other institutions as well as Yale. A more detailed analysis indicated that the period was effectively constant over the interval 1950–61, showing no drift exceeding one second. This determination of Jupiter's period of rotation established a definitive system of radio longitudes on the planet. (The Yale period was adopted by the IAU in 1962 but later, in 1976, revised by Warwick[18] to $9^h\ 55^m\ 29.71^s$.) *Second,* their analysis supported a suggestion previously advanced by Warwick[19] that decametric radiation on Jupiter is caused by auroral type of activity in Jupiter's atmosphere. *Third,* polarization measurements carried out at Bethany indicated relatively rare reversals in polarimitry at decametric wave lengths. Although this phenomenon was not yet explained, it was considered important toward the determination of acceptable models for the radiation source. Finally, taking into consideration that the intensities of the observed bursts of energy are often modulated by ionospheric diffraction and refraction effects, they estimated that the total energy generated during a strong storm must amount to 10^{20} to 10^{23} ergs. At about the same time, in the *Astronomical Journal,*[20] Douglas and Smith presented in greater detail their observations and analyses of the frequency and other characteristics of the Jovian storms. Their final report[21] on the investigations begun at Yale was published in 1967. Summarizing observations spanning more than four oppositions of Jupiter, they established the origin of the 1-sec component of time structure in Jupiter's decametric emission (called "L Pulses") as due to diffraction by inhomogeneities in the solar wind moving radially outward from the sun. An upper limit to the angular size of the decametric source was indicated to be about 5″. In a 1963 letter to *Nature*[22] Douglas and Smith reported on an apparent increase in the period of Jupiter by 0.8 sec sometime after 1961. This would not necessarily represent a change in the period of rotation of the planet; rather it was suggested to indicate a drift in the source of the radio emission.[22]

Years later Smith was to suggest that the work on Jupiter at Yale laid the early foundation for the discovery of pulsars. He commented,[23]

> Meanwhile Tony Hewish back in England had the insight to extend the idea of a tiny scintillating Jupiter source, casting its diffraction pattern on the earth through irregularities in the solar wind, to extra-solar-system sources, in order to use this technique to pick out those of very small angular size—the scintillators. To do so required building first a relatively large-aperture low-frequency array which would be able to record these relatively faint sources fast enough to permit the use of short enough time constants to look for scintillation effects in the solar wind. As I understand it, grad [*sic*] student Jocelyn Bell was assigned the job of building this antenna and putting it to use. Thus came into existence the first antenna which had a good chance of detecting the pulsars, and thanks to Jocelyn's skill and persistence (and obstinance?) as an observer, they were indeed discovered. Their properties were then tracked down through the combined efforts of the Cambridge astronomers including especially Tony Hewish. Here ends our version of the story of the background to the discovery of pulsars, and how in a real sense it began with an attempt at Yale to use Jupiter to measure the density of the interplanetary electron plasma.

Six Yale Ph.D. theses resulted from the leadership of Smith and Douglas. The first two with Smith as supervisor were by Douglas[15] and Philip,[24] who analyzed radio records obtained at Bethany on solar noise storms. Four Yale students completed their investigations at the University of Texas under the supervision of Douglas and received their degrees from Yale in 1968: F. A. Bozyan;[25] C. C. Brooks;[26] L. P. Pataki;[27–29] and G. F. Moseley.[30]

Meanwhile at Yale, Dr. Edward Lilley was associated with the Astronomy Department for a short time. He came to Yale from the U.S. Naval Research Laboratory, September 1, 1957. Like Smith, he had obtained his initial training in radio astronomy at Harvard under Bok. His 1954 thesis, A *Radio Study of the Association of Interstellar Gas and Dust,* dealt with observations at the 21cm wave length of hydrogen.[31] At Yale, Brouwer had hoped to obtain funding for a parabolic radio reflector and auxiliary apparatus[2] which would have been suitable for further such studies, but it never materialized. Lilley, however, was particularly successful in obtaining grants for studies of the distribution of electron densities above the earth's F2 layer and of radio emission intensities beyond the earth's atmosphere and for designing equipment for these purposes to be attached to artificial satellites.[32]

Douglas collaborated with Lilley on a project sponsored by the Air Force Cambridge Research Center including the Geophysics Research Directorate, and the Air Research and Development Command, whose objective was an investigation to be carried out from space vehicles[33,34] of the Earth's outer ionosphere, to be based on observations of extraterrestrial low-frequency noise. Under contract with the Air Force, radio telescopes for satellite transportation and lunar and planetary probes were being constructed. The first telescope was expected to be launched by 1960.[35] Lilley was also influential in making provision for qualified Yale staff and students to obtain access to the facilities of the National Radio Astronomy Observatory at Greenbank, Maryland, for cooperative projects.[2] But the radio project of greatest interest to Brouwer himself was that of obtaining the parallax of the sun from doppler shifts in the 21cm hydrogen line.[36] Higher accuracy was expected from this project than from earlier optical determinations. The project was expected to last several years. Thus the prospects for radio astronomy at Yale seemed bright. Alas, these great expectations remained unfulfilled.

In 1959 Lilley accepted a more attractive appointment at Harvard. Harlan Smith, besides his very time-consuming efforts to build the radio astronomy equipment, carried a heavy and successful teaching load; spent considerable time as associate editor of the *Astronomical Journal*; kept up his interests in optical astronomy, especially on variable stars; was acting secretary for the American Astronomical Society, and spent a considerable amount of time on administrative matters during Brouwer's extensive absences from New Haven (notably at the southern stations). Yet to Brouwer's dismay Smith was denied tenure because of the "publish or perish" policy of the University, none of his other major and essential contributions being considered equally eligible. He left in 1963. Several letters in the Archives indicate how shocked other astronomers were. The strongest disapproval was expressed by a western astronomer who wrote Brouwer,[37] "Harlan Smith is held in high opinion everywhere, and it just doesn't seem likely that his abilities and efforts would be regarded as insufficient at any institution. The fact that they were so regarded by Yale leads to the conclusion that something must be wrong with Yale, not with Harlan Smith."

Fig. 22.1. One Yagi Antenna at Bethany. (Photo by W. Cleaver)

Douglas, who succeeded Smith as head of the radio astronomy work, was considered among the most promising of young American radio astronomers,[38] and Kenelm Philip was appointed Research Staff Astronomer. With these appointments Brouwer felt that the immediate future of radio astronomy at Yale appeared to be assured.[39] However, Douglas left of his own volition in 1965, and literally all the students working in radio astronomy also followed Douglas to the University of Texas where Smith was head of the now flourishing Astronomy Department. All of Yale's radio astronomy equipment (Figs. 22.1 and 22.2) was dismantled and transferred to Texas.

In his report for 1963–64, in summarizing the status of the various branches of the Observatory, Brouwer remarked, "The future of radio astronomy at Yale is a separate problem that should probably be considered after the new status of astrophysics in the department has become more clearly defined." (He was referring to the need for an additional faculty appointment in astrophysics.) Then in 1964–65 Brouwer commented, "It is clear that a clean break with the past activity is unavoidable; it is equally evident that the Department needs someone on the staff with a strong background and a serious interest in radio astronomy." However, Brouwer, who had so optimistically expected this new branch of astronomy at Yale to attract many new graduate students, himself inadvertently sounded the death knell to any replacement of these pioneers. Provost Kingman Brewster (later President of Yale) had inquired what activity Brouwer would cut if his budget had to be reduced by 15 percent; and what new activities he would undertake were the budget to be increased by 15 percent.[40] In response, in a letter dated July 10, 1963, Brouwer noted that the Astronomy Department consisted of five major divisions, namely celestial mechanics, astrometry, theoretical astrophysics, optical astronomy, and radio astronomy, of which radio astronomy was the latest addition. Hence, he concluded,[41] "If we had to give up fifteen percent of our salary budget, I would recommend cutting out this activity altogether. This would leave a department sufficiently broad to provide a well-rounded instruction program to graduate students." He also remarked that if a fifteen percent increase in the salary budget were to be granted he would use it "to strengthen theoretical astrophysics."

Closely related to radio is radar. In the former, signals are received from distant radiating objects; in the latter, pulses are emitted and reflected back to the source by a distant object. From the time delay between the emission and the reception of the reflected wave, the distance of the reflector can be determined. At the Yale Observatory no experimental work with radar was carried out. However, such signals were bounced off the Moon by technicians at the U.S. Naval Research Laboratory in Washington, D.C., and Brouwer assisted in the analysis.[42,43] Previously, the best determination of the mean distance of the Moon had been determined from occultation data,[44] 384,407.6 ± 4.7km. By 1960 it was found that the radar result, 384.400.2 ± 1.1km, yielded higher accuracy although the uncertainty in the Moon's radius enters into the determination. (Since 1961 the American and the British Ephemerides have given 384,400km as the mean distance of the center of the Moon from the center of the Earth.)

Meanwhile V. A. Kotelnikov in the USSR was doing significant work with radar,[45] which won for him the Lenin Prize in 1964. In particular, from measurements of Venus in 1961 he determined the length of the astronomical unit as 149,599,300 ± 600km. This compares with values also determined from similar observations at various installations[46,47] ranging from 149,596,000 to 149,601,000km. The American and British Ephemerides for 1961 gave 149,600,000km

Fig. 22.2. An array of Yagi antennas at Bethany for studying radiations from Jupiter. (Photo. by W. Cleaver)

and for 1990, 149,597,871km. One of Brouwer's graduate students, Brian Marsden, in a paper, "An Attempt to Reconcile the Dynamical and Radar Determinations of the Astronomical Unit,"[48] concluded that although a solution could be found that satisfied both types of investigations, "It is deduced that the astronomical unit is determined more satisfactorily from radar bounces off Venus rather than by dynamical methods alone."

REFERENCES

1. H. J. Smith, May 1954, *Yale Obs. Archives.*
2. D. Brouwer, Memorandum, "Needs of the Department of Astronomy," presented to the Executive Committee of the Faculty of Arts and Sciences, Jan. 21, 1958. *Yale Obs. Archives.*
3. ———, *Ann. Report* for 1955–56.
4. B. F. Burke and K. L. Franklin, *Jour. Geophys. Research,* **60**, 213, 1955.
5. G. Westerhout, *Ann. N.Y. Acad. Sci.,* **198**, 218, 1972.
6. J. N. Douglas in appendix of letter from H. J. Smith to W. T. Sullivan, Feb. 19, 1979, of which a copy had been sent to R. Larson by Smith, Sept.3, 1981. *Yale Obs. Archives.*
7. D. Brouwer, *Ann. Report* for 1956–57.
8. H. J. Smith and J. N. Douglas, *Ast. Jour.,* **62**, 247, 1957.
9. D. Brouwer, *Ann. Report* for 1957–58.
10. ———, *Ast. Jour.,* **63**, 386, 1958.
11. H. J. Smith and J. N. Douglas, *International Ast. Union Symposium* No. 9, R. N. Bracewell, Ed., p. 53, 1959.
12. D. Brouwer, *Ast. Jour.,* **64**, 322, 1959.
13. ———, *Ann. Report* for 1959–61.
14. ———, *Ann. Report* for 1963–64.
15. J. N. Douglas, *A Study of Non-Thermal Radio Emission from Jupiter,* Ph.D. Dissertation, Yale Univ. Sept., 1960.
16. J. N. Douglas and H. J. Smith, *Inst. d'Astrophys., Liège, Collection in 8°,* Nr. 448, 551, 1963.
17. Unsigned memorandum on radio astronomy for 1963, *Yale Obs. Archives.*
18. *See* T. Gehrels, *Jupiter,* Univ. Arizona Press, p. 37, 1976.
19. J. Warwick, *Astrophys. Jour.,* **137**, 41, 1963.
20. J. N. Douglas and H. J. Smith, *Ast. Jour.,* **68**, 163, 1963.
21. ———, *Astrophys. Jour.,* **148**, 885, 1967.
22. ———, *Nature,* **199**, 1080, 1963.
23. H. J. Smith to W. T. Sullivan, Feb. 19, 1979, *Yale Obs. Archives.*
24. K. W. Philip. *Position Measurements on Solar Continuum Sources at Decameter Wavelengths,* Ph.D. Dissertation, Yale Univ., 1963.
25. F. A. Bozyan, *A Study of Spectral and Statistical Properties of Jupiter in the Decameter Region,* Ph.D. Dissertation, Yale Univ., 1968.
26. C. C. Brooks, *Accurate Declinations of 33 Radio Sources,* Ph.D. Dissertation, Yale Univ., 1968.
27. L. Pataki, Jr., *Properties of the Interplanetary Medium from Scintillations in Jovian Decameter Emission,* Ph.D. Dissertation, Yale Univ., 1969.
28. ———, *Ast. Jour.,* **73**, S112, 1968.
29. ———, *Bull. Amer. Ast. Soc.,* **2**, 213, 1970.
30. G. F. Moseley, *Precise Declinations and Optical Identification of Selected Radio Sources,* Ph.D. Dissertation, Yale Univ., 1969.
31. A. E. Lilley, *Astrophys. Jour.,* **121**, 559, 568, 1955.
32. D. Brouwer, *Ast. Jour.,* **63**, 384, 1958.
33. A. E. Lilley and J. N. Douglas, "Objective: Research directed toward investigation of the outer ionosphere by observation of extraterrestrial low frequency noise in space vehicles," *Quarterly Status Report* No. 1, SAR/AF (604) 3059, 1958.
34. J. N. Douglas, G. R. Huguenin, and A. E. Lilley, "A Very Low Frequency Radio Experiment for Satellite and Probe Transportation," *Final Report,* AF 19 (604) 3059, Jan., 1960.
35. D. Brouwer, *Ann. Report* for 1958–59.
36. E. Lilley and D. Brouwer, *Ast. Jour.,* **64**, 338, 1959.
37. J. D. Schopp to D. Brouwer, May 29, 1962, *Yale Obs. Archives.*
38. B. F. Burke, Head of the Radio Astronomy Section of the Carnegie Inst. of Washington, to D. Brouwer , Dec. 4, 1963, *Yale Obs. Archives.*
39. D. Brouwer, *Ann. Report* for 1961–63.
40. K. Brewster to D. Brouwer, June 10, 1963, *Yale Obs. Archives.*
41. D. Brouwer to K. Brewster, July 10, 1963, *Yale Obs. Archives.*
42. D. Brouwer, R. H. Bruton, A. C. Miller, and B. S. Yaplee, *Results of 10cm Lunar Radar Range Study,* London, 1960.
43. B. S. Yaplee, S. H. Knowles, A. Shapiro, K. J. Craig, and D. Brouwer, *I. A. U. Symposium,* **21**, 81, 1965. Also *NRL Report* 6134, 1964, and *Bull. Astronomique,* **25**, 81, 1965.
44. J. A. O'Keefe, *Ast. Jour.,* **57**, 108, 1952.
45. A. E. Salomonovich, in W. T. Sullivan, Ed., *The Early Years of Radio Astronomy,* Cambridge, U.K., p. 284, 1984.
46. D. O. Muhleman, *I. A. U. Symposium,* **21**, 155, 1965.
47. I. Shapiro, *Ibid.,* p. 214.
48. B. Marsden, *Ibid.,* p. 225.

23
THEORETICAL ASTROPHYSICS

In the *Annual Reports* of the Observatory the subtopic astrophysics did not appear until the 1941–42 report, under "Astrophysics Research." The term "Observational Astrophysics" was introduced in the 1963–64 report. However, nearly all of the categories included under this heading had previously been listed simply under "Photometry" or "Variable Stars." Here only the work in theoretical astrophysics is discussed.

Despite some criticism after Brouwer's death that it was he who effectively stymied the progress of theoretical astrophysics at Yale, there is considerable evidence to the contrary. It was true that astrophysics did not flourish as spectacularly as his own field of celestial mechanics. Although he made efforts to recruit more professional astrophysicists he was largely unsuccessful, primarily because tenure positions could not be offered. The only tenured positions in the Astronomy Department at the time of Brouwer's death were held by himself and astrophysicist Rupert Wildt (Fig. 23.1).

Astrophysicist Lyman Spitzer, Jr. (Yale B.A. 1935, Princeton Ph.D. 1938) spent his first year after obtaining his doctorate as National Research Fellow at the Harvard College Observatory. There he wrote several highly theoretical papers on the physical theory of line broadening:[1] "The Spectra of M Supergiant Stars":[2] "The Dissipation of Planetary Filaments,"[3] dealing with filaments of stellar matter produced by encounters between two stars; and "The Stability of Isolated Clusters,"[4] in which he computed that even in one hundred billion years [10^{11}] a globular cluster will not lose more than half its stars through evaporation. These were impressive contributions from a scholar only 25 years of age. He was appointed to the Yale Physics Department, starting in the fall of 1939. For the term 1941–42 Brouwer succeeded in having him transferred half-time to the Astronomy Department.

Spitzer's appointment appears to have marked the initiation of theoretical astrophysics at Yale. In his *Annual Report* for 1943–44 Brouwer remarked, "It is hoped and expected Mr. Spitzer's presence on the staff of the observatory will give the University a strong representation in the important field of theoretical astrophysics." Unfortunately World War II interrupted much of the work that had been started, and this field did not become revitalized again at Yale until 1946 when Spitzer returned to Yale as Associate Professor of Astrophysics. On being asked to recommend Spitzer, Harlow Shapley at Harvard wrote,[5] "It seems to me unnecessary to write in any detail about his qualifications. Apparently he will go further in astronomy than any of us who pass judgment on him." Spitzer himself responded to the proffered appointment[6] by proposing a Yale Astrophysical Bureau with close cooperation between the physics and astronomy departments, and he later[7] outlined plans for spectroscopic research using high-altitude rockets and coordinating with the military. At the close of World War II, Spitzer had been contacted by the RAND Corporation in regard to his thoughts on a then classified project for putting a large Earthcircling artificial satellite into orbit.[8] At Yale Spitzer hoped to entice Michigan Professor Leo Goldberg (Harvard Ph.D., 1938) to join him in projected solar ultraviolet spectroscopy—an offer (not accepted) that presumably expedited Goldberg's promotion as Director of the Michigan Observatory.[8] Brouwer wrote Dean W. C. DeVane,[9] "Astrophysics has become such an important part of astronomy that it is desirable for Yale University to have a strong representation in this field."

In his report for 1945–46 Brouwer reported that arrangements had been made for a considerably expanded program of astrophysical research calling for the appointment of an additional astrophysicist. Dr. Rupert Wildt, then at the University of Virginia, was appointed Assistant Professor of Astrophysics in the Department of Physics. His bibliography up to that time, amounting to 44 technical papers dealing with planets and stellar atmospheres, was impressive.[10] Wildt was particularly renowned for two discoveries: in 1932 at Göttingen he found methane and

Fig. 23.1. Rupert Wildt (1905–1976).

ammonia in the atmospheres of Jupiter and Saturn;[11] and in 1939 his work at Princeton led to his discovery that the negative hydrogen atom would account for the hitherto puzzling continuous absorption at 3650 Ångstroms in the spectra of the sun and cooler stars.[12,13] During the war he had been engaged in an underwater antisubmarine research project under the direction of Lyman Spitzer at Columbia University. Upon Wildt's appointment Brouwer commented,[14] "The prospects are bright for an era of fruitful cooperation between the two departments" (astronomy and physics). Within only a year, however, Spitzer left, having been offered the directorship of the Princeton Observatory, and Wildt was transferred as Associate Professor to the Department of Astronomy.

While at Yale, Spitzer published two series of erudite and innovative papers in the *Astrophysical Journal* on "The Dynamics of the Interstellar Medium,"[15] and "The Temperature of Interstellar Matter."[16] This latter series he continued over the following two years after he left Yale.[17] He considered these studies "to have important application to the general behavior of the interstellar medium, both in the condensation of atoms into dust and in the condensation of dust into stars."[18] Also in a paper in the *Physical Review* on the magnetic field of the galaxy,[19] Spitzer discussed the force of galactic radiation pressure on interstellar dust grains, and concluded that "The rotation of interstellar clouds containing ionized gas would tend to be slowed down by the eddy currents generated, thus facilitating the formation of stars. While these effects require further study, there is little question but that the possible presence of a galactic magnetic field must be taken into account in discussions of interstellar phenomena."

In view of his early pioneering investigations before the war and his abilities as a stimulating teacher and guide for graduate students, several people[20] expressed disappointment that this inspiring young astrophysicist and Yale alumnus had not been appointed Schlesinger's successor as Director of the Yale Observatory. In his report in 1951 as Chairman of the Yale Council's Committee on the Division of the Sciences, Spitzer remarked,

> "For several decades Yale has been internationally known as a center of research in the two fields of planetary motion and precise measurement of stellar positions. This work continues now under Professor Brouwer, whose election to the National Academy virtually coincides with the Committee's visit to New Haven. While these programs are too specialized to excite broad popular interest, they are basic to the progress of astronomy, and occupy an important place in a University dedicated to the advancement as well as the broad dissemination of knowledge."

It was thus apparent that these important older branches of astronomy would not have been neglected had Spitzer been appointed Director, whereas his enthusiasm for the modern fields might have attracted far more students to Yale. However, at the time of Schlesinger's retirement in 1939, Spitzer's administrative potentialities could not yet have been that well foreseen.

Before appointing Wildt as Spitzer's successor Brouwer solicited the opinions of several prominent astrophysicists. Chandrasekhar wrote,[21] "I do not feel there can be any doubt that Wildt has been one of the most successful astrophysicists in recent years." Henry Norris Russell called him "a man of distinguished record and outstanding ability."[22] Wildt was a dignified, reserved, and far less outgoing type of person than Spitzer, but similarly inspiring to the more able students. A. N. Vyssotsky[23] at Virginia described him as "a born teacher," precise and well organized. One of Wildt's outstanding pupils was W. C. DeMarcus (1924–82) whose obituary for Wildt[24] most clearly illustrates the best qualities of Wildt as a man, researcher, and teacher. Some students of only moderate ability, on the other hand, said they felt that Wildt rejected them as unworthy of his valuable time; he had little patience with mediocrity. While Spitzer was concerned largely with problems of stellar and galactic evolution, Wildt is best known for his work on the constitution of the planets, the greenhouse effect, and stellar atmospheres, particularly the role of the negative hydrogen atom (H-) in the interpretation of stellar spectra.

In 1953 Wildt was offered the directorship of the Observatory at Munich, Germany. Brouwer was concerned because the salary and pension benefits offered there were better than those any other German astronomer was getting and could not be matched at Yale. Aware that Wildt had no desire to leave the States, Brouwer solicited letters from several prominent astronomers to help convince the University administration that everything possible should be done to persuade Wildt to remain at Yale. Most of the replies were favorable but, surprisingly, two were not. Otto Struve, Director of the Yerkes Observatory, who in 1948 had earnestly hoped to be able to get Wildt to go either to the University of Chicago or its Yerkes Observatory,[25] now wrote,[26]

> Wildt has had a number of brilliant ideas which have greatly influenced the development of astrophysics. His mind is unique in its ability to discard old ideas and break through new paths. In this respect he surpasses all other astrophysicists in this country and abroad. His discoveries of the influence of negative hydrogen ions in stellar atmospheres is one of several examples.

On the other hand, Struve continued,

> Wildt has never been a fast worker, and I detect in him a tendency to become slower as he grows older. Nor has he ever done much to build up detailed theories upon the basis of those flashes of inspiration of which H- is an example. He either dislikes the important routine work needed to perfect astrophysical theory, or is incapable of carrying it to its end. In this respect several other astrophysicists surpass him. . . . Since the directorship at Munich would apparently give him a fine opportunity for inspiring others to do the routine work which he himself avoids, I am not certain that he should be especially urged to decline it.

Struve then gave a list of several young astronomers who, in his opinion, would in the long run contribute more to the success of astrophysical research and teaching at Yale than would Wildt.

Lyman Spitzer who was probably the astronomer most familiar with Wildt's background and recent work wrote,[27] in 1953, "Since WWII he published only two papers in the *Astrophysical Journal,* one work at Virginia, the other jointly with Spitzer and Schwarzschild ; . . . I fear that his days of really active research may have come to an end when he came to Yale."

The *Scientific Monthly* had already pointed out that mathematicians, by and large, achieved their most important results in their early twenties, physicists in their early thirties, and astronomers in their early forties.[28] Wildt at this time was already 48. The promotion Brouwer hoped to get for him at Yale, a full professorship, did not materialize until the spring of 1958, at which time Brouwer[29] indicated that Wildt still had to decide between continuing to serve Yale or accepting the post of Director of the Munich Observatory.

Despite these criticisms, Wildt's accomplishments continued to be impressive. His bibliography of articles written between World War II and 1953 includes six papers of which he was sole author, dealing with the constitution of the planets, the spectrum of the solar chromosphere, interstellar molecules, and emission lines in the spectra of dwarf M-stars. His earliest publication[30] in 1928, "Untersuchungen über den photographischen Diffusionslichthof" appeared in the *Zeitschrift für Wissenschaftliche Photographie* (the same journal, incidentally, in which E. Hertzsprung[31] first published his discovery of giant and dwarf stars in 1905). Between 1928 and 1972 Wildt published 72 papers.[10] Those since 1953 dealt with radiative transfer, molecules in stellar atmospheres, late-type stellar models, the greenhouse effect, planetary interiors, and especially a series on gray atmospheres.[32] DeMarcus[24] concluded his eulogy with this remark relative to the end of Wildt's career: "Finally, he was keenly aware of the defects in *fundamental theory* and the last years of his career were spent in an all-out assault on the theory of radiation fields not in thermodynamic equilibrium. Perhaps he bit off here more than he could chew. But is it not fitting that a man should end his career having passed always to ever more difficult problems?"

DeMarcus had earned his Ph.D. in 1951 in the Physics Department at Yale, but under the thesis guidance of Wildt. In that thesis they had come to the conclusion that the giant planets Jupiter and Saturn consist mostly of solidified hydrogen.[33] The following year Wildt was on leave of absence as visiting professor at the University of Hamburg; but for the 1953–54 term he was awarded a National Science Foundation grant for work on the equation of state of solid hydrogen with application to the Jovian planets.[34] As an obvious collaborator, DeMarcus was appointed Research Fellow for that year. At the December 1952 meetings of the American Astronomical Society DeMarcus had reported that Jupiter and Saturn must be covered by hydrogen oceans, Jupiter to a depth of 340 miles if the temperature were 150°K, and Saturn to 620 miles for a temperature of 130°K.[35] He pursued his theoretical models further,[36] finding that Jupiter and Saturn are composed of hydrogen in excess of 78 percent and 63 percent by mass, respectively. In all of his work on planetary theory DeMarcus gratefully acknowledged advice and encouragement from Wildt. The mutual respect and friendship between the professor and his former student lasted throughout their lives.

Meanwhile, in 1959 Wildt was again offered a tempting European post, this time at Basel, Switzerland, a post previously offered him in 1947.[37] Brouwer[38] wrote Provost Norman Buck persuading him to give Wildt assurance that his financial outlook would improve, adding, "It is hardly necessary for me to repeat how much I value Mr. Wildt's contributions to the department, nor how difficult it would be to replace him."

Although he was highly respected, comparatively few honors were conferred upon Wildt. The University of Göttingen, where he spent the happiest and most productive years of his youth, in 1960 conferred upon him the honorary title of Professor Emeritus. As is often the case, his highest honor came many years after the work for which it was awarded. The Eddington Medal of the Royal Astronomical Society, given for a single noteworthy achievement rather than a life's work,[39] Wildt received in 1966 for what he had accomplished in 1939, namely his identification of the contribution of the negative hydrogen ion to the opacity of stellar atmospheres. A book, *High Pressure Physics and Planetary Interiors,*[10] the proceedings of a conference held in Houston, Texas, in March 1972, was dedicated to him on the occasion of his retirement from Yale in 1973. The first paper in this volume contained Wildt's last publication, "Hydrogen Planets and High Pressure Physics," a brief account of his own early work on planets, as well as his complete bibliography.

After his retirement Wildt wrote Mrs. Albee, Executive Secretary of the Yale Astronomy Department,[40]

"There are moments when you feel like tooting your own horn and advertising what the astronomical establishment has ignored. My discovery of methane and ammonia in the atmospheres of the giant planets solved a spectroscopic puzzle of sixty years standing, which was no less conspicuous than the unidentified spectra in the solar corona and gaseous nebulae. Their unravelling was honored by medals [Lockyer, Rumford Medal 1874; Bowen, Draper Medal 1942, respectively] but planetary physics never rated with astronomers until the rocket cadets took over. A little recognition of my pioneering efforts in that field would have gone a long way had it come twenty years ago."

In the same letter, on the other hand, Wildt expressed pleasure in the honor AURA (the Association of Universities for Research in Astronomy) had bestowed upon him when he retired as its President in 1966. AURA had been incorporated in 1957 and Yale joined in 1958. Wildt became the Yale representative on its Board, and chairman of an ad hoc committee on formulating AURA policy with respect to scientific appointments and tenure, and chairman for three years of the Scientific Committee, followed by his election as President of AURA in 1965 and 1966.[41,42] Wildt remarked,

The send-off the AURA Board gave me left a flow of satisfaction. For one thing, the new Corporate Office building, occupied since January, is a lasting memorial to my administration. In its conference room my portrait, a large size photograph taken by the Kitt Peak photographer, will be displayed. . . . I received an elegant sterling silver tray engraved with the AURA seal. There was no precedent to these honors, which came as a complete surprise. All administrative Board members congratulated me warmly, while most scientists remained distinctly cool.

The astrophysicists on the Yale faculty before Professor Pierre Demarque became Director in 1968 are listed in Table 23.1.

Ludwig Oster had been transferred from the Yale Physics Department where his interests centered on plasma physics, especially radiation theory. He became mainly interested in the astrophysical applications of these theories, especially to solar phenomena. With his graduate students he investigated the interaction of radiation and matter in completely ionized gases, necessary to the understanding of astronomical problems of the solar atmosphere. He computed mathematical models describing the observed magnetic fields of sunspots, finding neutral points where the magnetic field strength vanishes. This phenomenon was considered to have bearing on the origin of solar flares.[43] Oster and graduate student C. C. Brooks investigated thermal radio emission from the transition layer between the chromosphere and the corona. Their models quite accurately predicted actually observed decimeter radio

Table 23.1
Astrophysicists Appointed Before 1968

Name	Terms	Title
Lyman Spitzer	1939–42	Instructor
	1946–47	Associate Professor
Rupert Wildt	1949–57	Associate Professor
	1958–73	Professor
Wendell C. DeMarcus	1953–55	Research Fellow
Ludwig Oster	1960–67	Associate Professor
Hollis R. Johnson	1961–62	Research Associate
Norman Baker	1963–64	Lecturer
Myron Lecar	1963–65	Lecturer
Martin D. Altschuler	1964–65	Instructor
James H. Hunter, Jr.	1965–70	Assistant Professor
D. W. N. Stibbs	1966–67	Visiting Professor from St. Andrews, Scotland
Sumner Starrfield	1967–71	Assistant Professor

emission from the sun.[44] Graduate student M. D. Altschuler collaborated with Oster on theoretical models of radio emission from the sun[45] and on electromagnetic radiation from plasma oscillations.[46] With S. Sofia, another graduate student, the emission of the quiet corona at meter wave lengths was studied.[47] Visiting astronomer S. Anand, with Oster and Sofia, also investigated the excitation of lunar luminescence by solar protons, a controversial topic.[48] During his seven years' association with the Yale Astronomy Department, Oster was author or co-author of about twenty technical papers.

Hollis Johnson worked with Wildt under a NASA grant on physical processes in stellar atmospheres. Before coming to Yale, Johnson had written an important paper, "The Equilibrium of NaI in the Solar Photosphere,"[49] in which he found that sodium in the solar photosphere demonstrates departures from LTE (local thermodynamic equilibrium) in the sense that excitation temperature differs from electron temperature. This work he pursued after coming to Yale.[50] He extended this type of analysis to other elements in a paper, "Abundance of Trace Elements and Non-Equilibrium Effects,"[51] in which he specifically investigated the solar abundance of beryllium, finding that departure from equilibrium cannot account for the difference between the abundances of beryllium in stellar atmospheres and in meteorites.

Myron Lecar earned his Ph.D. under Wildt in 1963 for the thesis, *Departures from Local Thermodynamic Equilibrium in an A0 Star at Atmospheric Pressure.* In the following two years at Yale he continued similar studies[52] and presented a paper, "A One-Dimensional Self-Gravitating Stellar Gas," at IAU Symposium 25 at Thessaloniki, Greece, in August 1964.[53]

James H. Hunter, Jr., a recent Ph.D. from the University of California at Los Angeles, was a prolific worker during his five years at Yale. In that interval he published no fewer than nineteen articles dealing with galaxy formation in a steady-state universe (an outgrowth of his UCLA dissertation),[54] star formation,[55–58] thermal stability and dynamic models of planetary nebulae (with Sofia),[59–61] the stability of interstellar gas,[62,63] the cosmology of Alfven,[64] thermal instabilities in the solar chromosphere,[65] and finally, numerous notes on quasars and their interpretation. With A. Wesselink, P. Lu, and J. Gibson, this work on quasars was one of the first applications of Yale's new 40-inch telescope.

The final astrophysicist employed under Wildt, in July 1967, was Sumner Starrfield, also from UCLA. His primary interest concerned the evolution of novae, particularly the accretion of matter onto the white dwarf component from the gaseous disk of the secondary.[66] However, all five of his publications of work accomplished at Yale came after 1968, the closing date of this history.

In the two and a half years between the death of Brouwer in January 1966, and the appointment of

Table 23.2
Numbers of Astronomers Doing Research in Three Branches of Astronomy

Year	Director	A	C.M	Astrophys.	Total
1940–41	Schlesinger	6	1	1	8
1964–65	Brouwer	2	9	6	17
1967–68	Wildt	1	6	5	12
1973–74	Demarque	1	2	12	15
1979–80	van Altena	3	1	11	15
1986–87	Larson	4	1	11	16

Demarque in 1968, nearly 53 per cent of the Astronomy faculty resigned, some voluntarily; some others Wildt urged to resign, evidently in order that the incoming Director not be handicapped in making his own appointments. Of those pursuing celestial mechanics, only Gerald Clemence (retired from the U.S. Naval Observatory) and Boris Garfinkel (retired from the Army's Ballistic Research Laboratories at Aberdeen Proving Ground, Maryland) remained. Garfinkel was the only one to continue teaching celestial mechanics to interested graduate students well into his eighties, voluntarily, without salary. He remained a loyal and devoted disciple of Brouwer as well as friend to the astrophysicists. Trained in celestial mechanics, E. Myles Standish, Ph.D. 1968, was appointed Assistant Professor from 1968 to 1972, but he specialized in the internal motions of star clusters, not in the classical solar-system problems.

Table 23.2 illustrates the change of emphasis in research over the years since Schlesinger's directorship. The numbers tabulated include professors, research associates, and research staff astronomers but do not include visiting professors, astronomers at the Southern Station, or part-time off-campus employees. Column 1 represents the final year of each directorship. The categories are A, astrometry; C.M., celestial mechanics; and Astrophys., astrophysics, including both observational and theoretical and radio astronomy (just 2, in 1964–65).

In recent years there has been a resurgence of activity in astrometry, largely because Yale has undertaken the determination of proper motions of southern stars relative to remote galaxies—the continuation of work at Lick Observatory for northern stars—and Yale's involvement in space sciences. But now, although graduate students at Yale can obtain thesis guidance in almost any branch of astronomy, astrophysics is, as Dr. Wildt would have wished, the predominant field.

REFERENCES

1. L. Spitzer, Jr., *Phys. Review,* **55**, 361, 699, 1939; **56**, 39, 1939.
2. ———, *Astrophys. Jour.,* **90**, 494, 1939; Abstract, *Pub. Amer. Ast. Soc.,* **9**, 231, 1938.
3. ———, *Astrophys. Jour.,* **90**, 675, 1939.
4. ———, *Monthly Notices Roy. Ast. Soc.,* **100**, 396, 1940.
5. H. Shapley to Dean DeVane, May 22, 1944. *Yale Obs. Archives.*
6. L. Spitzer to D. Brouwer, Dec. 16, 1945, *Yale Obs. Archives.*
7. ———, Aug. 7, 1946, *Yale Obs. Archives.*
8. L. Spitzer, *Ann. Review of Ast. and Astrophys.,* **27**, 1, 1989.
9. D. Brouwer to W. C. DeVane, March 5, 1946, *Yale Obs. Archives.*
10. S. K. Runcorn, Ed., *High Pressure Physics and Planetary Interiors,* Amsterdam, xviii + 209 pp., Bibliog. p. xv, 1972; *Lunar Science Inst. Contrib.* No. 89, 1972.
11. R. Wildt, *Veröffentlichungen Univ.-Sternwarte Göttingen,* **2**, Heft 22, 1932; Heft 26, 1932; **3**, Heft 40, 1934.
12. ———, *Astrophys. Jour.,* **90**, 616, 1939; Abstract, *Pub. Amer. Ast. Soc.,* **9**, 236, 1939.
13. A. Unsold, *Quarterly Jour. Roy. Ast. Soc.,* **17**, 523, 1976.
14. D. Brouwer, *Ann. Report* for 1945–46.
15. L. Spitzer, Jr., *Astrophys. Jour.,* **93**, 369, 1941; **94**, 232, 1941; **95**, 329, 1942; Abstracts *in Pub. Amer. Ast. Soc.,* **10**, 144, and 174, 1941.
16. ———, *Astrophys. Jour.,* **107**, 6, 1948.
17. ———, *Ibid.,* **108**, 276, 1948; **109**, 337, 1948.
18. ———, *Ast. Jour.,* **52**, 130, 1947.
19. ———, *Phys. Review,* (2), **70**, 777, 1946.
20. Private conversations mainly with faculty spouses, 1957.
21. S. Chandrasekhar to D. Brouwer, Feb.4, 1947, *Yale Obs. Archives.*
22. H. N. Russell to D. Brouwer, Feb. 4, 1947, *Yale Obs. Archives.*
23. A. N. Vyssotsky to D. Brouwer, Feb. 3, 1947, *Yale Obs. Archives.*
24. W. C. DeMarcus, *Icarus,* **30**, 441, 1977. Permission to quote granted by Academic Press.
25. D. Brouwer and O. Struve, Correspondence, November 1948, *Yale Obs. Archives.*
26. O. Struve to D. Brouwer, Nov. 27, 1953, *Yale Obs. Archives.*
27. L. Spitzer, Jr.,to D. Brouwer, Nov. 13, 1953, *Yale Obs. Archives.*
28. H. C. Lehman, *Scientific Monthly,* **61**, 127, 1945.
29. D. Brouwer, *Ann. Report* for 1957–58.
30. R. Wildt, *Zeitschrift für Wissenschaftliche Photographie,* **25**, 163, 1928.
31. E. Hertzsprung, *Ibid.,* **3**, 429, 1905; **5**, 86, 1907.
32. R. Wildt, *Astrophys. Jour.,* **140**, 1343, 1964; **143**, 363, 1966; **146**, 418, 1966.

33. D. Brouwer, *Ann. Report* for 1951–52.
34. ———, *Ann. Report* for 1952–53.
35. W. C. DeMarcus, *Ast. Jour.,* **58**, 36, 1953.
36. ———, *Ast. Jour.,* **59**, 116, 1954; **63**, 2, 1958.
37. D. Brouwer, *Ann. Report* for 1946–47.
38. D. Brouwer to N. S. Buck, July 15, 1959, *Yale Obs. Archives.*
39. T. G. Cowling, *Observatory,* **86**, 51, 96, 173; 1966.
40. R. Wildt to M. Albee, April 10, 1974, *Yale Obs. Archives.*
41. R. Wildt, *Ann. Report* for 1965–66.
42. D. Brouwer, *Ibid.,* for 1964–65.
43. L. Oster, *Ann. d'Astrophysique,* **27**, 14, 1964.
44. C. C. Brooks and L. Oster, *Astrophys. Jour.,* **134**, 948, 1961.
45. M. D. Altschuler and L. Oster, *Ast. Jour.,* **69**, 530, 1964.
46. L. Oster and M. D. Altschuler, "The Physics of Solar Flares," *NASA Symposium Proceedings,* p. 377, 1964.
47. L. Oster and S. Sofia, *Astrophys. Jour.,* **141**, 1139, 1965.
48. S. Anand, L. Oster, and S. Sofia, *Nature,* **202**, 1097, 1964.
49. H. R. Johnson, *Ann. d'Astrophysique,* **25**, 30, 1962.
50. ———, *Ast. Jour.,* **67**, 117, 1962; **68**, 76, 1963.
51. ———, *Astrophys. Jour.,* **137**, 858, 1963.
52. M. Lecar, *Ast. Jour.,* **68**, 77, 1963; **69**, 550, 1964.
53. ———, *International Ast. Union Symposium* No. 25, p. 46, 1966.
54. J. H. Hunter, Jr., *Monthly Notices Roy. Ast. Soc.,* **133**, 181, 1966.
55. ———, *Idem,* p. 239.
56. A. F. Aveni and J. H. Hunter, Jr., *Ast. Jour.,* **72**, 1019, 1967.
57. ———, *Ibid.,* **73**, S86, 1968.
58. ———, *Ibid.,* **74**, 1021, 1969.
59. S. Sofia and J. H. Hunter, Jr., *Ibid.,* **72**, 830, 1967.
60. ———, *Astrophys. Jour.,* **152**, 405, 1968.
61. J. H. Hunter, Jr., and S. Sofia, *Monthly Notices, Roy. Ast. Soc.,* **154**, 393, 1971.
62. J. H. Hunter, Jr., *Ast. Jour.,* **70**, 679, 1965.
63. ———, *Astrophys. Jour.,* **161**, 451, 1970.
64. ———, *Monthly Notices Roy. Ast. Soc.,* **137**, 271, 1967.
65. ———, *Icarus,* **5**, 321, 1966.
66. S. Starrfield, *Astrophys. Jour.,* **161**, 361, 1970.

24
MOVING, ESPECIALLY IN THE SOUTH

When Yale first acquired the Winchester Observatory in 1882 at the corner of Prospect and Canner Streets in New Haven, the grounds were spacious for future expansion of its facilities; the location reasonably close to the University yet sufficiently distant to have a clear horizon in nearly all directions; and at that time there was little trouble from city lights. When Schlesinger came in 1920, expansion began. He soon added a new wing to the Observatory administration building but, more importantly, he immediately began planning for the establishment of a southern observatory station. After considerable site testing he selected Johannesburg, South Africa. Schlesinger might aptly be described as the most successful Director in the years covered by this history, for it seemed he got almost everything he asked for, and was generally on cordial terms with the University Administration and the Observatory staff. His successor, Dirk Brouwer, for all his accomplishments, was far less successful in nearly all aspects of his administration, apart from celestial mechanics.

In a 1945 memorandum on desirable improvements in the equipment of the Observatory[1] Brouwer noted the obsolescence of the available telescopes and the need for more modern equipment in order to attract more graduate students and compete more favorably with other universities, notably Harvard and the Universities of California, Chicago, and Michigan. He added, "Plans for improving our telescopic equipment require an examination of the present site of the Observatory." City lights and large buildings close by were serious disadvantages to the astronomer. Brouwer quoted President Angell who in his 1929–30 Report had already commented on the deterioration of the location of the Observatory and stated that there were several sites within reasonable reach of New Haven "in which much more advantageous conditions could be secured at relatively small expense." But by 1945 no efforts seem to have been made to acquire a more favorable location. Now Brouwer remarked upon the difficulties that might arise were the entire Observatory to be moved some 15 miles from the center of the University. He advocated that only the instruments be moved out of town and that a simple building be erected there for sleeping quarters for the observers, a dark room, and a small workshop for minor instrument-repair jobs. If this were done, he advocated selling the property on Prospect and Canner Streets in order to fund a new building nearer the University, with more adequate office and laboratory space and a lecture room for classes. At the same time he cautioned that plans to build should keep in mind that maintenance costs must be kept to a minimum in order to ensure sufficient operating funds for the principal research projects of the department, namely celestial mechanics, theoretical astrophysics, and photographic astrometry. This early report outlined quite clearly what actually came to pass about a dozen years later, but by then the loss of the Prospect–Canner Street property seemed greater than the net gain.

Before that happened, however, similar problems arose at the Southern Station in South Africa. Late in 1943 the assistant observer resigned, and no suitable replacement could be found, leaving Dr. Alden to carry on alone, "operating a telescope that would require a staff of three for full-time operation."[2] The possibility of cooperative use of the equipment with Columbia University was already under discussion[3] and on January 1, 1946, a ten-year agreement between Yale and Columbia went into effect[4] whereby Yale would continue to use the equipment for astrometric work while Columbia would concentrate on photometric problems. The station was now renamed the Yale–Columbia Southern Station.

Meanwhile, in July 1946, Dr. Paul Skoberla of Breslau, a German astronomer interned in South Africa during the war, was temporarily employed at the South African Station. Thanks to his cooperation it was possible to make use of all possible observing nights, and 2,059 Yale plates were taken in the 1946–47 season in contrast to only 1,280 the previous year.[5] Besides plates for the parallax and proper-motion projects he began an investigation of southern star clusters. Although it was expected that Skoberla would soon be repatriated to Germany, he stayed until the closing of the Johannesburg Station in 1952.

Brouwer visited Johannesburg in 1948.[6] The Observatory, only a mile and a half from the center of Johannesburg and close to a railroad station, was suffering increasingly from city lights and smoke pollution. While possibilities for relocating in South Africa were being considered, British astronomer Dr. R. v.d.R. Woolley was establishing a new observatory at Mt. Stromlo, about ten miles from Canberra, Australia. Assuming that an influx of foreign astronomers would boost the progress of astronomy in Australia, he invited the Yale-Columbia Station to move to Mt. Stromlo, where the Australian government would provide a dome for the 26-inch refractor and housing (on a rental basis) for the observers.[6] The proposal seemed attractive except for the greater distance of Australia from the States. Hence further exploration for a possible new South African site in collaboration with either the Union Observatory or the University of Witwatersrand (both in Johannesburg, but not related) was being considered. In his Report for 1948–49, however, Brouwer stated, "The likelihood of obtaining adequate financial support for moving the telescope from Johannesburg to a more favorable site in South Africa has diminished. This is in part due to a general change in attitude by South African government officials and private individuals. A year ago a spirit of prosperity and optimism prevailed everywhere in South Africa; this has rather abruptly changed into an attitude of caution and lack of confidence in the future."

Meanwhile, Woolley's invitation to move the Yale-Columbia Station to the Commonwealth Observatory at Mt. Stromlo became definitive. Woolley visited the States for an international conference in 1950 when final plans for the transfer were worked out. Then in November—December 1950, Brouwer spent a month visiting Mt. Stromlo (Fig. 24.1). In a special report[7] Brouwer reported enthusiastically on conditions and prospects at Mt. Stromlo, anticipating close cooperation between the Australian and American astronomers and cooperative use of their respective telescopes. He was also impressed by the relationship between the Observatory staff and the Australian National University, predicting, "Within relatively few years the Australian National University may be expected to be an important graduate school and research center," a prediction fully verified by the number of graduates who have subsequently served on the Yale faculty. Brouwer concluded,[8] "There appears to be every reason to expect that our cooperation with the Commonwealth Observatory will be most cordial and fruitful and to the advantage of both parties." Unfortunately this happy relationship was to last hardly a decade.

Cyril Jackson, then in charge of the Southern Station at Johannesburg, began dismantling the 26-inch telescope on May 9, 1952 and escorted it to Durban, whence it was shipped to Australia on June 9. The Johannesburg Southern Station was finally closed June 30, 1952, ending 27 years of observations during which 66,462 plates were taken with the 26-inch.[9] Jackson followed to Australia in September, and Jan Schilt, Director of the Rutherfurd Observatory of Columbia University, spent June 1952 to April 1953 there as a Fulbright Fellow. Both men were therefore on hand for the construction of the dome for the telescope, which had originally been supposed to be ready when the telescope itself was to arrive. However, they encountered successive delays and the telescope did not resume normal operation until January of 1956. Seven years later, May 28, 1963, the last Yale plate was taken with the telescope, No. 69,950, before the Yale—Columbia Southern Station again moved, but the valuable veteran telescope was donated to the Commonwealth Observatory. It has been misnamed *The Yale-Columbia Telescope.* It would more appropriately have been called the *Schlesinger Telescope,* as only he was responsible for its construction and for the founding of the Southern Station in 1925. Moreover, the bulk of the plates taken were for Schlesinger's parallax project. Columbia had entered the scene, by invitation, twenty years later, because Yale could no longer find or afford sufficient staff to keep the telescope in operation full time.

Brouwer's happy initial expectations met numerous frustrations at Mt. Stromlo. Jackson who had given valuable service in Johannesburg, and gladly transferred to Mt. Stromlo, soon became disillusioned about the wisdom of the move. Among other problems, weatherwise the observing conditions at Mt. Stromlo were inferior to those at Johannesburg. Of this Brouwer had already been aware when he wrote Schilt on December 18, 1952, "It has bothered me a good deal that the weather on Mount Stromlo is not more satisfactory. Either the weather is exceptionally bad this year or we have been too optimistic on the basis of the reports that Woolley has furnished. It is no use lamenting. The only thing to do

Fig. 24.1. R. v. d. R. Woolley and D. Brouwer (right) at Mt. Stromlo.

is to make the best possible use of the available sky." At Mt. Stromlo a total of 400 hours of observing on 66 nights had been experienced in contrast to 2,000 hours on 220 observing nights at Johannesburg. (Woolley had reported that the average over five years at Mt. Stromlo had indicated a cloudless sky 26 percent of the time.) Jackson, moreover, did not get along well with the Mt. Stromlo staff; he expressed uncertainty about his future there[10] and seriously contemplated resigning. Finally he did resign and returned to South Africa in December 1954. He was replaced by Dr. Gerard de Vaucouleurs.[11]

For the period September 1955 through February 1956, Brouwer spent his sabbatical leave at Mt. Stromlo. There he used the 26-inch to photograph double stars. But he also visited the Sydney Observatory where he attached the Yale 8-inch catalogue camera to the astrographic telescope of the Sydney Observatory in order to complete the photography of the Yale southern zones. At the Commonwealth Observatory no suitable mounting was available, and Harley Wood, Director of the Sydney Observatory, graciously made his facilities available and completed the needed series of plates after Brouwer's departure.

While he was in Australia, Brouwer was invited to participate in the selection of a successor to Woolley as Director of the Commonwealth Observatory. Woolley had been named Astronomer Royal and was returning to England.[12] Dr. B. J. Bok of Harvard Observatory was chosen. The choice of a new Director was important to Yale as cordial relations between Yale and the Commonwealth Observatory were important. Brouwer wrote, "Professor Bok's personal and scientific qualifications give assurance of continued happy relations with the Commonwealth Observatory." Brouwer, at the conclusion of his visit, could report "with satisfaction that the Southern Station is operated with zeal and zest by Dr. de Vaucouleurs and Mr. Combas, and that their work is respected by their Australian colleagues." But soon de Vaucouleurs was offered an apparently better position at the Lowell Observatory and he left in March 1957.[13] Upon leaving, de Vaucouleurs was awarded an honorary D.Sc. degree from the University of Australia, and he reported to Brouwer, "This is a good way of leaving Canberra." Mr. Combas left for a more remunerative position at the Sydney Observatory.[14]

Cyril Jackson was invited to return to his former post as observer in charge of the Southern Station for a period of not less than two years. With the instruments now in operating order, his prospects appeared more favorable than before. As Jackson was unable to arrive before de Vaucouleurs' departure the latter wrote Bok, who was still at Harvard,[15] describing the housing arrangements that should be made for Jackson. After his arrival in May 1957, Jackson wrote Brouwer[16] that he found the conditions 500 percent better than when he had left. In the month of November he obtained a record number of 106 plates. On the other hand, in October he had three times experienced gale winds up to sixty miles an hour and feared the dome was in danger of flying off![17] Incidentally, he also commented that as the word *station* in Australia signified only "a sheep run, would it not be advantageous as well as more correct, for us to alter our title to *Yale–Columbia Observatory*?"

Brouwer had high expectations for carrying out in Australia not only the older astrometric projects, but especially an important new one. In about 1934 the Lick Observatory had begun construction of a 20-inch astrograph for the determination of proper motions relative to distant galaxies,[18,19] a project much delayed by World War II. The Lick plates could not reach far southern declinations. Hence in 1952[20,21] Brouwer began to explore the feasibility of Yale's extending this important work to the South Pole. Although several proposals were rejected, the National Science Foundation in 1957 did award a grant of $41,400 for exploratory studies.[14] As it had become apparent that Mt. Stromlo, so close to the national capital, Canberra, was not a suitable site, Dr. Isadore Epstein of Columbia spent nine months (September 1957 to May 1958), in Australia site-testing to find a better location for the Yale telescopes.[22] Meanwhile Dr. Bok, who had been asked to serve on a committee to look into problems relating to Yale's plans for a large astrographic instrument in Australia, had already initiated inquiries of his own a year before he came to Australia, regarding the desirability of establishing an observing station for the Commonwealth Observatory in an excellent climate somewhere else in Australia.[23] Cooperation among the various parties should have proved mutually beneficial.

Woolley had been highly supportive of foreign participation on Mt. Stromlo, providing not only housing, but also maintenance for the Yale—Columbia equipment. Bok, on the other hand, found so many urgent needs for funds for the Commonwealth Observatory that he could not justify the support of any foreign establishments. He estimated that the cost of the dome for the "antiquated 26-inch refractor" had "set back the research and development program in astronomy in Australia by a full year."[24] He therefore insisted that no help or support for the proposed new instrument or the moving of the 26-inch to a better location would be considered.

In 1957 in a confidential report to Brouwer, Jackson[25] commented on the disadvantages Yale had encountered by having accepted hospitality first at the University of Witwatersrand and then at the Commonwealth Observatory, without having provided the educational programs the host institutions had expected in return. (Jackson himself had hoped in vain to be given an opportunity to earn a Ph.D. at Witwatersrand.[26]) Keenly aware of the financial straits of the Yale Observatory, Jackson nevertheless strongly advocated a more independent course in the future. He also stressed that the objectives in locating a new site were different for Yale with its long-focus instrument than for the Australian short-focus instruments. Later, in 1959, in a letter to Professor Oliphant of The Australian National University,[27] Brouwer agreed with Jackson on some points but was more conservative on others, writing,

> If I am not mistaken, I see two distinct differences between the site survey by Mt. Stromlo observers and our own survey. From some reports I have the impression that the Mt. Stromlo observers are willing to compromise on a reasonably good site within a few hundred miles from Mt. Stromlo, while the Yale-Columbia survey has been carried out without any such thoughts of compromise. Secondly, I believe that we have paid more attention to quality of seeing [and] the absence of strong winds. After all we are thinking in part of a site for a long-focus telescope for which both seeing and wind are of greater significance than for instruments of shorter focal length.
>
> Regardless of these differences in emphasis I have, in principle at least, strongly favored a solution that would eventually result in a common site for a field station from [*sic*] Mt. Stromlo and Yale-Columbia. On the other hand, I have become increasingly aware of the difficulty of establishing fruitful contacts among individuals with widely different personalities. Enough has come to the surface while our site surveys were in progress to caution us that well intentioned schemes may fail due to clashes of personalities.

Brouwer cited a serious clash of personalities between Bok and Epstein, and considered Jackson's withdrawal from the life at Mt. Stromlo as "a move in self-defense to secure a position of independence of a small institution in its relation to the much larger Mt. Stromlo organization." In September 1961 Jackson had complained to Brouwer, "Since my stay at Mount Stromlo I have become allergic to humiliation." He was not being informed of impending changes at the Station,[28] and gradually the Director of Mt. Stromlo and Jackson were hardly on speaking terms.[29] For any future joint undertaking it would therefore be important to specify clearly in advance the rights and privileges of each party. Brouwer was not yet ready to break relations: "Together we could surely build better than if we do it independently."

By 1958 NSF support for the astrograph and moving the 26-inch had seemed assured by a pending grant of over a million dollars, already approved by the Bureau of the Budget but still awaiting congressional approval.[30] Disappointingly, this never materialized. Finally in mid-1960 the Ford Foundation awarded a grant of $750,000 for the purpose of establishing an astrographic telescope in the southern hemisphere.[31] The search for a satisfactory site in Australia, free from dust and summer cloudiness and at a suitable elevation, proved fruitless. Chile and Western Argentina in the eastern ranges of the Cordilleran mountains seemed promising and an area near San Juan, Argentina, was finally selected.[31] The Perkin-Elmer Corporation was to provide the optical parts for a double astrograph, two 20-inch telescopes (one for blue light the other for yellow) to be mounted in parallel, and Rademakers of Rotterdam would provide the mounting, all to be completed by the end of 1963. There was to be cooperation between the Yale–Columbia group and the University of Cuyo in San Juan.

While the final site was still under consideration, with Epstein looking over the possibilities both in Chile and the Argentine, Brouwer invited Jackson[32] to move to the new site, where the prospects were good that assistants might be supported by NSF. Although the Chilean site, where AURA would have its new large NSF-supported reflector, offered attractions, Brouwer was more inclined to favor Argentina because "the modest San Juan Observatory promises to develop into a center for astrometric research, and the people at San Juan are of a caliber that has impressed both Jan [Schilt] and me most favorably." Current cooperation between Yale and the Argentinian observatory proves the accuracy of those early impressions.

Between December 1960 and June 1962, site-testing observations were carried out by personnel of the Felix Aguilar Observatory. This resulted in the selection of El Leoncito, where approximately 70 percent of the nights were found suitable for astronomical observations.[33] The land selected consisted of about 100 acres some 100 miles from the city of

San Juan, at an elevation of about 8,000 feet.[34] In September 1962 an agreement between the Yale–Columbia Southern Observatory and the University of Cuyo was signed, providing for the lease of land at El Leoncito and in the city of San Juan by the University of Cuyo to the Yale–Columbia Southern Observatory, as well as for the plan of cooperation between the two parties in the operation of the Observatory. The optical parts of the astrograph were completed during the summer of 1963; the mounting was now expected early in 1964.[35,36]

Already in 1960 Brouwer had written Jackson[32] of the decision not to move the 26-inch to Argentina but to donate it to Mt. Stromlo, but without the 9-inch finding lens (originally the Sheffield Telescope lens) or the double-star camera. (The camera, however, remained at Mt. Stromlo on loan.) But arrangements for the transfer were not completed until after the approval of the Yale administration,[37] and Yale finally severed its connection with Mt. Stromlo. In accepting the telescope he had previously described as "antiquated"[24] Dr. Bok now announced in the local newspaper,[38] "The acquisition of this telescope by Mt. Stromlo Observatory means an important extension of activities by our section on positional astronomy." In particular, it would be used for the determinations of proper motions and the photography of double stars.[39] The telescope would also be used for public open nights. On the occasion of the transfer, M. J. Miller wrote a brief history of the telescope.[40]

Jackson arrived at El Leoncito (Fig. 24.2) in July 1964 to take up his duties as observer-in-charge. He and Arnold Klemola were to be on hand for the erection of the telescope in September 1964. Klemola, as a graduate student at the University of California, had already had some experience with the instrument at Lick Observatory that had served as the prototype for the new Yale double astrograph. He had also participated in the Yale Zone Catalogue measurements at Yale since the fall of 1961. The mounting for the 20-inch was delivered in October and dedication exercises were held on March 31, 1965.[41]

Almost the first use made of the new telescope was a search by Jackson for a possible remnant of the supernova of 1006. Bernard Goldstein of the History of Science Department, in 1964 searched for historical evidence for the supernova[42,43] which he inferred was, or was close to, the small nebula NGC 5882. Brian Marsden, a graduate student, determined its most probable search area.[44] Brouwer then wrote Jackson,[45] "It would be worth our while to secure some trial plates in this general area. A discovery of such an object during the early period of operation of the southern astrograph would not harm the reputation of the instrument." Jackson did indeed find a suspicious object within the designated area,[46] but this proved to be another planetary nebula some 3° from NGC 5882, and was not considered a likely supernova remnant.[47] Gardner and Milne[48] soon established that the SN remnant is radio source 1459–41 (at 15^h00^m—41° 42′ at the Lupus Centaurus boundary) in the catalogue of Mills, Slee, and Hill,[49] ultimately described[50] as "the largest and faintest of the five well-established remnants of historical supernovae in the last one thousand years," larger than the well-known Crab nebula, remnant of SN 1054. Because of the faintness of the SN 1006 remnant, it is not surprising that Jackson did not find it on the Yale plates.

In its first year at El Leoncito, Yale had welcomed a U.S. Naval Observatory facility (a 7-inch transit instrument) at this site.[51] Brouwer felt this would be beneficial to the Yale–Columbia operation by reducing the isolation of the small staff and bringing in additional competence of observers and technicians.[52] Moreover, the program to be carried out by the Naval Observatory personnel was of interest to Yale astrometrists, a program for the determination of accurate positions of *Southern Reference Stars* and *Fundamental Stars.* The installation and testing of the Naval Observatory equipment was completed in August 1967. In its first year of operation 9,000 observations were secured.[53] This association with the U.S. Naval Observatory survived until 1974 when the Naval Observatory completed its program and shipped its telescope back to Washington.[54] In all about 150,000 positional observations were accumulated.

Cyril Jackson resigned as Resident Director as of the end of June 1966, and Arnold Klemola took his place. Adriaan Wesselink was appointed to a newly created position as Executive Director, succeeding the late Dr. Brouwer, to oversee the programs for the telescope.[55] Klemola resigned in July 1967 to accept a post at the Lick Observatory, and was replaced by James A. Hughes of the Naval Observatory.[56] It was understood that the Hughes appointment was to be part-time only and for a period of about a year, until a permanent Resident Director could be appointed.[57,58] He was followed in 1967 by A. G. Samuel, and successive *Annual Reports* indicated changes every few years until C. U. Cesco of the Felix Aguilar Observatory at San Juan was appointed in October 1973.

Fig. 24.2. Yale Southern Observatory at El Leoncito.

That year the operation of the Yale–Columbia Southern Observatory, previously supported by NSF, was ended because of the lack of funds.[59] Of the primary program of 598 regions for stars south of −30°, 560 had been photographed; of the secondary program of 360 plates between 0° and −30°, intended for overlap with the Lick program for the analysis of possible relative errors, only 63 had been completed.[60] The following year van Altena reported that the operation of the station was being taken over by the University of San Juan until such time as second-epoch observations on the program were to begin (within a few years).[61] Meanwhile, the Argentinian Observatory continued with the necessary first-epoch plates. By 1982 most of the first-epoch plates for the Lick overlap regions had been taken and plans for the second-epoch plates were well underway.[62] Thus Brouwer's hopes and expectations were being fulfilled.

Jan Schilt had retired as Director of Columbia's Rutherfurd Observatory in 1962 and was succeeded in 1964 by Dr. Lodewyk Woltjer, who also took over the Columbia share of the directorship of the Southern Station.[63] This station continued to be operated jointly by Yale and Columbia until 1970 when Columbia revoked its commitment of joint support.[64] For about another decade, however, the station continued to be referred to in the *Annual Reports* as the *Yale-Columbia Southern Observatory*; since 1980 as the *Yale Southern Observatory.*

Plans for the 20-inch (Fig. 24.3) were primarily for the determination of proper motions. At first an interval between first and second epochs of 50 years was contemplated, later a minimum interval of 25 years. As it was assumed that the completion of the first-epoch plates would require only about seven years, the question naturally arose, "What use should be made of the instrument between the two epochs?" Brouwer[65] was emphatic that the instrument was sufficiently versatile that other projects could profitably be undertaken. Indeed, while Klemola was at El Leoncito he took photographs, when conditions were not suitable for the proper motion fields, for three other programs:[66] (1) fields of old novae, to try to identify them at minimum; (2) fields of clusters and clusters of galaxies; (3) observations of faint comets, asteroids, and satellites of Jupiter and Saturn. Plates were also taken for numerous investigations

Fig. 24.3. The Yale Southern double astrograph for determining proper motions of stars relative to distant galaxies.

of variable stars by Wesselink.[67] However, in the interval between the two proper-motion epochs, the equipment was temporarily in charge of the Argentinian observers for their own purposes as time for completing the first-epoch plates permitted.

Another instrument for El Leoncito, called the *camera cluster* or *cluster camera,* was designed according to Brouwer's specifications. Its primary purpose was to take plates for a continuation of Yale's proper motion Zone Catalogue projects. It offered as well high hopes for miscellaneous other projects. Unfortunately, it was not completed during Brouwer's lifetime. The cluster was to consist of the 8-inch Ross lens (the catalogue camera most recently used at Sydney, Australia); two 3-inch lenses, one for red the other for blue light, all taking 17x17-inch plates, the Sheffield 9-inch Clark to be used as a guide telescope; and a 4-inch finder formerly used with the 26-inch.

A memorandum, *Development at El Leoncito in period 1 April to 31 December 1965,*[68] noted that over five cubic meters of concrete had been poured in the smaller dome for the base pillar of the camera cluster, which was expected to arrive in early June 1966. The telescope was erected early in June and was reported to be ready for full-scale operation by the end of June.[69] Klemola wrote Wesselink[70] suggesting some ten possible observing programs, of which the highest priority was to go to the continuation of the Yale Zone Catalogues with the 8-inch, particularly the 0° to −30° zones. Other projects dealt with the photography of comets and asteroids, photometry of galaxies, nova and supernova search programs, and photoelectric observations with the

8-inch refractor. However, Klemola commented,[70] "Many objections to these suggested observing programs come up. Some are better done with other instruments. Others are very routine and of very low yield in the way of significant results when one considers the great effort involved. Also some will require auxiliary equipment not now available. Therefore few, if any, may actually be practical."

By November Klemola had not yet used the camera cluster but he planned to get Cesco started when the latter arrived. However, innumerable problems arose: the structure of the sliding roof of the shelter was defective and hampered the use of the telescope; the guiding telescope was so long that for elevations within 30° of the zenith the eyepiece was only a few inches from the floor; the pier had been so constructed that adjustment of the polar axis was difficult. After Hughes arrived he wrote[71] "Contrary to what you might have believed, the 'cluster camera,' including the Ross, is not really operative. . . . The guiding seems to be reasonable for a short exposure . . . but the major problem is focus, and in particular, tilt. . . . The mounting in general will not serve for any serious astrometric work. The idea of mounting four telescopes in a box, and hanging same on a modest sized axis system, appears to me ill conceived."

Another discouraging opinion came from Nicholas Mayall, Director of the Kitt Peak Observatory:[72] "While undoubtedly some things can be done to make it useful, I fear that, like so many such telescopes, it could become a white elephant and not find much interest for serious use by competent observers." One wonders how Brouwer himself would have coped with the execution of his plans for this instrument. Was it, for him, a blessing in disguise that he did not live to see the completion of his cluster of telescopes and cameras; or would he have succeeded in correcting the problems step by step to make the contraption work?

REFERENCES

1. D. Brouwer, Memorandum, July 24, 1945, *Yale Obs. Archives.*
2. ———, *Ann. Report* for 1943–44.
3. ———, *Ann. Report* for 1944–45.
4. ———, *Ann. Report* for 1945–46.
5. ———, *Ann. Report* for 1946–47.
6. ———, Report on a visit to the Yale-Columbia Southern Station, Oct. 26, 1948, *Yale Obs.Archives.*
7. ———, Report on a Visit to Mt. Stromlo, Canberra, Australia, Feb. 15, 1951, *Yale Obs. Archives.*
8. ———, *Ann. Report* for 1950–51.
9. ———, *Ann. Report* for 1951–52.
10. D. Brouwer to J. Schilt, Oct. 7, 1953, *Yale Obs. Archives.*
11. D. Brouwer, *Ann. Report* for 1954–55.
12. ———, *Ann. Report* for 1955–56.
13. G. de Vaucouleurs to D. Brouwer, May 15, 1956, *Yale Obs. Archives.*
14. D. Brouwer, *Ann. Report* for 1956–57.
15. G. de Vaucouleurs to B. J. Bok, March 8, 1957, *Yale Obs. Archives.*
16. C. Jackson to D. Brouwer, May 13, 1957, *Yale Obs. Archives.*
17. ———, Dec. 11, 1957, *Yale Obs. Archives.*
18. W. H. Wright, *Proc. Amer. Philosophical Soc.,* **94**, 1, 1950.
19. S. Vasilevskis, *Vistas in Astronomy,* **15**, 145, 1973.
20. D. Brouwer to C. Dollard, Feb. 23, 1953, *Yale Obs. Archives.*
21. D. Brouwer, NSF Proposal, June 26, 1956, *Yale Obs. Archives.*
22. D. Brouwer and J. Schilt to A. T. Waterman, July 28, 1958, *Yale Obs. Archives.*
23. B.J. Bok to D. Brouwer, May 25, 1956, *Yale Obs. Archives..*
24. B. J. Bok to R. Seeger, Dec. 24, 1957, *Yale Obs. Archives.*
25. C. Jackson to D. Brouwer, Nov. 16, 1957, *Yale Obs. Archives.*
26. ———, Feb. 4, 1953, *Yale Obs. Archives.*
27. D. Brouwer to M. L. Oliphant, Feb. 3, 1959, *Yale Obs. Archives.*
28. C. Jackson to D. Brouwer, Sept. 21, 1961, *Yale Obs. Archives.*
29. C. Jackson to B. J. Bok, Nov. 7, 1962, *Yale Obs. Archives.*
30. D. Brouwer, "Needs of the Department of Astronomy," Jan. 21, 1958, *Yale Obs. Archives.*
31. ———, *Ann. Report* for 1959–61.
32. D. Brouwer to C. Jackson, June 11, 1960, *Yale Obs. Archives.*
33. J. A. Lopez, *Inform. Bull. Southern Hemisphere* No. 4, p. 34, 1963.
34. *Yale Univ. News Bureau,* Jan. 15, 1962, *Yale Obs. Archives.*
35. D. Brouwer, *Ann. Report* for 1961–63.
36. ———, Ref. 33, p. 32.
37. D. Brouwer to K. Brewster, Apr. 19, , 1963, *Yale Obs. Archives.*
38. B. J. Bok, *The Canberra Times,* July 12, 1963.
39. ———, Ref. 33, p. 8.
40. M. J. Miller, *Idem,* p. 7.
41. D. Brouwer, *Ibid.,* No. 6, p. 2, 1965.
42. B. R. Goldstein, *Ast. Jour.,* **70**, 105, 1965.

43. B. R. Goldstein and Ho Peng Yoke, *Idem,* p. 748.
44. B. G. Marsden, *Idem,* p. 126, 1965.
45. D. Brouwer to C. Jackson, Nov. 12, 1964, *Yale Obs. Archives.*
46. D. Brouwer, *International Ast. Union Circular* No. 1893, 1965.
47. R. Minkowski, *Ast. Jour.,* **70**, 755, 1965.
48. F. F. Gardner and D. K. Milne, *Idem,* p. 754.
49. B. Y. Mills, O. B. Slee, and E. R. Hill, *Australian Jour. Phys.,* **13**, 676, 1960.
50. S. P. Reynolds and D. M. Gilmore, *Ast. Jour.,* **92**, 1138, 1986.
51. Anon., *Info. Bull. Southern Hemisphere* No. 20, p. 6, 1972.
52. D. Brouwer, *Ann. Report* for 1964–65.
53. K. A. Strand, *Pub. Amer. Ast. Soc.,* **1**, 118, 1969.
54. ———, *Ibid.,* **6**, 179, 1974; **7**, 213, 1975.
55. R. Wildt, *Ann. Report* for 1965–66.
56. ———, *Ann. Report* for 1966–67.
57. J. M. McDowell, Supt. U.S.Naval Obs., to Chief of Naval Operations, with *cc* to G. M. Clemence, May 27, 1966, *Yale Obs. Archives.*
58. J. A. Hughes to G. M. Clemence, May 27, 1966, *Yale Obs. Archives.*
59. P. Demarque, *Ann. Report* for 1973–74.
60. A. J. Wesselink, *International Ast. Union Symposium* No. 61, W. Gliese, C. A. Murray, and R. H. Tucker, Eds., p. 201, 1974.
61. W. van Altena, *Ann. Reports* for 1974–75 and 1975–76.
62. R. B. Larson, *Ann. Reports* for 1980–81 and 1981–82.
63. L. Woltjer, *Ast. Jour.,* **71**, 816, 1966.
64. P. Demarque, *Ann. Report* for 1970–71.
65. D. Brouwer, Research Proposal to NSF, Nov. 12, 1954, *Yale Obs. Archives.*
66. A. Klemola to N. U. Mayall, Nov. 1, 1966, *Yale Obs. Archives.*
67. A. Klemola to G. M.Clemence, Sept. 3, 1966, *Yale Obs. Archives.*
68. Anon., undated, *Yale Obs. Archives.*
69. C. Jackson to G. M. Clemence, June 7, 1966, *Yale Obs. Archives.*
70. A. Klemola to A. Wesselink, June 27, 1966, *Yale Obs. Archives.*
71. J. Hughes to F. P. Scott. May 26, 1967, *Yale Obs. Archives.*
72. N. U. Mayall to A, J. Wesselink, May 12, 1967, *Yale Obs. Archives.*

25
LOSS OF THE WINCHESTER SITE, AND ITS SEQUEL

The needs of the Observatory in New Haven were summarized by L. Spitzer[1] in 1951 at a meeting of the University Council, of which he was chairman of the Committee on the Division of the Sciences. The needs included additional teaching staff for the large undergraduate class; salary for a graduate-student research assistant; travel funds for professors to visit other observatories in order to obtain astrophysical observations not available at Yale; funds to remove the telescopes to a better location out of town; and funds for a new building closer to the center of the University. He did not mention the additional fact that overcrowding made it necessary to install IBM equipment in the basement of the Law School instead of in Observatory offices;[2] in the use of punched-card equipment for astronomy the Yale Observatory was a pioneer. With funding in scarce supply, little action was taken on Spitzer's recommendations. Finally in 1954, to alleviate crowded conditions, a second floor was added to the Observatory workshop to provide much needed room for the expanding library. Brouwer wrote,[3] "All concerned gladly accepted the temporary inconvenience that the construction entailed, and the expenditure of $10,500 from special Observatory funds that had slowly accumulated seemed well justified. This satisfaction was to last only a few months."

The Prospect Hill School and the Foote School, private secondary schools, had been interested in acquiring a portion of the Observatory grounds for a playing field. Assurance was given Brouwer[3] in March 1955, by Dean Pope, who was also a Trustee of the Prospect Hill School, that "nothing which would be against the interests of the Observatory was in his mind." Thus it came as a complete surprise when early in May Brouwer was informed that the University was considering moving the Observatory offices to 135 Prospect Street, the site of the Prospect Hill School for girls, and transferring the telescopes to a site outside New Haven. Offhand this might have seemed a good solution for all parties, but on closer examination, the Prospect Street building, originally a barn and then "a monkey house" (where Professor R. M. Yerkes conducted his psychological studies on monkeys) was both inconveniently arranged for office space and too small to offer any improvement over the existing quarters. Very costly alterations and additions would be necessary. Moreover, the advantages of transferring the existing telescopes out of town would also be minimal unless newer, modern telescopes could be added to the equipment. Brouwer pointed out, "Our telescopes are so modest that the cost and inconvenience of operating a station out of town would seem to be inappropriate in relation to the value of the instruments if there were no serious hope of acquiring new and improved equipment." In order for Yale to compete for graduate students with other major universities, better and more modern instruments were essential. Clearly, any move should be for the betterment of the Observatory, but Brouwer concluded "the proposal now being considered falls so far short of an ideal solution that it is by no means obvious that the move as a whole represents an immediate improvement for the Observatory."

It was while Brouwer was absent on his sabbatical in Australia from September 1955 to February 1956 that the most important decisions were reached, one might almost infer "behind his back." On September 29 Smith had written Brouwer[4] of confrontations with University officials in regard to the move these authorities were so strongly advocating. It turned out that the Foote School had acquired other property nearby and was no longer interested, whereas the Prospect Hill School alone could not carry the entire cost of the Observatory property. The officials were hence trying to cut Observatory costs in order to help the school. It was clear to them that the Observatory property was the *only* solution possible for the Prospect Hill School and therefore the move "really must go through." In December, Wildt[5] wrote Brouwer that Spitzer had expressed clearly the Astronomy Department's reluctance to move. Wildt added, "This brought forth a crisp statement from Dean Sinnott that the President attached great importance to the needs of the Prospect Hill School. It was apparent to the other department chairmen that great pressure is being put on us." One may well wonder about the allegiance of University officials who favored the needs of a private girls' school (indeed one whose students in that era would not be eligible for admission to masculine Yale) over the pressing needs of one of its own research departments.

The University adamantly decided to sell the Observatory property, move its offices to 135 Prospect Street, and move the telescopes to Bethany to an eleven-acre site owned by the New Haven Water Company, eleven miles northwest of the center of New Haven. This location had been suggested by Harlan Smith and secured with funds ($4,500) provided by Mrs. James E. Cooper and Mr. James W. Cooper. Originally a five-acre plot was intended, but this was increased to eleven to accommodate new radio-astronomy equipment being planned by Wildt and Smith. Construction of an addition to the building at 135 Prospect was to begin in March and be completed by mid-June. The original building was flush with the sidewalk and the addition was to extend likewise toward the west. But when a building permit was applied for, the plans were disapproved by the City of New Haven, which would approve a new addition only if it were set back at least 25 feet from the sidewalk. New plans had to be drawn up and construction could not even begin until after the originally specified completion date.

The Observatory was effectively evicted from its old quarters long before it was possible to move into the new. Although promises had been made by the Prospect Hill School that certain telescopes, equipment, and sundry possessions need not be moved until the Prospect Street property was ready for occupancy, events turned out otherwise.[6] The school had planned to add an auditorium which would not be connected to the old building, so construction could begin without discommoding the Observatory. However, without notifying Brouwer, the school's plans were abruptly changed and construction begun which did affect the Observatory. For example, only 12 hours' notice was given to remove heavy machines that would have been trapped in a basement if they were not suddenly removed. Again without previous warning, a building supervisor announced in April that the Reed telescope, which was being used extensively for student and open-night purposes, had to be dismantled immediately, and the many books stored in the building removed in order not to delay school construction. Equipment, books, and other belongings were moved into various dispersed buildings for storage. Office and working space was temporarily provided in the Osborn Zoology Laboratory, with the understanding that the astronomers would move into their new quarters before the beginning of the fall term of 1956—a deadline that proved impossible to meet because of further construction delays.

Brouwer was irritated that he had not been given sufficient notice of the moves to have time to initiate a fund drive for an appropriate building and a new astrographic telescope. He was keenly aware of the insurmountable deficiencies of the 135 Prospect Street building, even with a modest new wing. He would have preferred to raze the building and erect a properly designed new one on the premises. As for the move to Bethany, he was informed that it would be far easier to obtain funding for a new telescope if the Department already possessed a suitable site for it. Brouwer was forced to make the best of an undesirable situation.

Finally, in January 1957, the building at 135 Prospect (Fig. 25.1) was ready for occupancy. An Open House was celebrated on February 8, a time chosen to coincide with a meeting of the Yale Corporation, some of whom attended, as did officers of the University. Exhibits were set up illustrating the history and some of the achievements of the Department. At long last Brouwer's next Annual Report[7] expressed satisfaction:

> After having operated in the new headquarters of the Department for about a half year, we can express our general satisfaction with the facilities it provides. We have better housing of the library, a better workshop, a better classroom, and better storage facilities for publications and plate collections than the old Observatory building offered. The housing of the Computing Center in the same building, the proximity to other Science Departments and the relatively short distance to the center of the University are additional favorable features.

Brouwer also sincerely thanked the Sterling Library officials who had provided storage space for library materials and observatory publications for the past fifteen years, and the deans of the Law School for having provided space for the Observa-

Fig. 25.1. Observatory headquarters at 135 Prospect Street.

tory's punchcard operations for ten years in space they themselves needed.

This happy optimism did not last long. In a 1958 report to the Corporation, *Needs of the Department of Astronomy,* Brouwer stated,[8] "It is with some embarrassment that we find that the departmental building at 135 Prospect Street will soon be inadequate for our needs. At present all our office space is fully occupied. Next year one or two of the three rooms, now available to and occupied by graduate students, must be given up for microwave equipment. As a result, we shall be hard put to find room for graduate students." The Provost suggested that as a temporary solution to the space requirements for the growing numbers of graduate students, a few rooms in another building a few blocks distant might be made available. For a more permanent solution Brouwer recommended adding another wing to 135 Prospect, north of the present building, at an estimated cost of $90,000. The Computer Center also was outgrowing its allocated space. For its various needs the Department had already approached both the Ford and the Rockefeller Foundations, but unsuccessfully.

Again in 1960 Brouwer submitted plans for an addition,[9] this time on the corner of Prospect and Sachem Streets, at an estimated cost of $207,000, for which he was applying for funding.[10] But this plan was vetoed by the Provost[11] because now plans were underway for a computer center on Sachem Street.[12] The new Computer Center Building was completed and ready for occupancy in the summer of 1961, releasing some space for other activities at 135 Prospect.

In 1963 Provost Kingman Brewster, Jr., aware of the concern of the Budget Committee of the Corporation regarding various improvements at Yale, had sent an exploratory questionnaire to Brouwer asking, among other matters, what use he would make of a million dollars of new capital assuming such were to become available.[13] To this Brouwer replied:[14] "I would use it first for providing a large enough building, suitable for our needs, so that once again all our activities in New Haven could be housed under one roof. This is felt as the most important single major need of the Department." In his Report for 1963–64, perhaps inspired by Brewster's letter, Brouwer summarized and amplified the situation giving these details:

> When the department moved into its present quarters at 135 Prospect Street in 1957, the available space for offices and laboratories was adequate for the level of activity then existing. The expansion of staff and graduate student population of recent years could not then have been foreseen.

At the present time most of the activity in celestial mechanics is located at 334 Edwards Street. During the past academic year an overflow of graduate students was housed on the fourth floor of Winchester Hall, a solution that proved to be far from satisfactory. Notwithstanding these drastic steps, the building at 135 Prospect Street remains overcrowded and cannot accommodate a significant expansion in astrophysics. Thus the housing of the department remains a difficult problem that requires urgent attention.

This urgent need has hardly yet (1991) been met, despite two subsequent major moves.

Dr. Pierre Demarque came in 1968. The Computer Center had outgrown its quarters and moved out of the Watson building it had occupied at 60 Sachem Street, adjacent to the 135 Prospect St. building. In August the majority of the Astronomy Department personnel moved into the Watson building but kept only the first floor and basement of 135 Prospect, as well as the Reed telescope dome on the roof and the celestial mechanics headquarters in a house on Edwards Street. Soon the department also outgrew these allotted spaces. Retrieval of the second floor of 135 Prospect would have been the logical and the easiest solution to the problem from the standpoint of the Astronomy Department, especially as there was sufficient land for ultimate additions.[15] However, the University voted in favor of the social sciences and again the Astronomy Department had to move, this time, August 1975, into the equivalent of one floor of the Gibbs Physics Laboratory, where the new Director, Van Altena,[16] reported that the new location placed all faculty, staff, and students in the same building and greatly improved the facilities for the measuring equipment and the library. On the other hand valuable new plates from El Leoncito still had to be stored at the Beinecke Library, awaiting second-epoch plates. Yale Observatory publications, mountains of IBM cards rarely used but not yet obsolete, and sundry other materials had again to be stored in several other buildings (a special inconvenience to the secretaries who were in charge of distribution of publications). A department showing continuous scientific and personnel growth has never, since leaving the Winchester Observatory site, been given room for necessary and inevitable expansion.

In regard to the Bethany station (Fig. 25.2), construction began in November 1956 with the clearing of areas for roads and buildings. Things did not proceed too smoothly as the available funds, including a $22,000 NSF grant, would not suffice for even the barest requirements. However, the Yale Corporation voted an additional $30,000 for essentials,[7] and the following year, through the offices of Morris Hadley (son of a former President of Yale), a grant of $50,000 from the Rubicon Foundation was contributed for the completion of the central building at Bethany. This contribution actually came from Mr. Hadley himself, who wished to remain anonymous.[17]

Finally, the Opening Reception at the Bethany Observing Station was held on Sunday, December 14, 1958.[18] The telescopes then in place were the 10-inch Butler with photoelectric photometers, the 8-inch Ross Catalogue Camera for astrometric work and the Loomis polar telescope (Fig. 25.3) for determinations of the motion of the terrestrial pole. Two groups of Yagi antennas for planetary radio astronomy were also already installed. Yale President Whitney Griswold, who had been the first to propose the exchange of properties between the Prospect Hill School and the Observatory, was present at the dedication. Dr. Brouwer remarked to him that he had felt apprehensive for the future of the Observatory when the plans for moving were first presented, but that he was well pleased with what had been accomplished so far.

At its meeting of May 11, 1956, the Committee on the Sciences of the Yale University Council pointed out that the most urgent need of the Astronomy Department was a first-class instrument for positional astronomy, called an *astrograph*. The Report continued,

If such an instrument could be secured and installed, it would be of sufficient value to justify the operation of a site remote from New Haven, and would at the same time permit Yale to continue its present excellent program in the field of precision location of stars. This is one of the few types of astronomical work which is not especially dependent on ideal atmospheric conditions, so that the convenience of the site to the University far outweighs other geographical considerations. . . . The Committee strongly recommends that the administration give strong support and encouragement to the Department of Astronomy to obtain an outside grant for the purchase of a really first-class instrument around which the new observatory at Bethany could be centered.

However, it was a full ten years before Yale's 40-inch reflector was finally erected at Bethany.[19] In the meantime James Rodman,[12] a graduate student (but previously already a successful machine manufacturer), designed and helped build a 20-inch reflector (with optics by Tinsley Laboratories of California), which was completed in 1961 and immediately put to use by Harlan Smith in a search for evidence of

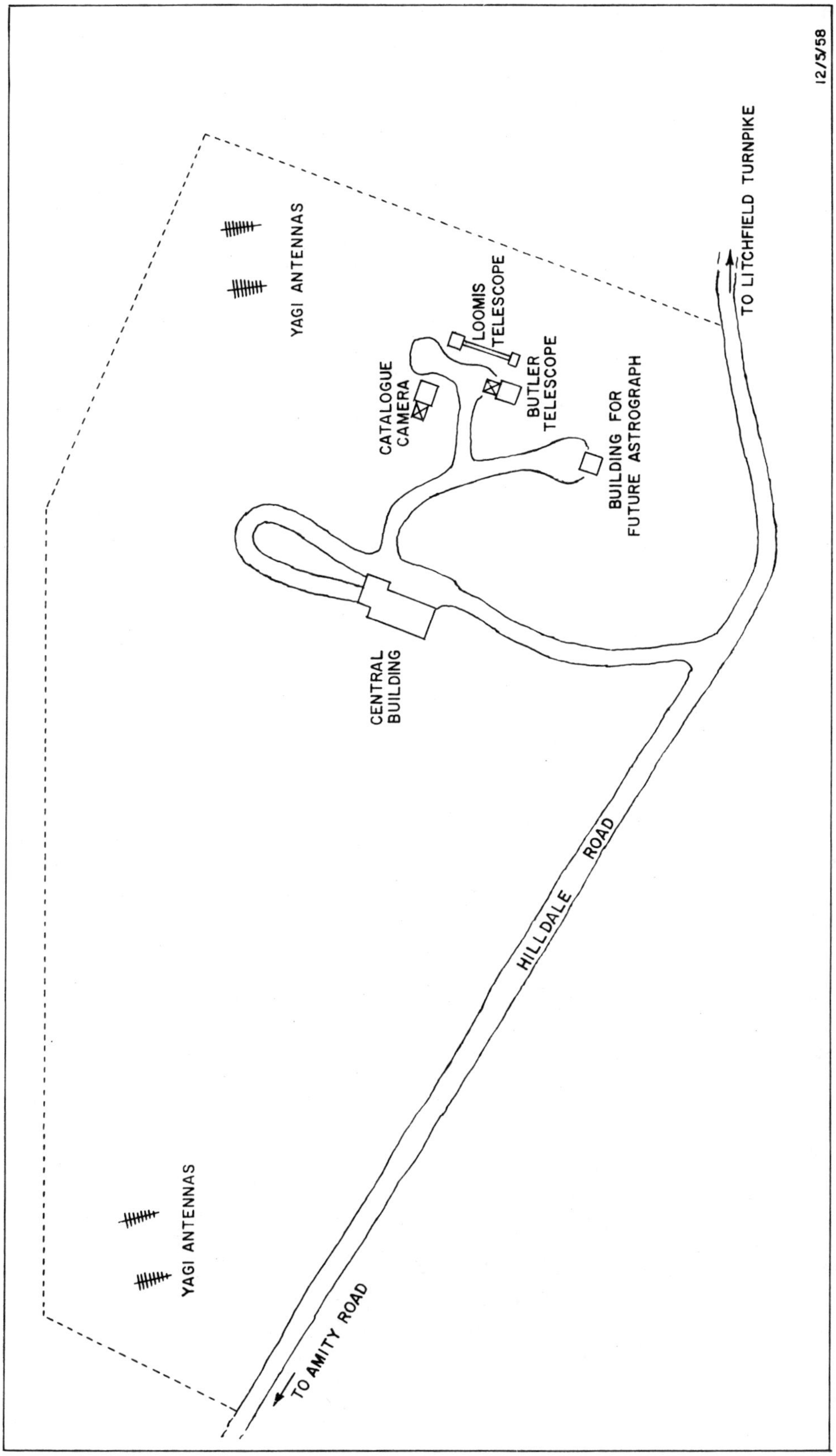

Fig. 25.2. Planned layout of the Bethany Station.

auroral activity in the atmosphere of Jupiter, and in studies of flare stars, using a photoelectric spectrophotometer also designed by Rodman and Smith under an ONR Grant to Smith.[20]

The National Aeronautics and Space Administration in the summer of 1962 provided a grant for a building and mounting for a 40-inch reflector. Funds for the optical parts, ordered from Perkin-Elmer in Connecticut, were not included in the grant but were provided from the Observatory's Higgins and other departmental funds.[20] The mounting was ordered from Boller and Chivens of California and the dome from Observ-a-Dome of Mississippi. All phases of the construction were scheduled for completion in the fall of 1964,[21] but the telescope was not installed until January 1966, and final adjustments were not completed until after the end of June.[19] Thus Brouwer did not live long enough to see the fruits of his long struggle to obtain a really fine modern telescope for Bethany. The telescope was used in the first few years under the general direction of Adriaan Wesselink, mainly for photographic studies of quasars and of variable stars in globular and galactic clusters.[22]

It soon became evident that Bethany was an unsuitable site for so fine a telescope as its 40-inch Ritchey-Chretien reflector. The telescopes of the National Optical Observatories at Kitt Peak and Cerro Tololo, supported by NSF and operated by AURA, were made available on a competitive basis to all member observatories; many sent their best telescopes to one or the other of these locations. Yale followed suit. By agreement with AURA the telescope would go to Cerro Tololo on loan for a period of 25 years. One-third of the observing time was to be reserved for Yale researchers.[23] Demarque commented that the relocation of the instrument to this superb site should enable the Yale observers to use it to pursue research in optical astrophysics. The telescope was shipped in 1973 and was ready for operation in 1974. At Bethany the 20-inch was refurbished and moved into the dome vacated by the 40-inch, where it has been used primarily for student-training purposes.

REFERENCES

1. L. Spitzer, Report of the Committee on the Division of the Sciences, University Council, May 1951, *Yale Obs. Archives.*
2. D. Brouwer, *Ann. Report* for 1947–48.
3. ———, *Ibid.,* for 1954–55.
4. H. J. Smith to D. Brouwer, Sept. 29, 1955, *Yale Obs. Archives.*
5. R. Wildt to D. Brouwer, Dec. 5, 1955, *Yale Obs. Archives.*

Fig. 25.3. The Loomis Telescope at Bethany. (Photo by W. Cleaver)

6. D. Brouwer, *Ann. Report* for 1955–56.
7. ———, *Ibid.,* for 1956–57.
8. ———, "Needs of the Department of Astronomy," Jan. 21, 1958, *Yale Obs. Archives.*
9. J. E. Sunblade, May 16, 1960, *Yale Obs. Archives.*
10. D. Brouwer to H. Sanford, Jr., May 5, 1960, *Yale Obs. Archives.*
11. N. S. Buck to D. Brouwer, May 18, 1960, *Yale Obs. Archives.*
12. D. Brouwer, *Ann. Report* for 1959–61.
13. K. Brewster to D. Brouwer, June 10, 1963, *Yale Obs. Archives.*
14. D. Brouwer to K. Brewster, July 10, 1963, *Yale Obs. Archives.*
15. P. Demarque, *Ann. Report* for 1968–69.
16. W. Van Altena, *Ibid.,* for 1975–76.
17. D. Brouwer, *Ibid.,* for 1957–58.
18. ———, Terminal Report, National Science Foundation Grant G-4172, "Relocation of the Yale Observatory," Jan. 2, 1959, *Yale Obs. Archives.*
19. R. Wildt, *Ann. Report* for 1965–66.
20. D. Brouwer, *Ibid.,* for 1961–63.
21. ———, *Ibid.,* for 1963–64.
22. R. Wildt, *Ibid.,* for 1966–67 and 1967–68.
23. P. Demarque, *Ibid.,* for 1971–74.

26
PROGRESS IN TEACHING ASTRONOMY

In the earliest days of the Collegiate School, its chief purpose was the training of Calvinist ministers. Indeed, 64 percent of the graduates between 1702 and 1707 entered the ministry, in contrast to only 3 percent of those graduated from 1901 to 1904.[1] Astronomy in the early days was taught primarily as an exemplification of the glory of God. Reports are somewhat conflicting as to how the Ptolemaic theory was actually taught. Some references state that it was presented as a statement of fact;[2] others that students could choose whether they preferred the Ptolemaic or the Copernican theory;[3] still others surmise that the Ptolemaic was taught, as at present, from the historical standpoint and for illustrating how completely different theories could equally well represent actual observations. In any case, the early tutors were not trained as professional astronomers, and it is more to their credit that they created an awareness of the subject than to their discredit that they were not currently well informed. How much worse an indictment of our own current teachings when we note how many college graduates, including those from Yale, still have faith in astrology even to the extent that one graduate advertised it in recent issues of the *Yale Alumni Magazine.*[4,5,6] With the acquisition of the books acquired by Dummer for the college in 1714, a greater understanding of science gradually developed.

Thomas Clap in addition to his strictly ecclesiastical and administrative functions was especially interested in mathematics and became a self-educated astronomer. In 1743, in the preface to his catalogue of the college library he recommended the following course of studies for students:[7]

> In the First Year to study principally the Tongues, Arithmetic, and Algebra; the Second, Logic, Rhetoric, and Geometry; the Third, Mathematics and Natural Philosophy; and the Fourth, Ethics and Divinity . . . Other less principal studies may be occasionally intermixed with these. Above all have an Eye to the great End of your Studies, which is to obtain the Clearest Conceptions of Divine things and to lead you to a Saving Knowledge of God in his Son Jesus Christ.

F. B. Dexter said of Clap, "his attainments in mathematics, astronomy and physics, placed him beyond doubt in the first rank of Americans in his generation." His stressing mathematics had a lasting effect on the Yale curriculum. Moreover, Dexter quotes Clap's contemporaries as considering him "preeminently an effective teacher." Although it has been stated that his theory on terrestrial comets smacked of Ptolemaic theory, he most certainly taught the Copernican—having been the first American to construct an orrery for teaching purposes, in 1740. At first it included only the six known planets, but later he added parabolas to represent a comet of 1743 and Halley's comet.

In 1781 Josiah Meigs was appointed Tutor in Mathematics, Natural Philosophy, and Astronomy, the first Yale officer to be appointed specifically to teach astronomy.[8] This appointment lasted only about three years, as Meigs's marriage (then frowned upon for faculty) and his wide interests in law and politics were thought to disqualify him. In 1794, he was appointed Professor of Mathematics and Natural Philosophy, but this lasted only until he resigned in 1801. The next University appointment specifically to include astronomy was that of Jeremiah Day as Professor of Mathematics, Natural Philosophy, and Astronomy, in 1801. With the strong emphasis on Latin and Greek, and to a lesser extent on mathematics, the curriculum for students was rigidly specified. In his *Annals of Yale College* (up to 1831), Ebenezer Baldwin indicates that astronomy was taught only to the Junior class, with additional lectures for Seniors.[9]

Olmsted, who was appointed Professor of Mathematics and Natural Philosophy in 1825 and additionally of Astronomy in 1839, was a strong advocate of teaching. In a Phi Beta Kappa oration[10] he quoted an analogy on bringing old and new scholars together:

Table 26.1
Relative Proportion of Sidereal System in Early Yale Texts

Author	Text	Date	% Stellar System
Olmsted	*Introduction to Astronomy*	1839	9
Norton	*Elementary Treatise on Astronomy*	1839	7
Norton	*Elementary Treatise on Astronomy*	1845	14
Olmsted	*Compendium of Astronomy*	1846	12
Olmsted	*Introduction to Astronomy*	1850	10
Norton	*Elementary Treatise on Astronomy*	1852	13
Olmsted	*Compendium of Astronomy*	1854	12
Olmsted	*Introduction to Astronomy*	1866	11
Norton	*Treatise on Astronomy, 4th ed.*	1867	15
Loomis	*Treatise on Astronomy*	1868	11
Loomis	*Treatise on Astronomy*	1871	10
Loomis	*Treatise on Astronomy*	1874	10
Loomis	*Elements of Astronomy*	1881	14
Loomis	*Treatise on Astronomy*	1888	11
Loomis	*Recent Progress of Astronomy*	1850	28
Loomis	*Recent Progress of Astronomy*, 3rd ed.	1856	46
Olmsted	*Outlines of a Course of Lectures on Astronomy*	1858	30

Thus stands an aged elm in ivy bound
Thus youthful ivy clasps the elm around.

Finding a deficiency of appropriate textbooks, Olmsted wrote many of his own. From the successive editions one can speculate on the advancement of knowledge of astronomy as such. In particular, the earliest textbooks contained very little besides the planetary system other than constellations and celestial coordinates, because relatively little was known about the stars—even though this was an era of important discoveries in Europe by Bessel, F. G. W. Struve, and the Herschels, on stellar parallaxes, double stars, and nebulosities. Progressively more text was added on the sidereal system: stars and nebulae. Table 26.1 gives the dates of some available Yale texts and the percentage of pages dealing with stars and nebulae. (Not included in these percentages are the pages dealing with celestial coordinates, time determination and other topics not specifically related only to either the solar or stellar systems.) Between 1839 and about 1850, both Olmsted and Norton increased their percentage of stellar-system pages by some 7 percent. Loomis, however, made relatively minor changes in his successive editions. The last three lines of Table 26.1 do not refer strictly to textbooks but also to treatises that do indicate a greater increase in the progress of knowledge of the stellar system than is indicated by the textbooks alone. After Loomis, no Yale faculty member wrote a general text on astronomy for undergraduates, although Paul Hodge, Yale 1956, Professor at the University of Washington, published two highly successful editions of his *Concepts of Contemporary Astronomy.* Table 26.2 gives a selected list of elementary American texts subsequent to those of Loomis. From 29 percent at the turn of the century the percentage of stellar-system coverage increased to an almost steady 50 percent from 1930 through Brouwer's regime. Recent texts may more appropriately be divided into three compartments: sun and solar system, stars and the Milky Way, and extragalactic systems. Approximately 10 to 20 percent of the current elementary texts are devoted to external galaxies.

Despite Olmsted's success as a teacher of astronomy, undergraduates were offered relatively few opportunities for studying it during the 19th century, when Latin, Greek and mathematics were the chief studies and nearly all courses were required, not elected by the student. Olmsted's courses were given mainly for the Junior class while Seniors were expected to take special selected subjects in Natural Philosophy and Astronomy. At the Scientific School (later renamed Sheffield) students were at first priv-

Table 26.2
Relative Percentage of Stellar System in Elementary American Texts

Author	Text	Year	% Extra-Solar
S. Newcomb and E. S. Holden	*Astronomy*	1883	29
C. A. Young	*Manual of Astronomy*	1902	29
E. A. Fath	*Elements of Astronomy*	1928	42
J. C. Duncan	*Astronomy*	1930	54
R. H.Baker	*Astronomy*	1938	45
C. H. Payne-Gaposchkin	*Introduction to Astronomy*	1954	53
O. Struve	*Elementary Astronomy*	1959	43
S. Wyatt	*Principles of Astronomy*	1964	52
R. H. Baker and L. Fredrick	*An Introduction to Astronomy*	1967	48
P. Hodge	*Concepts of Contemporary Astronomy*	1974	57
J. Pasachoff	*Contemporary Astronomy*	1977	57
P. Hodge	*Concepts of Contemporary Astronomy*	1979	62
M. L. Kutner	*Astronomy, A Physical Perspective*	1987	82

ileged to take the astronomy course in Yale College for noncredit and for a special fee. From 1860 to 1886 Professor Lyman taught astronomy at Sheffield, again, as in Yale College, only in the Junior year. Olmsted was reputed to be deficient in analytical mathematics,[11] hence his courses were largely what we now call Descriptive Astronomy. Lyman's were more strongly oriented to mathematics. Olmsted taught at Yale from 1825 to 1859; his students included future astronomers E. Loomis, F. A. P. Barnard, C. S. Lyman, E. P. Mason, H. L. Smith, W. Chauvenet, J. S. Hubbard, and H. A. Newton.

Elias Loomis, the most prolific of textbook writers, was Olmsted's follower in Yale College. He was a quiet, reserved man, most remembered by students and colleagues for his precise presentations, without wasting a single unnecessary word.[12] President T. Dwight said of him,

> As an instructor, ... he was characterized by great clearness of statement and equally marked conciseness in his expression of his ideas. The student who was attentive could not fail to understand his meaning and to gain from him the light which he wished for or needed. I have rarely, if ever, seen a man who, having distinctly grasped an idea, could set it forth in such aptly chosen words, or in so few of them. He had mathematical precision in its exactest measure. As an astronomer he must, it would seem, have also had an imaginative element in his nature. But he did not display it in his teaching, in any considerable degree. He was inexpressive ... he appeared before his classes in the lecture or recitation room as a man of intellect only, dealing with purely intellectual matters. There was, as a consequence, a certain strangeness or marvel about him in the thought of his students, as if science were personified, and its utterances were only of itself and were brief with the brevity of a definition.

The brevity and conciseness he himself practiced he required of his students. Upon his death lengthy obituaries appeared in newspapers all over the country, from as far west as Seattle, Washington.[13] He was described as "one of the most conspicuous educators and mathematical writers this country has produced." One obituary[14] commented that Loomis's lectures to Juniors in Natural Philosophy were always popular, and illustrated with relevant interesting experiments. On the other hand, in astronomy his teachings dealt largely with the mathematics of the subject, and he was apparently little interested in conveying the observational aspects of the subject, "a single view through the telescope being almost all the actual acquaintance with the heavenly bodies that the students under him were permitted to obtain." Some complained they would have received greater benefit from instruction that would have enabled them to tell one constellation from the other and enable them to pick out the leading objects in the heavens, than from the laborious assigned computations. Yet all respected and admired him, and his texts continued to be used both here and abroad.

Overlapping Loomis's tenure was astronomer H. A. Newton, Professor of Mathematics from 1855 to 1890. While primarily remembered for his researches on meteors, his biographer, the famous physicist J. Willard Gibbs, remarked,[15]

Table 26.3
Astronomy Courses

Years	Freshman	Sophomore	Junior	Senior
1822–43	None	None	2nd term	None
1850–51	"	"	2nd term	1st term, optional 2nd
1860–61	"	"	3rd term	1st and 2nd terms
1870–71	"	"	3rd term	1st term
1875–76	"	"	3rd term	1st term
1876–77	"	"	2nd term	1st term optional
1885–86	"	"	Optional	None
1894–99	"	"	None	None
1901–03	"	"	Optional	Optional
1903–08	"	"	None	None
1911–14	"	"	None	None

But the studies which have won for their author an honorable reputation among men of science of all countries form only one side of the life of the man whom we are considering. Another side, perhaps the most important, is that in which he was identified with the organic life of the College and University with which he was connected from a very early age. In fact, we might almost call the studies which we have been considering the recreations of a busy life of one whose serious occupation has been that of an instructor. If from all those who have come under his instruction we should seek to learn their personal recollections of Professor Newton, we should probably find that the most universal impression which he made on his classes was that of his enthusiastic love of the subject which he was teaching.

Until within a few years of his death H. A. Newton carried most of the teaching load, especially in mathematics. In his *Memories of Yale Life and Men* Timothy Dwight[16] wrote: "During the first twenty-five years of his professorial career, he carried on the regular instruction of entire classes in his department and was, as we may say, one of the routine workers of the Faculty. In the later period he limited himself to the teaching of smaller bodies of students, who made choice of courses which he offered as electives."

G. W. Pierson[1,17] has collected some statistics on the required and elective courses offered the different classes up to 1921. From these the offerings in astronomy are extracted and summarized in Table 26.3. Until 1876 the college year consisted of three terms; thereafter just two.

Pierson points out[18] that by 1892 "astronomy, which had been removed from the Junior year to give room for ethics, managed to escape being dropped entirely by getting included in Sophomore mathematics." But this "refuge" was also abandoned in 1894, when only physics, of the originally available sciences, was left, and then only as an alternative course to languages or mathematics in the Sophomore year. Taught since Rector A. Pierson's time in 1701, physics (though of course still offered) could from 1894 be completely avoided by the students.

Some efforts were made between 1921 and 1929 to introduce or expand studies in several less well represented subjects, including astronomy.[19] Yale's historian, G. W. Pierson remarked, "Yet for all the energy and enterprise of the times Yale College remains a chilly place for the strange or unconventional. It encouraged the study of politics and consented to the arts, but it was not interested in drawing students to the study of plants or the stars or the shape of the earth."

Yale's President Angell, on the other hand, was concerned with the advancement of science. He encouraged Schlesinger's work and urged more development in the teaching of astronomy. Between 1923 and 1930 only one course in astronomy was offered to upperclassmen. In 1930 Brouwer introduced a second, Advanced Astronomy, in which he included observational astronomy with the use of the Sheffield refractor. By 1935 the curriculum for Freshmen was no longer strictly prescribed, and qualified students could now elect astronomy.[20] In 1941 Lyman Spitzer first taught an undergraduate course in astrophysics. With the entrance of the United States into World War II, courses in navigation and naval science were added, and in 1945 an introduction to dynamical astronomy. Throughout Brouwer's regime general astronomy, astrophysics, and introductory celestial mechanics constituted the courses open to undergraduates.

Pierson[21] gives statistics on the percentages of Yale undergraduates studying astronomy between 1950 and 1976. In only one year (1966–67) did the proportion reach as high as one percent of the undergraduate body. No student could major in astronomy alone: between 1961 and 1966 four students graduated with a major in astronomy and mathematics (the end of the Brouwer era of mathematical astronomy); between 1956 and 1971, twenty-five graduated in astronomy and physics.

Although Masters degrees were awarded from the beginning of the Collegiate School, the Graduate School as such was not established for nearly another century and a half, in 1847. This was 25 years before a Graduate Department was founded at Harvard in 1872 under the inspiration of Charles W. Eliot in the third year of his presidency at Harvard. However the full-fledged Graduate School of Arts and Sciences at Harvard was not established until 1889. Although up to that time the numbers of graduate students kept gradually increasing, Kathleen Holden[22] points out that despite all its resources Harvard simply was not attracting graduate students: they were going to Yale or Johns Hopkins instead. (Incidentally, the first President of Johns Hopkins, founded in 1867, was Daniel Coit Gilman, a graduate of Yale Class of 1852.) For a topic in astronomy, the first Ph.D. at Yale and in America was not awarded until 1861.

In 1880 a program[23] indicating the materials on which a candidate for the Ph.D. degree should be expected to be examined included (1) Practical Astronomy and (2) Theoretical Astronomy. The first (1) covered the following:

(a) The theory, discussion of errors, and use of instruments including the transit, equatorial and accessories for spectroscopy and photometry; meridian circle; heliometer; surveying instruments; and instruments for the physical observation of total solar eclipses; and clock mechanisms.

(b) Practical application of formulae for the construction of astronomical tables, assignment of weights in star catalogues, and the methods of construction of independent and relative star catalogues.

(c) Mechanical construction of instruments and the design of instruments for special purposes.

(d) Considerable skill in the use of the meridian circle, to be demonstrated by investigations "which shall be difficult in themselves and lead to results of value in astronomical science."

(e) An acquaintance with the literature of Practical Astronomy.

The second (2) Theoretical Astronomy "with the necessary mathematics" was to include linear differential equations, determinants, dynamics, statics, theory of least squares, solid geometry, elementary celestial mechanics, orbit computation, the problem of stellar parallaxes, and double-star orbits.

These requirements in principle were subject to only relatively minor alterations well into the 20th century when some of the instruments had declined in usage, and astrophysics had in many other institutions gained the ascendancy and was finally introduced into the Yale curriculum.

Women were admitted to the graduate school (but not to the College as undergraduates) in 1892, and Margaretta Palmer in 1894 was granted the Ph.D. for a thesis in astronomy, the first to any woman from an American institution. From 1892 through 1971, when all the students who had started their graduate work during Brouwer's regime finished their theses, seventy scholars were awarded the Ph.D.degree for a thesis in some branch of astronomy. Among these were only four women (Margaretta Palmer in 1894, Sandra Schwartz in 1956, Jocelyn Gill in 1959, and Carol Williams in 1967). The small percentage in the later years cannot be attributed to discrimination against women in University entrance-requirements, but may be inferred as due either to less interest among women in the physical sciences or to a fear of later job discrimination.

Mathematics Professor E. W. Brown, probably more than anyone else at Yale in the early years of this century, concerned himself with improvements in the teaching of mathematics and science. As a disciple of George Darwin and G. W. Hill, he was keenly aware of the problems in celestial mechanics needing to be solved and the necessary background for solving them. Darwin, in his inaugural address as the first Plumian Professor of Astronomy and Experimental Philosophy at Cambridge University, had expostulated on the deficiencies in the system of tripos (honors) examinations at Cambridge, and more generally on the drawbacks in the system of teaching mathematics.[24] This was in 1883, the year before

Brown became a student in Cambridge. Darwin's opinions may well have influenced Brown's later efforts to improve the teaching of mathematics, especially applied mathematics, in America. Both Darwin and Hill were primarily theoretical research workers, but informally they were stimulating teachers for the few like-minded younger scholars. Brown realized his responsibilities in training future scientists to help solve the abstruse unsolved problems. Early training would be essential.

In a paper, "Mathematics for Freshmen and Sophomores,"[25] Brown discussed the differing needs of those who would go on to further studies either in pure or applied mathematics and those who would not. Should the teaching be aimed at the majority of the students in these early years? He concluded, "Mathematics for the majority is simply an aid to the other sciences. To continue to teach it solely as a discipline or as an art is to consign it to the same fate as the study of Greek. For the sake of the subject itself it must return to the earlier ideas from which it arose and in contact with other sciences become useful."

Before Brown was appointed Acting Director of the Observatory he had published several papers dealing especially with needs of the more gifted students who were being short-changed when teachers felt obliged to gear their teaching to the average student.[26] A survey by a committee of the American Association for the Advancement of Science, of which he was chairman, examined what other universities were doing to separate the gifted from the average. This can in general be done only in the larger institutions where classes for the average students are large. The report recommended that "for greater breadth of view" students should not do graduate work at their undergraduate institutions. For the *Yale Daily News,* Brown described the Reading Period at Cambridge University which had recently been emulated at Yale, and indicated its advantages for the motivated student.[27] Later,[28] in another article, "College Mathematics for Technical and Scientific Students and for General Mental Training—Concentration More Than Much Reading Needed," he outlines the types of courses available to fill the needs of students of differing aspirations.

The outbreak of World War I accentuated the potential need of men educated for scientific research. In an editorial, "Science in the War," for a New Haven newspaper,[29] Brown pointed out, "It is becoming evident that far more stress must be laid by the universities on their function as organizations for the advance of knowledge. Increased provision must be made for research. . . . Institutions of learning hold within their walls a large proportion of the scientific men of the country."

That same year (1917), in his address as retiring president of the American Mathematical Society, Brown spoke on "The Relations of Mathematics to the Natural Sciences" stressing the role of mathematical symbolism in the solution of scientific problems.[30] Here, incidentally, he used as one illustration the Fourier analysis in the study of variable-star periodicities where the light curves differ appreciably from sine curves. More to the point, he noted the natural bent of the nation to regard most of its affairs from a thoroughly practical point of view. Hence, "An incentive and the training once supplied, there seems no reason why a school of applied mathematics of the first rank should not flourish on this continent."

In a paper, "Efficiency in the Higher Education in Universities"[31] Brown, the researcher, was troubled by the amount of time required for teaching. He stressed the importance of small classes for advanced students if a university was to foster research and to educate men to do research. From the point of view of the teacher saddled with large undergraduate courses, however, he commented, "An active instructor with the education in his subject demanded at the present day is not generally content to spend his whole energies in teaching elementary subjects; he very properly asks for an opportunity to study his own special line, not only by investigation on his own account but also by expounding his subject to those who wish to learn." He considered the feasibility of sharing the teaching load among the eight or ten colleges that were giving first class higher instruction. Brown did not consider whether commuting could pose a problem. His proposal has now been implemented by the Five Colleges in Massachusetts (Amherst, Smith, Mount Holyoke, Hampshire, and University of Massachusetts). However, in 1919, daily transportation of students between colleges would not have been practicable.

Schlesinger's interests and activities in teaching have already been described in Chapter 20. In a letter to Professor Brown before coming to Yale, Schlesinger had written,[32]

> Again there appears to be at New Haven an unusual spirit of cooperation on the part of the Department of Mathe-

Table 26.4
Distribution of Astronomy Graduate Students by Specialization*

Year ending June	1963	64	65	66	67	68	69	70
Number of students	18	26	28	26	28	25	21	10
Celestial mechanics	12	17	16	13	19	17	14	1
Theoretical astrophysics	2	4	6	6	2	3	2	5
Observational astrophysics	2	2	1	1	3	1	2	4
Radio astronomy	2	3	5	4	4	4	3	0

*These numbers include students working on their theses *in absentia.*

matics and the Department of Physics, and such cooperation goes a long way toward successful attack upon many astronomical problems.

You have already emphasized the opportunity of a graduate school in astronomy, and I have told you how thoroughly this would fit in with my own ideas and ambitions. Here again the cooperation of the two departments named forms the unusual feature in the situation, for the education of a competent astronomer depends quite as much upon the instruction he gets in mathematics and physics as it does upon his preliminary training in astronomy itself.

No post–World War I undergraduate courses in astronomy were offered until 1923, although Schlesinger had already taught graduate courses in his first year at Yale. Subsequently, both undergraduate and graduate courses have been offered with little interruption, although the subject matter, particularly of the advanced courses, has fluctuated over the years, depending on the current interests of the professors and the demands of the students. During Brouwer's time, graduate courses were offered largely in celestial mechanics and, under the direction of Dr. Wildt, in astrophysics.

Starting with the *Annual Reports* of the Director of the Observatory for 1961–63, the numbers of graduate students are indicated who were enrolled or had completed their degrees in the Department of Astronomy and what their fields of specialization were. These are shown in Table 26.4. Most of the theses in radio astronomy were completed at the University of Texas under the guidance of James Douglas and Harlan Smith after all the Yale radio facilities had been transferred to Texas. This new field had attracted about as many graduate students to Yale as did astrophysics. The contrast between a student's choice of celestial mechanics or astrophysics before and after Brouwer's decease is striking. The preponderance of celestial mechanics was not only due to the demands of the early Space Age and Brouwer's eminence in his field, and that less emphasis was given to this topic at other institutions, but also to the fact that other universities were stronger than Yale in astrophysics, the brilliance of Yale's Professor Wildt notwithstanding. In recent years the breadth of subject matter from which students could choose dissertation topics has expanded significantly, including stellar evolution, stellar populations, galactic structure, interstellar matter, extragalactic studies, and cosmology.

In a report, *Astronomy at Yale,*[33] in 1965 Brouwer pleaded for the expansion of the department, concluding,

The number of graduate students in dynamical astronomy and in astrophysics that the department has attracted in recent years serves as a reliable indication that the expanded department will be among the leading departments of astronomy in the country. Even without a significant activity in radio astronomy, it will attract many graduate students because of its balance between dynamical astronomy and astrophysics.

Any significant expansion will require improvement of the housing of the department, preferably in a single building, large enough to house all of the department's activities in New Haven.

Those needs have hardly yet been fulfilled while the activities and size of staff and space requirements for new computers and instrumentation continue to expand.

The quality of both research and advanced teaching in a university may be inferred from the membership of its faculty and former students in the National Academy of Sciences (NAS). Founded in 1863, membership was limited to men distinguished for original research. Election has been considered as the most prestigious honor to be conferred upon

Table 26.5
Yale Professors Elected to NAS

B. Silliman 1863	C. S. Hastings 1889
B. Silliman, Jr. 1863	W. L. Elkin 1895
H. A. Newton 1863	F. Schlesinger 1913
E. Loomis 1873	E. W. Brown 1923
W. A. Norton 1873	D. Brouwer 1951
J. W. Gibbs 1879	L. Spitzer 1952
E. S. Dana 1884	G. M. Clemence 1952

a scientist in the United States. Of its fifty charter members[34] six had earned degrees from Yale in astronomy or related fields, namely F. A. P. Barnard (1828), W. Chauvenet (1840), J. S. Hubbard (1843), H. A. Newton (1850), Benjamin Silliman (1796), and Benjamin Silliman, Jr. (1837). Yale professors (including meteoriticists) who became members of the NAS and an inspiration to their astronomy students, included those listed in Table 26.5, elected in the years indicated.[35] Thus throughout an entire century from the founding of the NAS in 1863 through Brouwer's lifetime (1966), astronomy at Yale was guided by members of the National Academy of Sciences—possibly a unique continuous record, almost but not quite matched at Harvard Observatory.

Nearly 180 former Yale undergraduate or graduate students have been cited in this history of astronomy at Yale up to 1970. (These were the achievers, not necessarily all who had majored in astronomy.) Of these, 24 were deceased before the founding of the National Academy; of the remaining, 18 became members of the NAS. However, all but one (Lyman Spitzer, B.A. 1935) were elected prior to 1890. Of the 78 recipients of Yale Ph.D. degrees awarded for theses in astronomy (Appendix B), only A. W. Wright (1861) and C. S. Hastings (1873) were later elected to the Academy. The apparent decline in NAS membership by former Yale students can hardly be attributed to a decline in the capabilities of the best of the ever-increasing numbers of students relative to the older generations. Rather it may be related to the greater difficulty in getting accepted into a society whose total increase in membership was limited, while the scientific population at large proliferated at a greater rate than the numbers of astronomers. In 1863 the charter membership included few natural scientists, even though their total numbers already exceeded the physical scientists. This soon changed, with the concomitant relative decrease in the percentage of astronomers. Table 26.6 gives a sampling[34,36,37] of the total numbers of NAS members, the total number and percentage of astronomy members, and the number and percentage of the astronomers who were either Yale faculty or former Yale students.

In 1977, *American Men and Women of Science*[38] included approximately 30,000 members in the physical sciences alone, of whom about 1,200 or 4 percent were astronomers or astrophysicists. Inclusion of the natural sciences would have reduced to no better than 2 percent the number of astronomers among all scientists. Between 1863 and 1870 the total membership of NAS was limited to 50. In 1872 an additional 25 were inducted; but from then until 1915 only 10 new members a year could be elected.[39] This number was then increased from 15 in 1915 to 35 by 1958.[40] But 35 is only about 0.06 percent of all professional scientists in 1977, making the chances for the election of astronomers very slim in comparison with what they were in 1863. Thus the increases in the third column of Table 26.6 are better than might have been expected.

As early as 1827 Denison Olmsted[7] had pondered on the future of science, remarking, "Will some new instrument of investigation like logarithms, or fluxions, reduce beyond expression the labor of calcu-

Table 26.6
Membership in NAS

Year	Total	Astronomers		Yale Affiliates	
		No.	%	No.	% of All Astronomers
1863	50	11	22	6	55
1913	123	14	11	3	21
1963	637	20	3	1	5
1972	939	41	4	1	2
1977	1208	46	4	1	2

lation, and carry the researches of the mathematician into regions, where even the imagination is at present too slow to wander?" The advances in astronomy in the more than 160 intervening years have been beyond even Olmsted's keen powers of speculation. As the tools of the astronomer, including computers, space telescopes, and new physical theories, keep improving, the promise of future advances seems greater than ever. But advances increasingly require both able researchers and dedicated teachers. Yale continues to be in a position to help supply anticipated needs.

REFERENCES

1. G. W. Pierson, *A Yale Book of Numbers,* 633 pp., New Haven, 1983.
2. *E.g., Yale Alumni Weekly,* Feb. 5, 1937.
3. *E.g.,* J. F. Fulton, *Bull. Hist. Medicine,* **20**, 97, 1946.
4. P. Demarque, *Yale Alum. Mag.,* **53**, No. 2, p. 5, 1989.
5. S. G. Carlson, *Ibid.,* No. 6, p. 6, 1990.
6. M. Mallay, *Ibid.,* No. 8, p. 10, 1990.
7. *Cf.* F. B. Dexter, *New Haven Historical Society Papers,* **5**, 247, 1889.
8. L. W. McKeehan *Yale Science 1701–1801,* p. 53, 1947.
9. E. Baldwin, *Annals of Yale College from Its Foundation to the Year 1831,* New Haven, p. 226, 1838.
10. D. Olmsted, Phi Beta Kappa Oration, "On The Progressive State of the Present Age," Sept. 11, 1827.
11. M. Rothenberg, "The Educational and Intellectual Background of American Astronomy, 1827–1875," *Bryn Mawr Dissertation,* Chap. 2, 1974.
12. T. Dwight, *Memories of Yale Life and Men,* N. Y., p. 384, 1903.
13. Newspaper clippings, *Yale Obs. Archives.*
14. *The New York Evening Post,* Aug. 16?, 1889.
15. J. W. Gibbs, *Biographical Mem. Nat. Acad. Sci.,* **4**, 118, 1902.
16. Ref. 12, p. 309.
17. G. W. Pierson, *Yale College, an Educational History, 1871–1921,* 1952.
18. ———, *Idem.,* p. 86.
19. ———, *Yale: the University College, 1921–1937,* vol. **2** of Ref. 17, p. 188, 1956.
20. ———, *Idem.,* pp. 371, 520.
21. ———, Ref. 1. p. 271.
22. K. C. Holden, *Harvard University Graduate School of Arts and Sciences, Alumni Directory 100th Anniversary Edition,* pp. v-xii, 1989; *Newsletter,* Harvard Graduate Society, p. 1, Spring 1989.
23. L.Waldo, ms, June 30, 1880. *Yale Obs. Archives.*
24. G. H. Darwin, *Scientific Papers,* Cambridge, England, **5**, 1, 1916.
25. E. W. Brown, *The Yale Alumni Weekly,* **28**, 310, Dec. 13, 1918.
26. ———, *Science,* n.s., **42**, 368, 1915.
27. ———, *The Yale Daily News,* **38**, No. 189, p. 1, June 1, 1915.
28. ———, *Ibid.,* **39**, No. 150, 1, Apr. 7, 1916.
29. ———, *Journal and Courier,* New Haven, Apr. 21, 1917.
30. ———, *Bull. Amer. Math. Soc.,* **23**, 213, 1917.
31. ———, Copy of manuscript, March 25, 1919, *Yale Obs.Archives.*
32. F. Schlesinger to E. W. Brown, Oct. 25, 1919, *Yale Obs Archives.*
33. D. Brouwer, Manuscript copy, May 25, 1965, *Yale Obs.Archives.*
34. F. W. True, Ed., *A History of the First Half-Century of the National Academy of Sciences, 1863–1913,* Washington, p. 104, 1913.
35. R. C. Cochrane, *The First Hundred Years, 1863–1963,* National Academy of Sciences, Washington, p. 614, 1978.
36. Ref. 34, p. 337.
37. *Reports of the National Academy of Sciences* (annual) 1963–77.
38. J. Cattell Press, *American Men and Women of Science: Physics, Astronomy, Mathematics, Statistics, and Computer Science,* N.Y., 1977.
39. Ref. 34, p. 355.
40. Ref. 35, p. 525.

APPENDIX A
YALE TELESCOPES

(still in use, *)

Year Acquired

1734 Gregorian speculum reflector, diam. not specified, f.l. 2.5ft. Made by spectacle-maker Edward Scarlett. Deteriorated by 1789.

1789 Reflector, f.l. 3.5ft. Lost in shipwreck en route to England for repair in 1822.

1829 5-inch Dollond refractor given by Sheldon CLARK. On display in Yale's Peabody Museum.

1856 4-inch Troughton and Simms transit instrument, given by W. HILLHOUSE. (In 1927 Schlesinger noted that Yale had at least three transit instruments, none in use.)

1866 9-inch Alvan Clark refractor given by Joseph E. SHEFFIELD. 100 years later sent to El Leoncito as part of "camera cluster." At El Leoncito but not mounted.

1866 Meridian Circle with 5ft. f.l. lens presented by SHEFFIELD.

1882 6-inch Resold heliometer. Lens on exhibit at Department of Astronomy.

1882 * 8-inch Grubb refractor given by E. M. REED.

1889 Combination transit instrument and zenith telescope for the determination of latitude, made by C. S. LYMAN in 1853 and presented to the Sheffield School in 1889.

1894 6-inch Voigtländer portrait lens presented by C. S. BRAINERD.

1894 Meteorograph consisting of a battery of two 6-inch and two 5-inch portrait lenses on single mounting constructed by Warner and Swasey Co. (Dismantled in 1922.)

1896 Two more lenses added to meteorograph, the cost defrayed by BRAINERD.

1898 A second meteorograph with four cameras, mounted in Hamden. Two more cameras added by 1900, and moved from Hamden to Whitneyville.

1916 The LOOMIS Telescope, a coelostat designed by C. S. Hastings and built by Brashear of Pittsburgh, consisting of a 15-inch photographic refractor and a 10-inch Alvan Clark visual guide scope, tubes mounted parallel to the polar axis and with a 30-inch silvered flat to guide light up both tubes to the foci at the top of a 60-foot tower. Moved to Bethany 1957 but reversed (no longer a coelostat) for direct photography of North Pole only, for polar motion. Dismantled in 1982.

1922 5.6-inch Ross-McDowell camera for zone catalogue work.

1922 10-inch photographic zenith camera for determining change of latitude.

1925 26-inch photographic refractor for parallax work in Johannesburg, South Africa. Lens by Brashear, f.l. 36ft., with 10-inch visual guide scope, also 36 ft. f.l. Moved to Mt. Stromlo in 1952. Donated to the Mt. Stromlo Observatory in 1963.

1928 3-inch Ross camera for Johannesburg.

1933 5-inch Ross-Fecker, mounted on tube of 26-inch at Johannesburg.

1936 5-inch Ross for the northern hemisphere.

1949 * 8-inch Ross by Perkin-Elmer Corp. For north polar regions; now at Bethany.

1951 * The BUTLER 10-inch Alvan Clark refractor, presented by Mrs. Butler in memory of her husband, A. W. Butler.

1961 * 20-inch reflector designed by J. Rodman, optics by Tinsley Laboratories of California, mounting by Warner and Swasey.

1965 * 20-inch double astrograph for Yale Southern Station, one component for blue light the other for yellow. Optical parts by Perkin-Elmer Corp.

1966 * 40-inch Ritchey-Chretien-type reflector with Perkin-Elmer optics, for Bethany. Sent to Cerro Tololo, Chile, in 1974.

1966 Camera Cluster for El Leoncito, consisting of 8-inch "catalogue camera;" two 3-inch lenses, one for blue and one for red light; and the 9-inch Sheffield. The combination of four telescopes on one mounting proved inefficient and was abandoned.

APPENDIX B
Ph.D. THESES IN ASTRONOMY AWARDED THROUGH 1971

The theses listed here were not all awarded under the Department of Astronomy. Eleven were from the Department of Physics and a few others from Mathematics and History of Science. All, however, have direct bearing on astronomy. Although this history is intended to cover only through the administration of Dr. Dirk Brouwer (1902–1966) the list goes through 1971 as the final year in which the degree was awarded to students who began their graduate studies at Yale before the end of Brouwer's regime.

It is noted that although the first Ph.D. degree in Astronomy earned by a woman was awarded in 1894 to Margaretta Palmer, only three other women received the degree through 1971: Sandra Schwartz in 1956, Jocelyn R. Gill in 1959, and Carol Williams in 1967. This represents only 5 percent of the 78 theses listed here. (Subsequently, through 1990, there have been only four more women recipients, one each in 1980, 1981, 1988, and 1990.)

Of the 78 theses awarded the Ph.D. through 1971, 40 (51%) are in celestial mechanics or related topics, 20 (26%) in astrophysics, 8 (10%) in photometry, 6 (8%) in radio astronomy, and 4 (5%) on miscellaneous topics.

1861 Arthur Williams Wright (Physics), *Having Given the Velocity and Direction of Motion of a Meteor on Entering the Atmosphere of the Earth, to Determine its Orbit About the Sun, Taking into Account the Attraction of Both These Bodies.*

1866 Francis Englesby Loomis (Physics), *Periodic Stars.*

1873 Charles Sheldon Hastings (Physics), *Comparison of the Spectra of the Limb and of the Center of the Sun, Made at the Sheffield Scientific School.*

1877 Thomas Alexander Smith, *Computation of the Orbit of Electra.*

1880 Thomas Marcus Blakslee, *The Orbit of Swift's Comet of July 1878.*

1885 Jefferson Engel Kershner (Mathematics), *The Determination of Longitude Between New Haven and Harvard College Observatory, Cambridge.*

1889 Asaph Hall Jr., *The Mass of Saturn and the Orbit of Titan.*

1891 Frederick Lincoln Chase, *Triangulation of the Victoria Comparison Stars.*

1894 Margaretta Palmer, *Determination of the Orbit of the Comet 1847 VI.*

1901 Joshua Larson, *Orbit of S 3062.*

1904 Oliver Clarence Lester (Physics), *On the Oxygen Absorption Bands of the Solar-Spectrum.*

1912 Archibald Lamont Daniels, Jr., (Mathematics), *The Librations of Bodies Whose Periods Are One-Third That of the Disturbing Body.*

1912 Harley Richard Willard (Mathematics), *A Family of Oscillating Orbits of Short Period.*

1913 Theodore Henry Brown (Mathematics), *The Effect of Radiation on a Small Particle Revolving about Jupiter.*

1917 William Leonard Crum (Mathematics), *The Perturbations Caused by a Close Approach of Two Asteroids.*

1923 Carl Leo Stearns, *Trial of the Loomis Memorial Telescope for Stellar Photometry.*

1928 Paul Slavenas, *The Stellar Case of the Problem of Three Bodies.*

1931 Wallace John Eckert, *The General Orbit of Hector.*

1933 John Scoville Hall, *Photo-Electric Photometry in the Infra-Red with the Loomis Telescope.*

1935 Theodore Grefe Mehlin, *Trial of a Photoelectric Method for Measuring Astronomical Photographs.*

1935 Robert Ingersoll Wolff, *First-Order Perturbations of the Minor Planet (3) Juno by Jupiter.*

1941 Hans Georg Hertz, *On the Theory of the Trojan Asteroids.*

1943 Boris Garfinkel, *An Investigation in the Theory of Astronomical Refraction.*

1948 Harold Hooker Lane, *Photographic Magnitudes of Stars in the Vicinity of the North Celestial Pole.*

1950 Morris Schuyler Davis, *A Study of the Method of Rectangular Coordinates in General Planetary Theory.*

1951 W. C. DeMarcus (Physics), *Density Relations with Applications to the Constitution of the Planets.*

1955 Alvan H. Davis, (Physics), *The Kinetic Temperature of an Atmosphere Supported by Radiation.*

1956 Raynor Lockwood Duncombe, *The Motion of Venus, 1750–1949.*

1956 Sandra Schwartz, *The Inhomogeneous Milne Equation.*

1959 Jocelyn Ruth Gill, *The Motion of Neptune's Satellite Triton.*

1959 Mahendra Singh Vardya, *Convection in Late-Type Stellar Atmospheres and Interiors.*

1960 James N. Douglas, *A Study of Non-Thermal Radio Emission from Jupiter.*

1963 Robert Goldman, *Coulomb Interactions in a Strong Magnetic Field.*

1963 Myron Lecar, *Departures from Local Thermodynamic Equilibrium in an A0 Star at Atmospheric Pressure.*

1963 Kenelm Winslow Philip, *Position Measurements on Solar Continuum Sources at Decameter Wavelengths.*

1963 James P. Rodman, *An H-Alpha Spectrophotometer and Some Astronomical Applications.*

1964 Martin D. Altschuler, *Interaction of Fast Charged Particles with the Coronal Plasma.*

1965 Giorgio E. O. Giacaglia, *Evaluation of Methods of Integration by Series in Celestial Mechanics.*

1965 William H. Jefferys, III, *Some Dynamical Systems of Two Degrees of Freedom in Celestial Mechanics.*

1965 Claus Oesterwinter, *The Motion of a Lunar Satellite.*

1966 William D. Arnett (Physics), *Physical Processes and Supernova Dynamics.*

1966 Carl J. Hansen (Physics), *Neutrino Losses from Dense Stellar Interiors.*

1966 Brian G. Marsden, *The Motions of the Galilean Satellites of Jupiter.*

1966 Normal Paul Patterson, *Predicted Ultraviolet Radiation from Stellar Chromospheres.*

1966 David A. Pierce, *Singularities in General Perturbations.*

1966 Prativadibhayankaram Krishnaswamy Raju, *Formation of Solar Prominences.*

1966 William A. Sloan, *Rocket Observations of the Sun in the Light of Lyman Alpha.*

1966 Sabatino Sofia, *The Origin of Condensations in Planetary Nebulae.*

1966 James W. Truran (Physics), *Thermonuclear Reactions in Supernova Shock Waves.*

1966 Peter Hermann Ulmschneider, *The Structure of the Outer Atmosphere of Cool Stars.*

1967 John Phillips Britton (History of Science), *On the Quality of Solar and Lunar Observations and Parameters in Ptolemy's Almagest.*

1967 Eugene Frank Milone, *Photoelectric Photometry of Some Selected Variables.*

1967 Douglas A. O'Handley, *Determination of the Mass of Jupiter from the Motion of 65 Cybele.*

1967 Julian Ivanhoe Palmore, *Numerical Experimentation into Effects of Varying the Mass Ratio on Periodic Solutions of the Restricted Problem of Three Bodies.*

1967 Carol Ann Williams, *An Analysis of the Performance of the Loomis Polar Telescope.*

1968 Frank Arakel Bozyan, *A Study of the Spectral and Statistical Properties of Jupiter in the Decameter Region.*

1968 Alan D. Fiala, *Determination of the Mass of Jupiter from a Study of (57) Mnemosyne.*

1968 George S. Greenstein (Physics), *The Helium Problem in Cosmology.*

1968 Stephen Howard Knowles, *Determination of the Astronomical Unit from Hydrogen-Line Radial-Velocity Measurements.*

1968 Jay Henry Lieske, *A Dynamical Determination of the Solar Parallax from the Motion of (433) Eros.*

1968 Paul Edward Nacozy, *Validity of Hansen's Method of Partial Anomalies.*

1968 C. Frederick Peters, *Treatment of Encounters in Stellar Dynamics.*

1968 Jeffrey David Rosendhal, *Evolutionary Effects in the Rotation of Supergiants.*

1968 Erland Myles Standish, *Numerical Studies of the Gravitational Problem of n Bodies.*

1969 Dale Gaylard Bettis, *Stabilization of Finite Difference Methods of Numerical Integration in Celestial Mechanics.*

1969 Clinton Childs Brooks, *Accurate Declinations of 32 Radio Sources.*

1969 William J. Klepczynski, *The Motion of the Minor Planets 10 Hygeia, 24 Themis, 31 Euphrosyne and 52 Europa and the Mass of Jupiter.*

1969 William D. Langer (Physics), *Effects of Hyperons on the Variations of Neutron Stars.*

1969 Charles F. Martin, *A Study of the Rate of Rotation of the Earth from Occultations of Stars by the Moon—1627–1860.*

1969 Gerard Franklin Moseley, *Precise Declinations and Optical Identification of Selected Radio Sources.*

1969 Louis Peter Pataki, Jr., *Properties of the Interplanetary Medium from Scintillations in Jovian Decametric Emission.*

1969 Thomas Charles van Flandern, *A Discussion of 1950–1968 Occultations of Stars by the Moon.*

1970 Kaare Aksnes, *A Second-Order Solution for the Motion of an Artificial Earth-Satellite Based on an Intermediate Orbit.*

1970 William Jackson Breedlove, *A Solution of a Low Thrust Lawden Problem.*

1970 Pasquale Bernard Esposito, *An Analytical Comparison of Newtonian and Relativistic Planetary Orbits.*

1970 Roy Edward Laubscher, *The Motion of Mars 1751–1969.*

1971 David Waring Dunham, *The Motions of the Satellites of Uranus.*

1971 Wayne H. Osborn, *The Positions of Globular Cluster Stars in the Physical HR Diagram.*

APPENDIX C
YALE IN THE ASTEROID BELT

The first asteroid, in the belt between Mars and Jupiter, was discovered by Giuseppi Piazzi in Italy on January 1, 1801, and named Ceres. Since then over 4,500 asteroids (also known as minor planets) have been discovered and verified. Nowadays, new discoveries are not named immediately but only after an orbit has been determined, which takes into account observations at several oppositions; or in rare instances is otherwise unambiguously verified. (Sometimes "new" discoveries are later found to be rediscoveries of previous apparitions.) The discoverer usually has the privilege of naming the object once it has been accepted as a verified new object. The names were originally assigned to mythological characters. Later they could be named for almost anything: famous astronomers or other scientists, musicians, artists, authors, often members of the discoverer's family; flowers and towns abound; once even a cat was honored ("Mr. Spock," belonging to and named by James Gibson, the discoverer of the asteroid). Eventually, objections were raised to naming asteroids after animals.[1]

Discoverers do not always name their asteroids, and often the discoverer has died before his discovery has been adequately verified. Now a small committee in the International Astronomical Union (IAU) judges on the suitability of names proposed by the discoverers, or by others if the discoverers are no longer living. Over the years no fewer than thirty Yale astronomers or students now or formerly associated with Yale have had objects named in their honor, although in many instances the honors were conferred for accomplishments not related to the person's work at Yale. Table C.1 lists the serial number of the discovery of the asteroid, its name, the name of the discoverer, the dates of discovery and of naming, the astronomer so honored, and the years when he or she was actively associated with Yale. (According to the convention adopted by the IAU Commission on Asteroids, they are always tabulated in alphabetical order of the first letter of the name, regardless of whether that is a first name or a surname; thus the asteroid Beatrice Tinsley is correctly listed in Table C.1 under B, not T.)

In a few instances the asteroid was named for two individuals with the same surname (Irja and Arnold Klemola; R. P. and C. U. Cesco). In these cases the names of the Yale affiliates are italicized in Table C.1.

J. W. Gibbs (1839–1903), the physicist, is not generally considered an astronomer, but he is included here because his citation notes that "he made significant contributions to minor planet studies through his work on orbits."

Asteroid Osita is named after Ursula Gibson, the wife of James Gibson, the discoverer. (Osita is the Spanish equivalent of Ursula). She was a volunteer assistant at El Leoncito, 1971–73.

The youngest person honored is P. M. Tamblyn, Yale class of 1990, who was a volunteer assistant at the Harvard–Smithsonian Center for Astrophysics in the summer of 1987, working on asteroid identifications under the direction of Brian Marsden, the current Director of both the International Astronomical Union's Central Bureau for Astronomical Telegrams and its Minor Planet Center.

In addition to the above, asteroid Cunningham might be mentioned, discovered by Delporte in 1930 and named after Leland E. Cunningham in 1976. Cunningham, as a graduate student at Harvard, had spent part of 1941 at Yale obtaining Ph.D. thesis guidance in celestial mechanics under Professor Dirk Brouwer, recognized world authority in this field.

Another asteroid, discovered at the Felix Aguilar Observatory in Argentina in 1974, was named El Leoncito in 1984 after the location where, in collaboration with the Argentinian observatory, Yale had established its Southern Station in 1963.

Besides the Yale astronomers for whom asteroids were named, those who actually discovered new as-

Table C.1
Asteroids Named For Yale Astronomers

No.	Name	Discoverer	Dates	Astronomer	At Yale
2067	Aksnes	Vaisala	1936–78	K. Aksnes	PhD 1970
2023	Asaph	Indiana U.	1952–77	Asaph Hall	LLD 1879
2157	Ashbrook	Reinmuth	1924–80	J. Ashbrook	1946–53
3087	Beatrice Tinsley	Gilmore + Kilmartin	1981–85	B. Tinsley	1975–81
1746	Brouwer	Indiana U.	1963–68	D. Brouwer	1927–66
1643	Brown	Reinmuth	1951–76	E. W. Brown	1907–38
1571	Cesco	Itzigsohn	1950–82	R. P. + *C. U. Cesco.*	1966–87
1919	Clemence	Gibson + Cesco	1971–76	G. Clemence	1963–74
3415	Danby	Reinmuth	1928–86	J. M. A. Danby	1961–67
3416	Dorrit	Reinmuth	1931–87	D. Hoffleit	1956–
3368	Duncombe	Bowell	1985–87	R. Duncombe	1948–49 PhD 1956
3123	Dunham	Bowell	1981–86	D. Dunham	PhD 1971
1750	Eckert	Reinmuth	1950–76	W. Eckert	PhD 1931
1761	Edmondson	Indiana U.	1952–71	F. K. Edmondson	1968–69 Freshman 1929
2937	Gibbs	Bowell	1980–84	J. W. Gibbs	1863–1903
2742	Gibson	Shoemaker	1981–83	J. B. Gibson	1971–74
3299	Hall	Shoemaker	1980–85	J. S. Hall	PhD 1933
3842	Harlansmith	Bowell	1985–89	H. J. Smith	1953–63
2193	Jackson	H. E. Wood	1926–80	C. Jackson	1945–66
1723	Klemola	Vaisala	1936–80	Irja + *A.Klemola*	1961–67
3454	Lieske	Bowell	1981–86	J. Lieske	PhD 1968
1877	Marsden	van Houten	1971–75	B. Marsden	1963–64 PhD 1966
1691	Oort	Reinmuth	1956–70	J. Oort	1922–23
1971	Osita	Gibson	1971–71	Ursula Gibson	1971–73
1953	Rupertwildt	Indiana U.	1951–82	R. Wildt	1946–76
2308	Schilt	Cesco + Klemola	1967–81	J. Schilt	1926–31
1770	Schlesinger	Cesco + Klemola	1967–76	F. Schlesinger	1920–43
1911	Schubart	P. Wild	1973–76	J. F. Schubart	1965–66
2160	Spitzer	Indiana U.	1956–84	L. Spitzer	1939–51
3420	Standish	Bowell	1984–87	M. Standish	PhD 1968 1968–72
2035	Stearns	Gibson	1973–78	C. L. Stearns	PhD 1923 1920–25
3417	Tamblyn	Reinmuth	1937–87	P. M. Tamblyn	Yale '90
2370	vanAltena	Klemola	1965–85	W. F. van Altena	1974–
1945	Wesselink	van Gent	1930–85	A. Wesselink	1964–

teroids deserve to be recorded. Chief among these is Cyril Jackson of South Africa. Yale established its first Southern Station at Johannesburg, South Africa, in 1925, and in 1952 moved it to Mt. Stromlo, Australia; and finally in 1963 to El Leoncito, Argentina. At all three of these stations Jackson was chief observer from 1945 to 1966, except for an absence between September 1954 and March 1957. Between 1929 and 1939, before becoming associated with Yale, he was at the Union Observatory, also in Johannesburg. There he discovered 65 asteroids, most of which he named after sites in South Africa. This is the largest number of asteroids discovered by any one individual in the southern hemisphere, although K. Reinmuth of Germany discovered vastly more (287 named asteroids) in the northern skies in a span of 43 years (1914–57).

Table C.2 lists the numbers of named asteroids discovered by Yale observers while they were associated with Yale. The last column indicates the span of years represented by these discoveries, all made in the southern hemisphere.

Of the more than thirty Yale people honored in the asteroid belt, twenty are still living (as of 1991), of whom Hoffleit and van Altena continue to be active at Yale. I am indebted to Brian Marsden, Yale Ph.D. 1968, for considerable help in the compilation of Table C.1.

Table C.2
Asteroids Discovered by Yale Observers

Observers	Number	Location	Years
Klemola	2	El Leoncito	1965
Cesco	3	" "	1970–71
Cesco and Klemola	5	" "	1967
Cesco and Samuel	1	" "	1968
Cesco and Gibson	2	" "	1971
Gibson	4	" "	1971–73
TOTAL	17		

REFERENCE

1. B. Marsden, *Trans. I.A.U.,* **19B**, 179, 1985.

APPENDIX D
MEDALS
Awarded before 1970
(Not necessarily complete)

1882 Silver Medal, Tottenridge Park School, England, to E. W. Brown, a Sophomore.

1884 J. Lawrence Smith Medal of the National Academy of Sciences, to H. A. Newton.

1887 Gold medal for excellence in reading the Scripture Lessons in Chapel, Christ's College, Cambridge, England, to E. W. Brown.

1888 Belden Mathematical Prize, The College of the City of New York, to Frank Schlesinger, a Sophomore.

1889 Belden Mathematical Prize, The College of the City of New York, to Frank Schlesinger, a Junior.

1890 Ward Medal, The Free Academy of the City of New York, to the student of most proficiency in Natural Philosophy, to Frank Schlesinger.

1898 Donohoe Medal of the Astronomical Society of the Pacific, to F. L. Chase.

1907 Medal of the Royal Astronomical Society, to E. W. Brown.

1908 Prix Lalande, French Academy of Sciences, to W. L. Elkin.

1914 Royal Medal, Royal Society, London, to E. W. Brown.

1920 Catherine Wolfe Bruce Medal, Astronomical Society of the Pacific, to E. W. Brown.

1926 Prix Valz, Académie des Sciences, Institut de France, to Frank Schlesinger.

1927 Medal of the Royal Astronomical Society, to Frank Schlesinger.

1929 Catherine Wolfe Bruce Medal, Astronomical Society of the Pacific, to Frank Schlesinger.

1935 Townsend Harris Medal, the Associate Alumni of the College of the City of New York, to Frank Schlesinger '90.

1936 James Craig Watson Medal, National Academy of Sciences, to E. W. Brown.

1948 Donohoe Medals, to J. Ashbrook and C. Jackson.

1952 James Craig Watson Medal, National Academy of Sciences, to H. R. Morgan.

1955 Gold Medal of the Royal Astronomical Society, to Dirk Brouwer.

1957 Rittenhouse Medal, to Lyman Spitzer.

1962 Conrad Medal, USN, to Gerald M. Clemence.

1962 Bronze Medal, Université de Liège, to Rupert Wildt.

1963 USN Distinguished Service Award and Medal, to Gerald M. Clemence.

1964 Graduate Society Medal, Radcliffe College, to Dorrit Hoffleit.

1965 Royal Association Gold Medal, to Gerald M. Clemence.

1966 Catherine Wolfe Bruce Gold Medal, to Dirk Brouwer.

1966 Eddington Gold Medal, to Rupert Wildt.

1966 James Craig Watson Medal, to Wallace Eckert.

APPENDIX E
CONTRIBUTIONS TO THE *ASTRONOMICAL PAPERS OF THE AMERICAN EPHEMERIS AND NAUTICAL ALMANAC* STEMMING FROM A JOINT ONR CONTRACT WITH YALE, THE U.S. NAVAL OBSERVATORY, AND THE IBM WATSON SCIENTIFIC COMPUTING LABORATORY

1. G. M. Clemence, "First Order Theory of Mars," **11**, Part 2, 278 pp., 1949.
2. H. R. Morgan, "Primary Reference Stars for Pluto," **11**, Part 3, 19 pp., 1950.
3. P. Herget, G. M. Clemence, and H. G. Hertz, "Rectangular Coordinates of Ceres, Pallas, Juno, Vesta, 1920–1960," **11**, Part 4, 67 pp., 1950.
4. W. J. Eckert, D. Brouwer, and G. M. Clemence, "Coordinates of the Five Outer Planets 1653–2060," **12**, xxxi + 327 pp., 1951.
5. A. J. J. van Woerkom, "The Motion of Jupiter's Fifth Satellite, 1892–1949," **13**, Part 1, 77 pp., 1951.
6. D. Brouwer and A. J. J. van Woerkom, "The Secular Variations of the Orbital Elements of the Principal Planets," **13**, Part 2, 29 pp., 1950.
7. H. R. Morgan, "Catalog of 5,268 Standard Stars, 1950.0, Based on the Normal System N30," **13**, Part 3, xxviii + pp. 109–321, 1952.
8. G. M. Clemence, "Coordinates of the Center of Mass of the Sun and of the Five Outer Planets, 1800–2060," **13**, Part 4, viii + 41 pp., 1953.
9. G. M. Clemence, "Perturbations of the Five Outer Planets by the Four Inner Ones," **13**, Part 5, 42 pp., 1954.
10. P. Herget, "Solar Coordinates 1800–2000," **14**, xi + 735 pp., 1953.
11. E. W. Woolard, "Theory of the Rotation of the Earth Around its Center of Mass," **15**, Part 1, pp. 1–165, 1953.
12. H. G. Hertz, "The Mass of Saturn and the Motion of Jupiter 1884–1948," **15**, Part 2, pp. 167–215, 1953.
13. P. Herget, "[Heliocentric] Coordinates of Venus 1800–2000," **15**, Part 3, x + pp. 217–523, 1955.
14. R. L. Duncombe, "Motion of Venus," **16**, Part 1, pp. 1–258, 1958.
15. G. M. Clemence, "Theory of Mars (Completion)," **16**, Part 2, pp. 261–333, 1961.
16. P. Herget, "Rectangular Coordinates of Ceres, Pallas, Juno, Vesta 1960–1980," **16**, Part 3, pp. 337–95, 1962.
17. E. S. Jackson, "Determination of the Equinox and Equator from Meridian Observations of the Minor Planets," **20**, pp. 1–131, 1968.
18. R. L. Duncombe, "Heliocentric Coordinates of Ceres, Pallas, Juno, Vesta 1928–2000," **20**, Part 2, pp. 135–309, 1969.
19. A. D. Fiala, "Determination of the Mass of Jupiter from a Study of the Motion of 57 Mnemosyne," **21**, Part 2, pp. 49–163, 1972. (An expansion of the author's Yale Ph.D. Dissertation, 1968.)
20. D. A. O'Handley, "Determination of the Mass of Jupiter from the Motion of 65 Cybele," **20**, Part 3, pp. 311–68, 1969.
21. D. A. Pierce, "Star Catalogue Corrections Determined from Observations of Selected Minor Planets," **22**, Part 3, pp. 205–360, 1978.
22. R. E. Laubscher, "The Motion of Mars 1751–1969," **22**, Part 4, pp. 361–490, 1981. (Also the author's Yale Ph.D. Dissertation, 1970.)

APPENDIX F
YALE AND USNO PERSONNEL

There is a long record of cooperation between the United States Naval Observatory and Yale.[1] On numerous occasions USNO sent some of its employees to Yale to further their education in celestial mechanics and astrometry, fields of great importance especially to its Nautical Almanac department. More Yale Ph.D. graduates gained employment at USNO than at any other observatory, Yale itself not excepted. And upon early retirement from USNO, several prominent astronomers came to Yale as research associates in order to collaborate with the Yale celestial mechanicians whom they highly respected. Table F.1 indicates the dates when these astronomers were associated with USNO before and/or after working or receiving a degree at Yale.

REFERENCE

1. D. Hoffleit, *Comments on Astrophysics,* **16**, 17, 1992.

Table F.1
Yale and USNO Personnel

Name	USNO	Yale	USNO
J. S. Hubbard		Class of 1843	1845–63
Asaph Hall	1862–91	Hon LL.D. 1879	
Asaph Hall, Jr.	1882–85	Ph.D. 1889	1889–92, 1905–08
H. B. Hedrick	1886–1910	1910–19	
H. R. Morgan	1901–05		
	1907–44	1947–53	
C. B. Watts	1930–59	1959–66	
G. M. Clemence	1930–63	1963–74	
W. Eckert		Ph.D. 1931	1940–45
John S. Hall		Ph.D. 1933	1948–58
H. Hertz	1941–43	Ph.D. 1941	1943–58
R. Duncombe	1942–48	1948–49, Ph.D 1956	1949–75
D. A. O'Handley	1961–67	Ph.D. 1967	
W. J. Klepczynski	1961f	Ph.D. 1969	1961f
A. Fiala	1963f	M.S. 1964, Ph.D. 1968	1963f
T. C. van Flandern	1963f	Ph.D. 1969	1963f

APPENDIX G
EVOLUTION OF THE BRIGHT STAR CATALOGUE
and Other Yale Catalogues

The Bright Star Catalogue, starting with E. C. Pickering's "Revised Harvard Photometry," *Annals of Harvard College Observatory,* **50**, 1980, known as the HR catalogue, has gone through six editions:

1908 E. C. Pickering, *Harvard Revised Photometry*
1930 F. Schlesinger, 1st Yale ed., *The Catalogue of Bright Stars*
1940 F. Schlesinger and L. Jenkins, 2nd ed., *The Catalogue of Bright Stars*
1964 D. Hoffleit, 3rd ed., *The Catalogue of Bright Stars*
1982 D. Hoffleit and C. Jaschek, 4th ed., *The Bright Star Catalogue*
1988 D. Hoffleit and W. Warren, Jr., 5th ed., *The Bright Star Catalogue* (available on tape only)

When Schlesinger compiled the first edition of the *Yale Catalogue of Bright Stars,* he chose the HR as his starting point, assigning the HR numbers to the stars of his catalogue, and advocating that henceforth they be referred to as BS numbers. From the HR catalogue he incorporated sundry cross-reference columns, and especially the Harvard visual magnitudes, the only homogeneous such collection available at the time. He gave more precise equatorial coordinates; added precessional terms; updated the spectral types, the *Henry Draper Catalogue* having meanwhile been published (*Harvard Annals,* **91–99**, 1917–24); and most importantly, added proper motions, parallaxes, and radial velocities as available. These data were revised in the 1940 edition, but no new categories were added.

After a lapse of 24 years Hoffleit compiled a much-enlarged edition, giving numerous cross-references to other major catalogues and adding columns for the identification of variable stars, double or multiple stars and their principal components, positions for both 1900 and 2000, galactic coordinates, and (B-V) colors. The tabulated information now required a double-page spread. The fourth edition added a column, I, to identify which stars were infrared sources, and columns for the additional colors (U-B) and (R-I), and rotational velocities, $v\sin i$.

Bright stars have continued to be widely observed, as these stars have generally been the only ones sufficiently bright for precise observations of astrophysical properties essential for physical interpretations. Consequently, new data are continuously becoming available. Table G.1 indicates the percentage growth of data in the various categories represented from one edition to the next. The apparent decline in the parallaxes between the second and third editions needs clarification. In his first two editions Schlesinger included parallaxes of all types: trigonometric, dynamic, moving cluster, and spectroscopic, without differentiating which belonged to which category, whereas the later editions included only trigonometric and dynamic parallaxes. The spectroscopic were the most numerous but were later omitted because the spectra are correlated with absolute magnitudes, not directly with distances. The final column, Δn, indicates the actual increase in numbers between the final two editions for cases where both the percentages and percentage increases are small. The radial velocities are essentially complete, lacking for only six stars.

During the Schlesinger and Brouwer eras four general catalogues of parallaxes were published. Table G.2 indicates how the numbers have progressively increased, though the average annual rates have steadily declined from over 200 between 1910 and 1924, to under 50 per year between 1963 and 1987.

Relatively few proper motions had been determined before Schlesinger, shortly after coming to

Table G.1
Percentage of HR Stars in Various Categories of Information

Edition	1	2	3	4	5	Δn
	1930	*1940*	*1964*	*1982*	*1988*	
V			50	95	96	
B-V			50	94	97	
U-B				73	80	
R-I				28	28	24
MK Lum			40	87	89	
P.M.	84	100	100	100	100	
Par.	(41)	(60)	30	32	32	16
RV	47	63	75	93	100 –	
*v*sin*i*				43	43	19
D			25	32	33	86
SB			10	19	19	27
Var + suspects			8	22	24	176

Yale, inaugurated his programs for reobserving photographically the *Astronomische Gesellschaft* zones for which high-accuracy earlier-epoch positions had been determined. Table G.3 shows which zones were measured and reduced, and the years of publication.

All the northern zones and to −30° were carried out mainly by Ida Barney under the direction of F. Schlesinger. From −30° south, the plates were taken by and under the direction of D. Brouwer and the measurements and reductions were carried out mainly under the supervision of D. Hoffleit. Among numerous others, Phillip K. Lu participated in a large part of the work on the southern zones.

The zones +60° to +85°, +30° to +50°, and −50° to −60° were not covered by the Yale programs (*see* Chapter 18). For all northern zones the AGK3 Catalogues, and for those south of −30° the various catalogues of the Cape Observatory are available. Indeed, the Yale programs for stars south of −30° had been planned by Brouwer in order to strengthen the Cape proper-motion determinations.

Table G.2
Trigonometric Parallaxes

1910	365	Mostly Yale heliometer
1924	1679	1870 parallax stars of which 191 not trigonometric
1935	3800:	7534 entries including trigonometric, spectroscopic, and dynamic parallaxes. Only about 50% are trigonometric
1952	5822	All trigonometric
1963	6399	" "
1988	7607	" "

Table G.3
Proper Motions

Zones		*The Zone Catalogues* Vol. Yale Trans.	Years
+85° to	+90°	26 Part I	1954
+50	+60	4, 7, 26 Part II, 27	1925, 1930, 1959
+30	−30	3 Part IV, 5, 9–14, 16–25	1926–1954
−30	−40	28, 29	1967–1968
−40	−50	30	1970
−60	−70	32 Part II	1983
−70	−90	31	1971
Total no. vols.		27	

APPENDIX H
WOMEN ASTRONOMERS AT YALE THROUGH 1968

In a strictly men's university it is hardly surprising that women astronomers played no role until the last decade of the nineteenth century. At the women's colleges, on the other hand, young women were often better trained in astronomy than were the undergraduate men at their more research-oriented male universities. Matthew Vassar opened his Female College in 1865, equipped it with a 12-inch refracting telescope, and hired Maria Mitchell, America's first woman astronomer—a lady from Nantucket Island, initially trained by her father, and then mostly self-educated. She proved to be an excellent and stimulating teacher. Several women followed in her footsteps and also became famous astronomers.[1] Other women's colleges followed suit, but graduate schools were seldom opened to these women. The first Ph.D. degree awarded to any woman in America was in 1877 from Boston University for a thesis on Greek drama.[2] Yale opened its doors to women graduate students in 1892 and awarded three Ph.D. degrees in 1894, one to a woman, Margaretta Palmer. Previously, one American woman, Dorothea Klumpke (Mrs. I. Roberts) had already in 1893 received the equivalent *Docteur-ès-Sciences* in Paris. Through 1900 only three Ph.D. degrees had been awarded to women astronomers in the United States[2] (one to Anna Delia Lewis from Carleton College in 1896, another to Caroline Furness from Columbia in 1900). Professional jobs for well-trained women were generally hard to find; employment other than teaching in a women's college was seldom open to them.

MARGARETTA PALMER,[3] Vassar 1887, one of Maria Mitchell's last students, became an assistant to H. A. Newton at Yale in 1889, and in 1892 was one of the first women in any field admitted to the graduate school, earning in 1894 the first Ph.D. in astronomy ever awarded a woman in America. She was an expert in the computation of the orbits of comets. Her first publication, her thesis, was on the comet discovered by Maria Mitchell in 1847, the comet that catapulted Maria Mitchell to international fame. She then computed numerous other orbits. However, as was so typical of women's work in astronomy, she was assigned a necessary but highly routine task. Under the direction of W. L. Elkin, H. A. Newton's daughter Miss J. S. Newton had in 1897 begun an index of all available sources on the precise positions of the nearly 460,000 stars north of $-23°$ in the *Bonner Durchmusterung* and its southern supplement. This massive undertaking, a forerunner of the German *Geschichte des Fixsternhimmels,* was turned over to Miss Palmer with whom Miss Newton continued to collaborate. The Yale compilation was never published, but for many years Miss Palmer provided (on request) other astronomers with the data they needed on any specified star.

With the perfection of dry-plate celestial photography[4] by the early 1880s, other observatories, particularly Harvard under the direction of Professor E. C. Pickering, obtained hundreds of plates showing vast numbers of faint stars. Thus the need arose for assistants to examine the plates, measure the positions and magnitudes of the stars, search for variable stars, study stellar spectra, etc. This opened the door to many aspiring young women astronomers, especially at Harvard College Observatory. There, from 1880 to 1900 no fewer than 26 women were employed to aid in computations and plate measurements.[5] In this interval Yale had employed only the Misses Palmer and Newton. Photography was not yet an important asset to Yale astronomy; indeed it did not play an important role (except in meteor photography) until Frank Schlesinger became the Director in 1920. Then, as we have seen, Schlesinger valued the services of women, not for fundamental original research, but for carrying out meticulously accurate measurements and computations according to his own precepts.

IDA BARNEY[6,7] (Yale Ph.D. in mathematics, 1911) and LOUISE JENKINS[8] exactly fulfilled Schlesinger's preconceptions of what women in astronomy should do, beliefs in complete agreement with those of Pickering. The famous work of Annie J. Cannon at Harvard on the classification of the spectra of a quarter-million stars was after all largely routine, as was Miss Barney's determination of the proper motions of about 150,000 stars. All this work required complete understanding of what was involved in their assignments: knowledge, ability, intelligence, and especially concentration. But they were expected not to introduce innovations; at best they made minor improvements based on practical (not theoretical) experience.

In 1924 Schlesinger, with the assistance of Margaretta Palmer and Alice Pond had compiled the first *General Catalogue of Stellar Parallaxes.* Miss Jenkins collaborated in his second edition in 1935, and was the sole author of the third edition in 1952 and a *Supplement* in 1963. Until a new parallax catalogue by W. van Altena (to come out *ca* 1992), Jenkins's is still the one most often consulted. She also measured the parallaxes of many stars, again according to Schlesinger's precepts; was co-author of Schlesinger's second edition of the *Catalogue of Bright Stars*; and acted as both the observatory librarian and executive secretary to the department.

In 1956 I became successor to Miss Barney to take charge of the zone catalogue work for the determination of the proper motions of southern stars; but I also inherited the updating of the *Catalogue of Bright Stars*; and for a time had charge of the departmental library when it was moved from the old observatory on Canner Street to its new location at 135 Prospect Street; and I shared in some of the teaching.

Dr. Brouwer, who was chairman of the search committee for a new director of the Nantucket Maria Mitchell Observatory, found that this small observatory had funding for only a half-yearly appointment. Hence he hired me for the months November through April at Yale, while Nantucket employed me May through October to specialize on variable stars. At Yale I was expected to do the standard woman-astronomer routine jobs while at Nantucket I enjoyed freedom of choice. There I inaugurated a summer research participation-training program on variable stars for women students, mainly undergraduates. This program was later made coeducational when vital government grants would be in jeopardy if we discriminated against men, and we were therefore finally provided with more than one dormitory room for the students.

Sharing joint appointments between the Yale and the Maria Mitchell Observatories paid dividends for both institutions. During my tenure at Nantucket (1957–78), three Yale students participated in the program: Jocelyn Gill (1957), Susan Hess (1966), and Harriet Dinerstein (1973). Four Maria Mitchell participants found temporary employment at the Yale Department of Astronomy: Nancy Gregg (Y 1968, N 1969) Barbara Hatfield (Y 1971, N 1972), Wendy Whiting (N 1978, Y 1980–1981), and especially Katherine Wood Paranya (N 1966, Y 1968–1970), an expert computer-programmer and co-author of my research on common proper motion stars.[9]

Schlesinger enjoyed cooperation with the astronomy department at Columbia University for his work on the Yale Zone Catalogues. There Sarah J. Hill, Columbia Ph.D., assisted Jan Schildt in the determination of magnitudes for the stars in some of the Yale zones.[10,11] Later, Brouwer and Wallace Eckert at IBM collaborated extensively on numerous projects. Eckert designed and had built a semiautomatic plate-measuring machine to determine the coordinates of stars in the Yale zones.[12] This machine was first used by Rebecca B. Jones (former assistant at Lick and Harvard) for measuring the plates for Miss Barney's final catalogues.[13] Then Mrs. Dorothy Eckert, assisted by Mrs. Meredith B. Weddle (Maria Mitchell Observatory participant 1960), measured the plates for the southern zone -40 to $-50°$.[14]

During Schlesinger's directorship, no Ph.D. degree in astronomy was awarded a woman. The first after 1894 was earned by Sandra Schwartz in 1956 for a thesis in astrophysics written under the guidance of Rupert Wildt. Jocelyn R. Gill in 1959 received her degree for work in celestial mechanics under the direction of Dirk Brouwer, and Carol A. Williams in 1967, also earned her degree in celestial mechanics, initially under Brouwer's and finally, after Brouwer's death in 1966, under Gerald Clemence's direction.

In all generations professional women have complained about discrimination, both in rank and in salary relative to men of similar ability and education. During the Great Depression (1929) such discrimination actually proved a blessing in disguise in some quarters. For example, Harlow Shapley at Harvard College Observatory could hire two women for the price of one man. We women therefore kept our beloved jobs while few men could be afforded.

REFERENCES

1. D. Hoffleit, *Maria Mitchell's Famous Students,* Published by the AAVSO, Cambridge, MA, 1983.
2. W. C. Eells, "Earned Doctorates for Women in the Nineteenth Century," in *Bull. Amer. Assoc. Univ. Professors,* **42**, p. 644, 1956.
3. *See* Chapters 12, p. 79; 14, pp. 93–95; 17, p. 122; and 26, p. 190.
4. D. Hoffleit, *Some Firsts in Astronomical Photography,* Harvard College Observatory, p. 32, 1950.
5. S. Bailey, *History and Work of the Harvard Observatory,* Cambridge, p. 274, 1931.
6. *See* Chapters 17, p. 117; 18, pp. 127–28; and 20, p. 142.
7. D. Hoffleit, *Status,* p. 1, June 1990.
8. *See* Chapters 17, p. 122, and 18, p. 129.
9. D. Hoffleit and K. Paranya, *I.A.U. Colloquium No. 7,* p. 80, 1971.
10. J. Schilt and S. J. Hill, *Contr. Rutherfurd Obs.,* Columbia Univ., Nos. 30, 1937, and 31, 1938.
11. S. J. Hill and J. Schilt, *Ibid.,* No. 32, 1952.
12. W. J. Eckert and R. B. Jones, *Ast. Jour.,* **59**, 83, 1954.
13. I. Barney, D. Hoffleit, and R. B. Jones, *Trans. Ast. Obs.,* Yale Univ., **26**, Part II, 1959; **27**, 1959.
14. D. Hoffleit, D. Eckert, P. Lu, and K. Paranya, *Ibid.,* **30**, 1970.

APPENDIX I
CHRONOLOGY

1492 Fall of meteorite at Ensisheim, Alsace, the first known preserved meteorite.

1511 Raphael commissioned to paint Madonna di Foligno, depicting 1511 fireball at Crema.

1638 Colony founded by John Davenport at the mouth of the Quinnipiac River, later known as the New Haven Colony. Davenport's first (unfulfilled) hopes included the establishment of a college, like Harvard.

1641 First Grammar School in the Colony, conducted by Ezekiel Cheever until 1650; supported by taxes.

1660 Hopkins Grammar School in New Haven founded. Still active.

1662 Connecticut and New Haven Colonies merged by Royal Charter.

1701 The Collegiate School founded, the forerunner of Yale.

1702 First Commencement, at Saybrook; five Masters Degrees awarded.

1714 Books from England acquired through Jeremiah Dummer, including a copy from I. Newton of his *Principia,* the first copy to reach America.

1718 Gifts from Elihu Yale; the College named in his honor.

1721 Death of Elihu Yale; his intended benefactions for the college not honored by his heirs.

1740 Thomas Clap's construction of a planetary orrery. Later he added parabolas to represent the comets of 1743 and Halley's.

1742 November 24. Brilliant detonating fireball, which inspired Clap's theory of "terrestrial comets."

1745 Trustees of the College renamed "The President and Fellows." Rector Clap became the first President.

1746 The College of New Jersey, later (1896) named Princeton, founded by one Harvard and six Yale graduates, with the Rev. Jonathan Dickinson, Yale 1706, as its first President.

1750 President Clap enunciated his theory of *Terrestrial Comets.*

1750 Peter Dollond in England started his business as maker of optical instruments.

1752 Peter's father, John Dollond, joined his son's company, Dollond and Sons, and became its most brilliant member. Peter's nephew George and son John also became members.

1758 Dollond's produced the first successful achromatic telescopes.

1758 Return of Halley's Comet. Whittelsey and Clap predicted perihelion for September 14, 1758. Clairault predicted April 15, 1759. First seen in Europe at Christmas 1758. John Winthrop at Harvard saw it April 3, before the Yale astronomers. Perihelion actually occurred March 12, 1759.

1761 John Dollond experimentally proved Newton's belief wrong that chromatic aberration could not be avoided.

1778 Ezra Stiles the first Yale graduate to be elected President of Yale.

1781 President Stiles published Clap's *Terrestrial Comets.*

1790 Fall of meteorites in Provence Gascoigne, France, proved extraterrestrial by Chladni, but his proof was not yet accepted.

1798 Göttingen students, Brandes and Benzenberg, first to obtain meteor heights from observations at two stations.

1802 H. W. M. Olbers postulated that asteroids are fragments of a disrupted planet.

1803 Fall of meteorites in l'Agile, France, proved by Biot to be extraterrestrial, but belief slow in acceptance.

1805 J. C. D. Wildt postulated that meteorites might have arisen from the disruption of a planet.

1807 Fall of meteorites at Weston, Connecticut, investigated by Silliman and Kingsley. Thomas Jefferson said to have been incredulous.

1810 Part of the George Gibbs Cabinet of Minerals deposited on loan at Yale, the remainder in 1812.

1817 Jeremiah B. Day author of a highly esteemed text, *Navigation and Surveying*. In 1814 he had also published *An Introduction to Algebra*.

1818 B. Silliman founded the *American Journal of Science and Arts*.

1822 The college telescope lost in shipwreck of the *Albion* on way to England for repairs.

1824 Yale purchased the entire George Gibbs Cabinet of Minerals for $20,000. It contained a few meteorites, including the largest segment of the Weston fall.

1829 Sheldon Clark donated $1,200 for the purchase of a new telescope.

1830 The 5-inch Dollond telescope erected in the remodeled tower of the Athenaeum (built in 1761), an unsatisfactory arrangement that lasted 40 years.

1833 Spectacular Leonid Shower. Olmsted and Twining the first to discover the radiant. Olmsted proved that the meteors were of cosmic, not terrestrial origin.

1834 Olmsted published his theory on the origin of hailstones and meteors.

1835 Return of Halley's Comet. First seen in America by Olmsted and Loomis at Yale with the 5-inch Dollond.

1835 The Red River, Texas, iron meteorite, weighing 1,635 lbs, donated to Yale by Mrs. George Gibbs.

1837 James Dwight Dana published *A System of Mineralogy,* including a brief section on meteoritic iron.

1839 Student E. P. Mason was the first in America to observe telescopic meteors systematically.

1840 William Chauvenet, Yale 1840, graduated B.A. with high honors. Future author of *Spherical and Practical Astronomy,* 2 vols, Lippincott & Co., Philadelphia, 1863, and of several mathematical texts.

1843 First *Weather Maps,* by E. Loomis, played important role in establishment of the U.S. Weather Bureau in 1891.

1844 Israel Munson, physician and merchant, bequeathed to Yale $15,000, the largest endowment up to that time. Used to establish the Munson Professorship of Natural Philosophy and Astronomy. The first recipient was D.Olmsted, 1844–59, followed by E. Loomis, 1860–89, and D. Brouwer, 1944–66. P. Demarque is the current Munson Professor.

1847 Beginning of a Yale Scientific Department.

1847 The Graduate School established.

1854 William Hillhouse presented to the College a transit circle, 4-inch aperture, 5 ft focal length.

1854–55 The Chemistry and Engineering Schools combined as the Yale Scientific School.

1858 Mrs. James A. Hillhouse (Cornelia) and daughters Mary and Isaphine gave six acres of land for the site of an observatory, in memory of her father-in-law, James Hillhouse (Yale Treasurer 1782–1832).

1858–60 Joseph Sheffield bought the old Yale Medical School at the corner of Grove and Prospect Streets and converted it for the use of the Scientific School. The building was named Sheffield Hall, later South Sheffield when North Sheffield was built in 1872/3.

1860 The degree of Ph.D. established.

1860 J. Homer Lane's first suggestion of rotating shutters for photographic determinations of meteor velocities.

1861 The Scientific School renamed the Sheffield School.

1861 The first Ph.D. degree in America awarded A. W. Wright of the Sheffield School for a thesis on meteor orbit determination.

1863 Founding of the National Academy of Sciences. Of the 50 charter members 11 were astronomers. Included were six Yale graduates cited in this history of astronomy at Yale: F. A. P. Barnard, W. Chauvenet, J. S. Hubbard, H. A. Newton, B. Silliman Sr., and B. Silliman, Jr.

1866 Return of the Leonids correctly predicted by H. A. Newton who also advanced the theory of their capture by Jupiter.

1866 Sheffield gave the Scientific School a 9-inch Alvan Clark refractor and a small meridian circle. The refractor was first mounted in a Monitor gunboat turret atop Sheffield Hall where it remained until 1927 when the housing was no longer considered suitable.

1866 Francis E. Loomis received Ph.D. at Yale.

1869 Francis E. Loomis received Ph.D. from Göttingen University which published his thesis, *Periodic Stars*.

1869 First catalogue of meteorites in the Yale collection, by George J. Brush, Curator of the Gibbs Mineral Cabinet.

1870 A cylindrical dome added to the Athenaeum for better use of the Dollond telescope.

1871 Oliver Winchester gave 32 acres at a cost of $90,000 for an observatory for the College, the first gift to Yale for the endowment of research rather than for educational facilities.

1872 Future President A. T. Hadley, then a Yale undergraduate, by pure chance was the first at Yale to observe the Bielid meteors.

1872 J. Homer Lane elected to the National Academy of Sciences.

1873 Elias Loomis and W. A. Norton elected to the National Academy of Sciences.

1874 G. W. Hill elected to the National Academy of Sciences.

1879 The Horological Bureau for checking timepieces and providing standard time, established under the direction of Leonard Waldo. Exact time to be furnished by telegraph to all railroads in Connecticut. Checking of clocks and watches ceased in 1887, but a *Time Service* continued until 1918.

1880 The Thermometric Bureau similarly established by Waldo, mainly for checking clinical thermometers (active at least until 1900).

1880 H. A. Newton ordered a 6-inch heliometer from Repsold, the only heliometer in America, delivered in the spring of 1882.

1880 An anonymous gift, later identified as from Edward M. Reed, enabled H. A. Newton to order from Howard Grubb of Dublin an 8-inch equatorial refractor, and a Troughton and Simms transit instrument.

1881 Gift of clock from William Hillhouse.

1881 A. W. Wright, Yale's first Ph.D. recipient, elected to the National Academy of Sciences.

1882 H. A. Newton named Director of the Observatory in May.

1882 Exchanges of telegraph signals between Harvard and Yale resulted in an accurate determination of the longitude difference between the meridian circle at Harvard and the transit instrument at North Sheffield Hall at Yale.

1882 Erection of the Winchester Observatory, soon renamed the *Observatory in Yale College* at the request of the Winchester family.

1882 Transit of Venus, December 6, 1882, observed with the new heliometer, the Reed Telescope, and other equipment.

1883 Subscription among faculty to raise funds in order to employ the heliometer expert, W. L. Elkin.

1884 H. A. Newton resigned as Director on May 8, but continued on the Board of Managers. No new Director appointed for more than a decade, shortly before Newton's death.

1884–87 W. L. Elkin's first use of the heliometer for the triangulation of stars in the Pleiades. Bessel in 1841 had also determined the relative positions of the Pleiades stars with the historic Königsberg heliometer. Elkin compared the two results to determine the relative proper motions.

1885–1910 Stellar parallaxes determined by Elkin and assistants from observations with the heliometer.

1885 Observatory threatened with deficit by sudden cancellation by Connecticut of its contributions for Time Service to the State.

1887 Waldo persuaded to resign.

1888 Elkin's first published results on stellar parallaxes for ten bright stars based on Yale heliometer measurements.

1889 Elkin observed the total eclipse of the sun at Winnemucca, Nevada, on January 1.

1889 Elias Loomis bequeathed his entire estate to Yale for support of the Astronomy Department, two-thirds subject to life interests. The full income became available in 1939.

1889 Asaph Hall, Jr., from heliometer measurements determined an orbit for Titan and the mass of Saturn, agreeing with Bessel but not with A. Hall, Sr.

1889 C. S. Hastings, who later designed the Yale 26-inch and the Loomis telescopes, elected a member of the National Academy of Sciences.

1892 February 13, death of Edward M. Reed, who willed two-thirds of his estate to the Yale Observatory; by 1920 income from the entire fund became available.

1892 Margaretta Palmer computed a definitive orbit for Maria Mitchell's Comet 1847VI.

1893 Amateur J. E. Lewis's photograph of a fireball, analyzed by H. A. Newton. First indication of meteoroid rotation.

1893 Initial experiments by H. A. Newton and Elkin on meteor photography.

1894 Margaretta Palmer the first woman to receive a Ph.D. in astronomy.

1894 Grant from the J. Lawrence Smith Fund of the National Academy of Sciences in support of meteor photography. An equatorial mount for a 6-camera meteorograph ordered from Warner and Swasey.

1895 W. L. Elkin elected to the National Academy of Sciences.

1896 Elkin appointed Director of the Observatory in June.

1896 Death of H. A. Newton on August 12.

1897–1908 E. B. Brown, *Theory of the Motion of the Moon,* published in five installments in the *Memoirs* of the Royal Astronomical Society.

1898 An additional grant from the J. Lawrence Smith Fund for meteor photography.

1898 A second meteor station, four cameras on a fixed mounting, erected on estate of O. F. Treadwell in Hamden in preparation for the approaching Leonids.

1898 F. L. Chase discovered a faint new comet on one of his meteor plates taken at Hamden on November 21. For this he was awarded a Donahoe Medal of the Astronomical Society of the Pacific.

1899 A. T. Hadley elected the first lay President of Yale.

1899 First Yale experiments for photographically determining velocities of meteors. First segmented trail, July 31, 1899.

1899 Expected Leonid display a disappointment.

1900 Second meteor station moved to estate of J. H. Webb in Whitneyville where a new equatorial mounting by Grubb was erected for the now also 6-camera meteorograph.

1900 F. L. Chase observed the May 28 total solar eclipse at Pinehurst, North Carolina, and Mason F. Smith at Norfolk, Virginia.

1906 Parallaxes determined by the heliometer for 163 stars published in Volume 2 of the *Transactions of the Yale Astronomical Observatory.*

1909 Much of the land donated by Winchester was sold and the proceeds used to establish the Oliver F. Winchester Fund.

1910 Elkin resigned because of poor health. Chase appointed Acting Director.

1912 C. S. Hastings designed a coelostat (also called siderostat or heliostat) intended for parallax work. The plans were submitted to Schlesinger, then at Allegheny, and met with his enthusiastic approval.

1912 Remainder of Yale heliometer parallax determinations published in Volume 2 of the *Transactions.*

1914 On November 16 the Corporation voted to expend not over $10,000 for a building for the coelostat.

1914 Anson P. Stokes published *Memorials of Eminent Yale Men,* including some 30 of the Yale men cited in this history.

1915 Siderostat building erected.

1916 F. Schlesinger elected to the National Academy of Sciences.

1918 In July Mathematics Professor E. W. Brown appointed Acting Director for the duration of WWI or until a new Director could be appointed.

1918 Observatory closed and valuable instruments stored in Sloane Physics Laboratory.

1919 Publication of *Tables of the Motion of the Moon,* by E. W. Brown, 3 vols., Yale Univ. Press.

1920 On April 1 Frank Schlesinger assumed the Directorship.

1920 The new photographic polar coelostat designed by Hastings installed and named the *Loomis Telescope.*

1920 Miss Palmer and Schlesinger began compilation of a catalogue of stellar parallaxes.

1920 Schlesinger started discussion meetings called *The Neighbors*. No women admitted.

1921 First use of Loomis Telescope by C. L. Stearns for variable stars.

1921 Work on a *Bright Star Catalogue* begun.

1921 Ida Barney measured and discussed spectrograms of Algol obtained by Schlesinger at Allegheny.

1921 Meteorograph cameras dismantled.

1922 Plans underway for Zone Catalogue work.

1922 Schlesinger attended the Rome meetings of the International Astronomical Union as head of the American delegation.

1923–25 Publication of *The Theory of the Trojan Group of Asteroids,* by E. W. Brown, in vol. 3, *Transactions of the Yale University Observatory.*

1923 E. W. Brown elected to the National Academy of Sciences.

1923 On February 12 the Corporation authorized purchase of a 26-inch photographic refractor for the determination of stellar parallaxes in the southern hemisphere.

1923 Site-testing for a southern station began in New Zealand, but the final site selected was at Johannesburg, South Africa.

1924 First *General Catalogue of Stellar Parallaxes,* by F. Schlesinger, containing 1,870 stars.

1924 Jan Oort, at Yale, investigated relations between globular clusters and high-velocity stars.

1925 First Yale Zone Catalogue, by Schlesinger and Ida Barney, for +50° to +55°, based on Allegheny plates.

1925 On January 25, total solar eclipse visible from Yale. Many visiting astronomers as well as local specialists carried out observations.

1925 On February 18, ground broken at University of Witwatersrand, Johannesburg, for the 26-inch telescope building. On June 22 the building was formally dedicated by the Prince of Wales.

1925 On May 1, Harold Alden of the University of Virginia, appointed in charge of the Southern Station at Johannesburg, South Africa.

1925 E. W. Brown was chairman of the American delegation to the International Astronomical Union in Cambridge, England, in July, and Schlesinger was elected one of five vice-presidents of the Union.

1926 E. W. Brown found evidence for changes in the rate of rotation of the Earth.

1926 J. Schilt determined photometric magnitudes using a photometer he earlier designed at Leiden.

1927 On March 11, Schlesinger delivered the first George Darwin Lecture of the Royal Astronomical Society.

1927 Dirk Brouwer, Leiden Ph.D., appointed assistant to E. W. Brown.

1927 The 39th meeting of the American Astronomical Society, held in New Haven, December 29–30, with the Yale Observatory as host organization.

1927–28 Schilt investigated rotation of the galaxy.

1928 Schilt noted that there are two classes of Cepheid variables, the high luminosity classical Cepheids, and less luminous variables with periods greater than 10 days having a more general distribution. This discovery precedes by about 15 years Baade's discovery of Populations I and II.

1930 Celestial mechanics computations by Brown and Brouwer were being expedited by the use of Hollerith machines at Columbia University.

1930 Graduate student John S. Hall made pioneering use of the Loomis telescope for infrared photometry, with a photometer he designed and built himself.

1930 Publication of the first edition of the *Catalogue of Bright Stars* by F. Schlesinger.

1930 The 45th meeting of the American Astronomical Society, held in New Haven in December.

1931 Jan Schilt resigned to become Director of the Rutherfurd Observatory of Columbia University.

1931 Schlesinger one of two delegates of the National Research Council to meetings of the International Research Council and the International Union for Physics, in Brussels in July.

1931 Sheffield telescope moved to roof of Bingham Hall where it and a transit instrument were used by undergraduates.

1932 On May 14 the Corporation approved making the Observatory an official Department of the University. The Board of Managers was replaced by the Faculty of the Department.

1932 At the International Astronomical Union meeting at Harvard, Schlesinger was elected President of the Union.

1933 Publication of the first Yale Zone Catalogue (+25° to +30°) for which only Yale photographs were used, made with a 5-inch Ross Catalogue Camera.

1933 Death of W. L. Elkin on May 30.

1935 Publication of the second edition of the *General Catalogue of Stellar Parallaxes,* by F. Schlesinger and L. Jenkins.

1935 Mrs. Elkin turned Professor Elkin's books over to the Observatory, duplicates to be sold to establish a William Lewis Elkin Fund for general use of the Observatory.

1935 Schlesinger attended Paris Meetings of the International Astronomical Union as its President. Brown was Chairman of the Lunar Commission and Brouwer presented plans which were approved for determining systematic errors of fundamental star positions by intensive observations of asteroids.

1937 Reductions of Elkin's important pioneering photographic meteor observations finally completed and published by C. P. Olivier at the University of Virginia.

1937 With the cooperation of W. J. Eckert at Columbia voluminous computations relevant to the Zone Catalogues were being carried out on the Columbia Hollerith machines.

1937 The amateur New Haven Astronomical Society was founded and received Schlesinger's encouragement.

1938 Arthur Bennett completed photoelectric measurements in the far infrared with equipment of his own design. He observed some 400 stars in addition to 30 low-temperature variable stars.

1938 Completion by W. J. Eckert, under the supervision of E. W. Brown, of Brown's lunar theory, a task that occupied Brown for more than 20 years.

1938 On July 22, death of E. W. Brown, whose will made the Observatory his chief legatee. He also left $5,000 to the American Astronomical Society, which did not yet own the *Astronomical Journal,* for support of a journal devoted to astronomical research.

1938 The 8th Satellite of Jupiter, lost since 1935, rediscovered on the basis of calculations by Brown and H. Hertz.

1940 Publication of the second edition of the *Catalogue of Bright Stars,* by F. Schlesinger and L. Jenkins.

1941 On June 30 Schlesinger resigned as Director, and was succeeded by D. Brouwer.

1941 Lyman Spitzer, Jr., appointed Instructor in Physics and Astronomy starting July 1, but he left for war work in January.

1941 In September Brouwer assumed the office as Editor of the *Astronomical Journal,* a post he held until his death in 1966.

1942 On June 12–14 the American Astronomical Society held its 68th meeting in New Haven at the invitation of Yale.

1942 Members of the amateur New Haven Astronomical Society began successful occultation observations at the Reed telescope for Dr. Brouwer's researches.

1943 V. Goedicke gave the first radio address, "The Place of Astronomers in the War," in the series *Yale Interprets the News.*

1943 On July 10 Frank Schlesinger died. At a Memorial, President Angell remarked that Schlesinger's achievements "will long stand as one of the most brilliant chapters in the history of our Observatory."

1943 The Sheffield telescope dismantled and removed from roof of Bingham Hall on September 23.

1944 Publication of *Spherographical Navigation,* by D. Brouwer, F. Keator, and D. McMillen, The Macmillan Co., N.Y., xxiii + 200 pp.

1945 Harold Alden left Johannesburg to become the Director of the Leander McCormick Observatory of the University of Virginia. Replaced by Cyril Jackson.

1945 The Loomis telescope reversed, now a fixed polar refractor designed to photograph only the north polar region for determining the motions of the Earth's pole of rotation.

1946 The University entered into a ten-year agreement with Columbia University whereby the Yale and Columbia Observatories would jointly operate the Southern Station, a measure primarily of economy. Yale would concentrate as before on astrometry, Columbia on photometry.

1946 Initial support by the Office of Naval Research for long-term work on the motions of the major planets; collaboration with the U.S. Naval Obs. and the Watson Scientific Laboratory.

1947 Resignation of Lyman Spitzer, a Yale graduate who had been associated with the Departments of Physics or Astronomy 1939–42 and 1946–47, to assume the duties of Director of the Princeton University Observatory. Succeeded by Rupert Wildt.

1948 The Astronomy Department acquired its first IBM equipment, housed in the basement of the Law School, for work in celestial mechanics; R. L. Duncombe in charge.

1948 Brouwer visited the Southern Station in South Africa. As observing conditions at Johannesburg had been deteriorating, plans were considered for moving the station to a more favorable site, either in South Africa or Australia.

1948 In April Yale was host to the first North-East Regional Convention of the Astronomical League, attended by both amateur and professional astronomers.

1948 In August Joseph Ashbrook of Yale, observing at Flagstaff, Arizona, and Cyril Jackson at the Southern Station in Johannesburg, within 12 hours independently discovered a telescopic comet, named Comet Ashbrook-Jackson.

1948 The 80th meeting of the American Astronomical Society held in New Haven in December.

1950 Decision to move the Southern Station from Johannesburg to Mt. Stromlo, Australia, at the invitation of Sir R. Woolley, Director of the Commonwealth Observatory at Canberra. In November–December Brouwer visited Mt. Stromlo.

1950 A 10.75-inch refractor with housing donated by Mrs. Arthur W. Butler.

1951 Brouwer elected a member of the National Academy of Sciences.

1951 Publication by Brouwer, Eckert, and Clemence of the coordinates of the five outer planets at 40-day intervals from 1663 to 2060. (*Ast. Papers of the Amer. Ephemeris and Nautical Almanac.*)

1952 Publication of new edition of the *General Catalogue of Trigonometric Stellar Parallaxes,* by Louise Jenkins.

1952 Dismantling of the telescopes at Johannesburg.

1952 L. Spitzer and G. M. Clemence elected to the National Academy of Sciences.

1954 D. Kimball was group leader to solar eclipse in Greenland, June 30, sponsored by Air Force Cambridge Research Center.

1955 Brouwer awarded the Gold Medal of the Royal Astronomical Society and appointed George Darwin Lecturer for 1955.

1955 The University unexpectedly informed Brouwer that the Observatory must soon move to accommodate a girls' school unrelated to the University. The Observatory to acquire the girls' school building at 135 Prospect St.

1955 Brouwer awarded a Fulbright Grant to carry out research at Mt. Stromlo, Australia, from September 11, 1955 to February 29, 1956.

1955 H. J. Smith and D. Kimball observed the annular solar eclipse in Formosa, on December 14, making use of new photoelectric equipment.

1956 Land acquired in Bethany for an observatory, the cost met by funds from James W. Cooper and Mrs. James E. Cooper.

1956 Delays in the renovation of 135 Prospect Street building; temporary quarters provided at Osborn Zoological Laboratories.

1956 Initiation of plans for radio astronomy at Yale.

1956 Routine observations with the 26-inch were resumed at Mt. Stromlo.

1956 Morris Davis appointed head of Observatory computing facilities.

1957 In January the Department finally moved to 135 Prospect Street. Dedication exercises February 8.

1957 Research Corporation Grant of $10,000 to Wildt and Smith to begin research in planetary radio astronomy.

1957 National Science Foundation Grant of $41,400 for studies related to the acquisition of a large astrographic telescope for the southern hemisphere.

1957 Gift from the Rubicon Foundation of $50,000 for completion of the central building at the Bethany Station.

1958 P. Hodge and R. Wildt published their analysis of meteoritic dust, indicating a worldwide accretion rate of $5x10^8$kg/year.

1958 Additional $21,000 funds from the National Science Foundation to H. J. Smith for support of radio astronomy.

1958 A. E. Lilley received several funds totalling over $61,000, supporting two years' work in radio astronomy.

1958 A. E. Lilley the recipient of the first Bart J. Bok Prize from Harvard Observatory for his studies of absorption spectra of radio stars.

1958 Yale joined AURA, the Association of Universities for Research in Astronomy, created to establish and operate a National Optical Astronomy Observatory in Arizona. Wildt made a member of the Board of Directors of AURA.

1958 The annual report indicates that the Observatory headquarters at 135 Prospect Street were already becoming over crowded.

1958 Opening reception at the Bethany Station, December 14.

1959 NASA Grant of $113,000 for determination by radio observations of the distance of the Earth from the Sun.

1959 A three-year NASA grant of $149,580 in support of research on planetary and satellite motions. This supplemented an annual grant of $30,000 from the Office of Naval Research.

1959 First NSF Grant, $40,000, to support a Summer Institute in Dynamical Astronomy.

1959 Ida Barney retired after having served the Observatory for 37 years, during which she was author or co-author of 22 volumes of Zone Catalogues covering the sky from $+85°$ to $+90°$, $+50°$ to $+60°$ and $-30°$ to $+30°$.

1959 A. E. Lilley resigned to accept an appointment at Harvard.

1959–83 Zone Catalogue measurements and reductions of some 70,000 southern stars, both at Yale and under Yale supervision at the IBM Watson Laboratory in New York, supported by the Army Map Service.

1960 Ford Foundation Grant of $750,000 for establishing a 20-inch double astrographic telescope in the southern hemisphere. Its purpose, to extend to the South Pole the Lick Observatory survey of proper motions relative to distant galaxies. Optics to be supplied by Perkin-Elmer, mounting by Rademachers of Rotterdam.

1960 Site testing in Australia and in South America led to the selection of El Leoncito near San Juan, Argentina, for the Yale–Columbia Southern Observatory.

1960 Loomis telescope moved to Bethany and equipped with new semiautomatic plate drive.

1961 Publication of *Methods of Celestial Mechanics* by D. Brouwer and G. M. Clemence, Academic Press, N.Y., 528 pp.

1961 Yale astronomers H. J. Smith and D. Hoffleit and their students visited Harvard where they examined plates and were the first to discover that the quasar 3C273 is variable.

1961 An NSF Grant enabled construction of an addition to the administration building in Bethany for the benefit of radio astronomy.

1962 A Research Center in Celestial Mechanics was started with an annual budget of $350,000, supported by the Office of Naval Research and the Air Force, half this amount being contributed the initial year.

1962 NASA and other funds made available for the acquisition of a 40-inch reflector for Bethany.

1962 An agreement signed in September between the Yale–Columbia Southern Observatory and the University of Cuyo providing lease of land at El Leoncito and in San Juan by the University of Cuyo to the Yale–Columbia Southern Observatory.

1962 Publication of *Fundamentals of Celestial Mechanics,* by J. M. A. Danby, Macmillan, N.Y., 348 pp.

1962 The American Astronomical Society held its 111th meeting at Yale, August 26–29.

1963	The Yale 26-inch refractor at Mt. Stromlo donated to the Australian National University.
1963	The lenses for the southern astrograph completed by Perkin-Elmer in the summer of 1963.
1963	Harlan J. Smith left Yale to become Director of the McDonald Observatory and head of the Department of Astronomy at the University of Texas.
1963	Cyril Jackson appointed Observer-in-Charge of the Yale-Columbia Southern Observatory.
1963	Gerald M. Clemence resigned as head of the U.S. Nautical Almanac Office and joined the newly established Yale Observatory Research Center for Celestial Mechanics.
1963	Publication of *Supplement to the General Catalogue of Trigonometric Stellar Parallaxes,* by Louise Jenkins.
1964	Publication of third edition of the*Yale Bright Star Catalogue,* by D. Hoffleit.
1964–67	R. Wildt served as President of Commission 16 (physical studies of planets and satellites) of the International Astronomical Union.
1965	Erection of the southern astrograph completed, and dedication of the Yale–Columbia Southern Station at El Leoncito, Argentina, March 29—April 2.
1965	Erection of U. S. Naval Observatory facilities at the Yale–Columbia El Leoncito site.
1965	G. M. Clemence received the Gold Medal of the Royal Astronomical Society. He was Chairman of the Division of Physical Sciences of the National Research Council, and was elected Chairman for three years of the Section on Astronomy of the National Academy of Sciences.
1965	J. N. Douglas, radio astronomer, resigned to join Smith in Texas. This resignation spelled the end of radio astronomy at Yale.
1965	R. Wildt elected President of AURA. Served in this capacity 1965–68 and 1971–72. Chairman of the Board in 1972.
1966	Dr. Brouwer died January 31. Dr. Wildt appointed Acting Director.
1966	The Dirk Brouwer Memorial Prize of $1000 established by friends of Dr. Brouwer to be awarded at irregular intervals to a student in the Astronomy Department "for a contribution of unusual merit in any branch of Astronomy."
1966	40-inch reflector installed at Bethany. In 1973 it was dismantled and transferred to the AURA station at Cerro Tololo, Chile.
1966	Photography begun at El Leoncito on plates for proper motion determinations relative to faint galaxies.
1966	Cyril Jackson resigned at the end of June and Arnold Klemola assumed the duties of Acting Director of the Yale–Columbia Southern Observatory. A. J. Wesselink appointed Executive Director in charge of the programs for the Southern Observatory.
1966	R. Wildt awarded the Eddington Medal of the Royal Astronomical Society. Re-elected President of AURA for another year.
1966	Publication of *Methods in Astrodynamics and Celestial Mechanics,* R. Duncombe and V. Szebehely, Eds., Academic Press, N.Y., 436 pp.
1967	A. Klemola resigned to accept a position at the Lick Observatory.
1967	Publication of *Theory of Orbits,* by V. Szebehely, Academic Press, N.Y., 668 pp.
1968	Pierre Demarque chosen Chairman of the Department of Astronomy, effective July 1.
1968	Headquarters of the Department moved to 60 Sachem Street, formerly the Watson Computing Center. The department retained the ground floor and basement of 135 Prospect St., and the Reed Telescope on the roof. In 1976 the Department was evicted from both these buildings, except for the telescope dome on the roof.
1970	Columbia University revoked its commitment to partial funding of the Yale–Columbia Southern Observatory.
1970	K. Aksnes was the first recipient of the Brouwer Memorial Prize, for his Ph.D. thesis in celestial mechanics.
1973	U.S. Naval Observatory discontinued use of its facility at El Leoncito.
1973	R. Wildt retired.

INDEX OF PROPER NOUNS

Y indicates either an undergraduate or graduate degree from Yale. Honorary degrees are cited only from Yale and for non-Yale graduates on the Yale astronomy or related faculty. NAS indicates member of National Academy of Sciences; c, charter member; f, foreign associate. Dates of birth and death given if available.